Functionalized Nanomaterials and Structures for Biomedical Applications

Functionalized Nanomaterials and Structures for Biomedical Applications

Editors

Paul Cătălin Balaure
Alexandru Mihai Grumezescu

Basel • Beijing • Wuhan • Barcelona • Belgrade • Novi Sad • Cluj • Manchester

Editors

Paul Cătălin Balaure
"C.D. Nenițescu" Department
of Organic Chemistry
National University of
Science and Technology
Politehnica Bucharest
Bucharest
Romania

Alexandru Mihai Grumezescu
Department of Science and
Engineering of Oxide
Materials and Nanomaterials
National University of
Science and Technology
Politehnica Bucharest
Bucharest
Romania

Editorial Office
MDPI AG
Grosspeteranlage 5
4052 Basel, Switzerland

This is a reprint of articles from the Special Issue published online in the open access journal *Materials* (ISSN 1996-1944) (available at: www.mdpi.com/journal/materials/special_issues/nano_biomedical_application).

For citation purposes, cite each article independently as indicated on the article page online and as indicated below:

Lastname, A.A.; Lastname, B.B. Article Title. *Journal Name* **Year**, *Volume Number*, Page Range.

ISBN 978-3-7258-1800-6 (Hbk)
ISBN 978-3-7258-1799-3 (PDF)
doi.org/10.3390/books978-3-7258-1799-3

© 2024 by the authors. Articles in this book are Open Access and distributed under the Creative Commons Attribution (CC BY) license. The book as a whole is distributed by MDPI under the terms and conditions of the Creative Commons Attribution-NonCommercial-NoDerivs (CC BY-NC-ND) license.

Contents

About the Editors vii

Preface ix

Ana M. Díez-Pascual
Surface Engineering of Nanomaterials with Polymers, Biomolecules, and Small Ligands for Nanomedicine
Reprinted from: *Materials* 2022, *15*, 3251, doi:10.3390/ma15093251 1

Hajra Ashraf, Davide Cossu, Stefano Ruberto, Marta Noli, Seyedesomaye Jasemi, Elena Rita Simula and Leonardo A. Sechi
Latent Potential of Multifunctional Selenium Nanoparticles in Neurological Diseases and Altered Gut Microbiota
Reprinted from: *Materials* 2023, *16*, 699, doi:10.3390/ma16020699 41

Guillermo Tejada, Natalia L. Calvo, Mauro Morri, Maximiliano Sortino, Celina Lamas, Vera A. Álvarez and Darío Leonardi
Miconazole Nitrate Microparticles in Lidocaine Loaded Films as a Treatment for Oropharyngeal Candidiasis
Reprinted from: *Materials* 2023, *16*, 3586, doi:10.3390/ma16093586 60

Byron Mubaiwa, Mookho S. Lerata, Nicole R. S. Sibuyi, Mervin Meyer, Toufiek Samaai, John J. Bolton, et al.
Green Synthesized sAuNPs as a Potential Delivery Platform for Cytotoxic Alkaloids
Reprinted from: *Materials* 2023, *16*, 1319, doi:10.3390/ma16031319 74

Vladimer Mikelashvili, Shalva Kekutia, Jano Markhulia, Liana Saneblidze, Nino Maisuradze, Manfred Kriechbaum and László Almásy
Synthesis and Characterization of Citric Acid-Modified Iron Oxide Nanoparticles Prepared with Electrohydraulic Discharge Treatment
Reprinted from: *Materials* 2023, *16*, 746, doi:10.3390/ma16020746 90

Rajaram Rajamohan, Chaitany Jayprakash Raorane, Seong-Cheol Kim and Yong Rok Lee
One Pot Synthesis of Copper Oxide Nanoparticles for Efficient Antibacterial Activity
Reprinted from: *Materials* 2023, *16*, 217, doi:10.3390/ma16010217 102

Morgane Valentin, Damien Coibion, Bénédicte Vertruyen, Cédric Malherbe, Rudi Cloots and Frédéric Boschini
Macroporous Mannitol Granules Produced by Spray Drying and Sacrificial Templating
Reprinted from: *Materials* 2023, *16*, 25, doi:10.3390/ma16010025 118

Shashiraj Kariyellappa Nagaraja, Shaik Kalimulla Niazi, Asmatanzeem Bepari, Rasha Assad Assiri and Sreenivasa Nayaka
Leonotis nepetifolia Flower Bud Extract Mediated Green Synthesis of Silver Nanoparticles, Their Characterization, and *In Vitro* Evaluation of Biological Applications
Reprinted from: *Materials* 2022, *15*, 8990, doi:10.3390/ma15248990 134

Mohammad Mahdi Eshaghi, Mehrab Pourmadadi, Abbas Rahdar and Ana M. Díez-Pascual
Novel Carboxymethyl Cellulose-Based Hydrogel with Core–Shell Fe_3O_4@SiO_2 Nanoparticles for Quercetin Delivery
Reprinted from: *Materials* 2022, *15*, 8711, doi:10.3390/ma15248711 154

Hissa F. Al-Thani, Samar Shurbaji, Zain Zaki Zakaria, Maram H. Hasan, Katerina Goracinova, Hesham M. Korashy and Huseyin C. Yalcin
Reduced Cardiotoxicity of Ponatinib-Loaded PLGA-PEG-PLGA Nanoparticles in Zebrafish Xenograft Model
Reprinted from: *Materials* **2022**, *15*, 3960, doi:10.3390/ma15113960 . . . . **175**

Jyoti Dhatwalia, Amita Kumari, Ankush Chauhan, Kumari Mansi, Shabnam Thakur, Reena V. Saini, et al.
Rubus ellipticus Sm. Fruit Extract Mediated Zinc Oxide Nanoparticles: A Green Approach for Dye Degradation and Biomedical Applications
Reprinted from: *Materials* **2022**, *15*, 3470, doi:10.3390/ma15103470 . . . . **200**

About the Editors

Paul Cătălin Balaure

Assoc. Prof. Dr. Paul Cătălin Balaure received his Ph.D. in chemistry in 1996 from the Romanian Academy's "Costin D. Nenitzescu" Institute of Organic Chemistry, under the scientific supervision of Senior Researcher Dr. Eng. Petru Filip. His thesis, "Dimerization and ring opening reactions in benzocyclobutenic systems", explored fundamental aspects of organic chemistry, such as annulene aromaticity and stereochemical aspects of concerted ring-opening reactions. His work contributed to the synthesis of complex chemotherapeutic compounds with sterane or taxane carbon skeletons via palladium-complex-mediated cyclization reactions.

In 2001, Dr. Balaure joined the Department of Organic Chemistry at the National University of Science and Technology Politehnica Bucharest, expanding his research to include core–shell magnetic nanoparticles for biomedical applications, modified release pharmaceutical forms, smart drug nanocarriers, essential oil encapsulation, and enzyme-based electrochemical sensing. These topics align with his teaching in pharmaceutical technology, biochemistry, and environmental toxicology.

In 2007, he gained expertise in electroanalysis during a research stage at the University of Modena and Reggio Emilia, Italy.

Dr. Balaure has published 46 papers, 23 as the first author in ISI journals, and presented over 40 communications at conferences.

Alexandru Mihai Grumezescu

Alexandru Grumezescu is a renowned scientist with significant contributions to the field of nanotechnology and biomedical engineering. He is known for his pioneering research in developing innovative nanomaterials and advancing their applications in medical sciences.

Grumezescu completed his undergraduate and postgraduate studies in Chemical Engineering at the University Politehnica of Bucharest, where he also obtained his Ph.D. His doctoral research focused on the synthesis and characterization of nanostructured materials for biomedical applications.

Dr. Grumezescu has authored and edited numerous scientific books and papers that have been highly influential in his field. His work primarily revolves around the development of nanosystems for drug delivery, antimicrobial treatments, and medical device coatings. Some of his notable publications include works on nanostructured materials, bioactive compounds, and advanced therapeutic systems.

Throughout his career, Alexandru Grumezescu has received various awards and honours recognizing his contributions to science and technology. His research has not only advanced the field of nanotechnology but also provided practical solutions for medical challenges, leading to improved patient outcomes and healthcare innovations.

Preface

Nanomedicine, encompassing the use of nanotechnology for the prevention, diagnosis, and treatment of diseases, is one of the most exciting and fastest-growing fields in modern healthcare. The systemic administration of bulk drugs suffers from a series of major drawbacks such as poor bioavailability, rapid degradation within the body, improper biodistribution, and a lack of targeted delivery to injured tissues, resulting in important side effects, the development of multidrug resistance, and low therapeutic efficacy.

Nanoscale formulations of drugs can circumvent the above disadvantages due to a series of unique features derived from their small dimensions, which are size-correlated with the biological systems on which they act. Smart engineering of nano drug delivery systems (NDDSs) endow them with multifunctional capabilities such as stealth properties to evade the patient immune system and prolonging circulation time, specific ligand-guided drug delivery to targeted diseased areas without damaging healthy tissues, stimuli-responsive drug release allowing specific spatiotemporal controlled release, the ability to penetrate cells, the ability to improve drug solubilization in biological fluids and bioavailability, and the ability to modulate drug pharmacokinetics.

Nanotechnology also helps with imagistic diagnosis and clinical analyses. Theranostic nanomedicine uses nano formulations that integrate both drugs and imaging agents into a single platform. Such nano formulations are used for monitoring drug accumulation at the targeted site versus off-target localization (non-invasive monitoring of drug biodistribution, which is in strict correlation with the magnitude of the harmful side effects), for monitoring drug release after intracellular uptake of the nanocarrier, and for assessing therapeutic outcome, for instance, malign tumor regression in cancer disease. Therefore, theranostic nano formulations have great potential to predict which individual patients have the best chance of respond ingappropriately to a particular nanomedicine treatment, especially in the case of heterogenous diseases such as cancer. Thus, theranostic nanomedicine broadly opens the way to personalized medicine, shifting the therapeutic paradigm to a more holistic approach of "treating the patient, not just the disease".

Through appropriate surface or bulk functionalization, the physicochemical and pharmacological features of nanocarriers used for drug delivery and theranostic purposes can be finely tuned, given the large diversity of molecular and supramolecular structures of such nanovehicles.

The aim of this Special Issue is to highlight the newest and most significant achievements in developing novel functionalized nanomaterials and nanostructures to be applied in the biomedical field.

Paul Cătălin Balaure and Alexandru Mihai Grumezescu
Editors

Review

Surface Engineering of Nanomaterials with Polymers, Biomolecules, and Small Ligands for Nanomedicine

Ana M. Díez-Pascual

Universidad de Alcalá, Facultad de Ciencias, Departamento de Química Analítica, Química Física e Ingeniería Química, Ctra. Madrid-Barcelona, Km. 33.6, 28805 Alcalá de Henares, Madrid, Spain; am.diez@uah.es

Abstract: Nanomedicine is a speedily growing area of medical research that is focused on developing nanomaterials for the prevention, diagnosis, and treatment of diseases. Nanomaterials with unique physicochemical properties have recently attracted a lot of attention since they offer a lot of potential in biomedical research. Novel generations of engineered nanostructures, also known as designed and functionalized nanomaterials, have opened up new possibilities in the applications of biomedical approaches such as biological imaging, biomolecular sensing, medical devices, drug delivery, and therapy. Polymers, natural biomolecules, or synthetic ligands can interact physically or chemically with nanomaterials to functionalize them for targeted uses. This paper reviews current research in nanotechnology, with a focus on nanomaterial functionalization for medical applications. Firstly, a brief overview of the different types of nanomaterials and the strategies for their surface functionalization is offered. Secondly, different types of functionalized nanomaterials are reviewed. Then, their potential cytotoxicity and cost-effectiveness are discussed. Finally, their use in diverse fields is examined in detail, including cancer treatment, tissue engineering, drug/gene delivery, and medical implants.

Keywords: functional nanomaterials; nanomedicine; polymers; tissue engineering; cancer therapy; drug delivery; medical implants

Citation: Díez-Pascual, A.M. Surface Engineering of Nanomaterials with Polymers, Biomolecules, and Small Ligands for Nanomedicine. *Materials* **2022**, *15*, 3251. https://doi.org/10.3390/ma15093251

Academic Editor: P. Davide Cozzoli

Received: 18 March 2022
Accepted: 28 April 2022
Published: 30 April 2022

Publisher's Note: MDPI stays neutral with regard to jurisdictional claims in published maps and institutional affiliations.

Copyright: © 2022 by the author. Licensee MDPI, Basel, Switzerland. This article is an open access article distributed under the terms and conditions of the Creative Commons Attribution (CC BY) license (https://creativecommons.org/licenses/by/4.0/).

1. Introduction

Nanomaterials with unique physicochemical properties have recently attracted a lot of attention since they offer a lot of potential in many fields, particularly in biomedical sciences, including drug and gene delivery systems [1–3], cancer treatment [4,5], monitoring systems [6], tissue engineering [7], and so forth. New generations of engineered nanostructures, also known as designed and functionalized nanomaterials, have opened up new possibilities in the applications of biomedical approaches such as biomolecular sensing, drug delivery, biological imaging, and therapy. A wide number of nanomaterials have great potential to be used in biomedicine, including nanotubes, nanoparticles, nanoplates, and nanowires, to mention but a few [8–10]. Nonetheless, they must meet specific characteristics to be used in biomedical applications [11]. In this regard, their potential cytotoxicity, which can be induced by their structure, chemical content, or features, for example, as well as their biocompatibility, have to be assessed [12]. Their colloidal stability should also be maintained under physiological conditions, ideally across a wide pH range [13]. As a result, it is critical to consider these criteria to ensure the safety, nontoxicity, and biocompatibility of the nanomaterials. Specific interactions with polymers, natural biomolecules, and synthetic ligands of interest are required to modify and functionalize the nanomaterial surface in order to meet these criteria [14–16].

The methods for creating and manipulating functionalized nanomaterials (FNMs) open up exciting new opportunities for developing novel multifunctional biological devices [17]. Furthermore, functionalization prevents nanoparticles from agglomeration and makes them compatible in subsequent phases. As a result, FNMs can transport more efficiently after systemic injection and have better pharmacokinetic characteristics in vivo. FNMs can be deeply driven into tissues through narrow capillaries and epithelial coating, leading to improved therapeutic agent delivery to the targeted location [18]. Furthermore, the small size of FNMs enhances exceptional physicochemical features such as solubility, diffusivity, immunogenicity, and the capacity to target the designated region with minimum diffusion to its surrounding [19,20].

The nanomaterial interface can be designed and applied in different ways. These approaches are classified as replacement, noncovalent, and covalent conjugations based on the primary concept of the type of functionalization interaction [21]. The interface between nanoparticles (NPs) and the attached molecules is modified via the replacement approach, which comprises ligand exchange and ligand addition [22]. Noncovalent techniques rely on many interactions, most of them weak, such as electrostatic, Van der Waals, hydrophobic, and hydrogen bonds, and it is particularly useful with metallic nanoparticles [23]. They are straightforward and do not modify the molecular structure nor their interaction with targets. However, these modifications are strongly dependent on parameters such as ionic strength and pH. On the other hand, covalent attachment techniques have been proposed to alter the external functionalization of nanomaterials to bind molecular entities for biomedical purposes, hence giving the nanoparticles additional functionality [24].

This study aims to provide particular examples to cover the different ways of nanomaterial functionalization using polymers, natural biomolecules, and small ligands (Figure 1), via covalent and noncovalent conjugation. Before highlighting specific examples of each type of functionalization, the basis of the functionalization will be summarized. Although some studies on nanoparticle surface modification for medical and nanotechnological application have been reported [8,10–12,25], most of them are not updated, deal only with nanoparticles rather than nanomaterials in general, and focus only on either the nanoparticle synthesis or on certain biomedical applications. Thus, the current paper reviews recent studies on nanomaterial surface engineering, divided by nanomaterial type and specialized uses. Besides, the cytotoxicity, cost effectiveness, and use of FNMs as a versatile tool in nanomedicine will be discussed. Due to their beneficial characteristics such as biodegradability and biocompatibility in physiological mechanisms, wide availability, suitability for chemical treatment, and wide range of potential synthesis process from different sources, nanomaterials have been extensively explored in the literature. This article offers novel insights on surface functionalization of nanomaterials, focusing on their therapeutic, diagnostic, tissue-engineering, and medical-implant applications. Following a brief overview of the different surface modification strategies and different types of functionalized nanomaterials, a summary of the most relevant biomedical applications is presented.

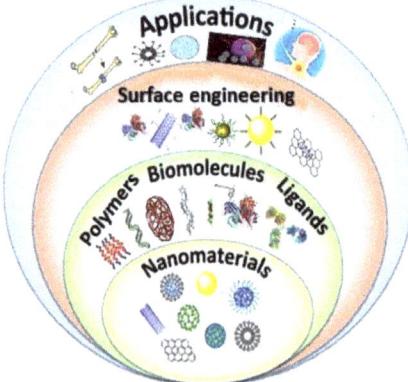

Figure 1. Schematic representation of the functionalization of different types of nanomaterials by polymers, natural biomolecules, and synthetic ligands, and their applications in nanomedicine.

2. Strategies for Surface Functionalization of Nanomaterials

The exclusive properties of nanomaterials compared to their microsized counterparts, such as big, specific surface areas and nanometer sizes, have involved huge attention in the scientific community. Depending on the desired final properties, the composition of nanomaterials can vary from metals or metal oxides to carbon or polymers (Figure 2). Metallic nanoparticles (like gold or silver) are beneficial for designing drug delivery and imaging systems, but their safety has to be investigated in detail to prevent undesirable side effects in humans [26]. Iron oxide, with outstanding magnetic properties, is the most common selection as the core of functionalized nanoparticles. Silica NPs are frequently used in drug delivery applications. Mesoporous silica nanoparticles (MSN), with tunable pore size, are widely used to load small molecules, including amino acids, nucleic acids, drugs, and so forth [27]. However, due to reactive surface silanol groups, there are biocompatibility issues regarding the use of silica nanoparticles for nanomedicine. Concerns regarding the toxicity of carbon-based nanomaterials such as carbon nanotubes (CNTs), quantum dots (QDs), and graphene have also been reported [28]. Polymeric nanoparticles are a widespread option for biomedical applications owed to their tailorable physicochemical properties, excellent biocompatibility, and capability to liberate molecules in a continued way [29]. Numerous polymeric micelles such as thiomers, pluronic, polysaccharides, and polyethylene glycol (PEG) have been investigated [14]. In addition, other colloidal nanostructures such as dendrimers, liposomes, polysomes, and cyclodextrins have been designed for targeted applications [30].

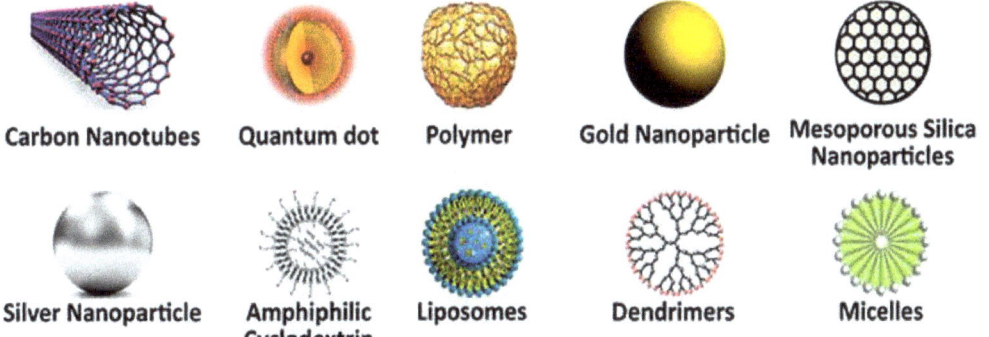

Figure 2. Representation of different types of organic and inorganic nanomaterials used in nanomedicine.

Various types of targeting agents have been implemented to be incorporated on the surface of nanoparticles, especially peptides [31], aptamers [32], antibodies [33], polyethylene glycol (PEG) [34], cationic molecules, folic acid [35], drugs, and fluorescent probes, as depicted in Figure 3. It should be noted that biomolecular interactions rely on the chemical modification of the nanoparticle surface when using NPs for in vitro or in vivo applications [36]. Through a ligand–receptor interaction, such targeting moieties can allow nanoparticles to be embodied into cancer cells and tissues. To facilitate active targeting of NPs to receptors, which are located on the surface of the membrane, the nanoparticle surface can be tailored with targeting ligands, resulting in increased cellular internalization and/or selective absorption via receptor-mediated endocytosis [37]. Researchers are particularly interested in discovering new biomarkers and their relevant ligands in targeted medication administration. The binding of NPs to analytes, pathogens, and biomarkers might cause their signal to be amplified, making it easier to detect and image [38]. When the scaffold surface is decorated with bioactive cues to allow FNPs to interact with cells and the extracellular matrix (ECM) to elicit tissue-specific phenotypes, this is referred to as functionalization [39]. Chemists can easily make the suitable functionalities for use in clinics thanks to the easiness of such functionalization. For example, cell surface molecules have been used to identify nanoparticles functionalized with ligands that show a varied affinity for proteins [34].

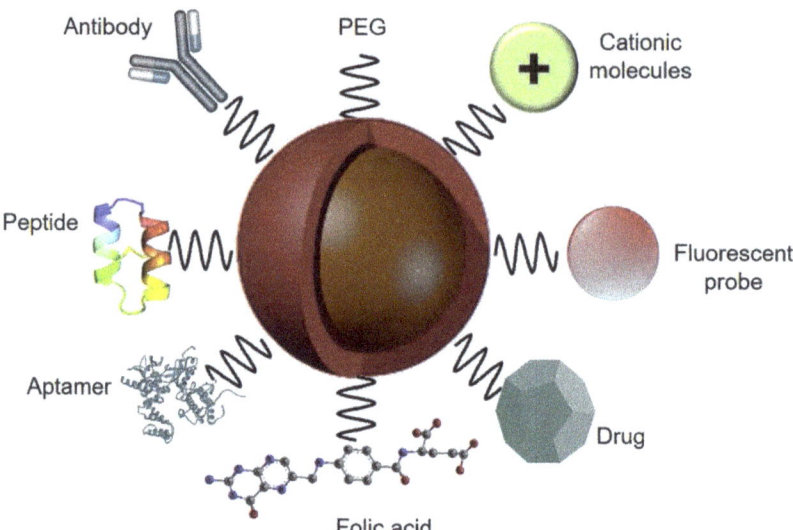

Figure 3. Representation of a nanoparticle functionalized with different types of ligands, polymers, therapeutic compounds, and biomolecules. Adapted from Ref. [40], copyright 2020, with permission from Impact Journals LLC.

Furthermore, functionalization has been proven to protect NPs against agglomeration and make them biocompatible materials in other application stages [41]. Functionalization improves the NPs' physical, chemical, and mechanical characteristics, resulting in synergetic effects [42].

2.1. Functionalization by Covalent Conjugation

The covalent conjugation comprises the reaction of a conjugator (also named linker) with a certain species or chemical group, in a way that the molecules are attached on the nanomaterial surface [43]. Carboxylic acids, amines, thiols, disulfides, phosphates, nitriles, and so forth have been used for covalent conjugation via chemical reactions [44]. Amine groups are the most widely used for functionalization in the biomedical field. The strategy consists in anchoring small molecules or proteins on the nanoparticles. Further, amine functionalization can be used with the aid of n-hydroxysuccinimide (NHS) and different carbodiimides such as EDC. Similarly, carboxylic groups can form ester or amide bonds with alcohol or amine groups on the NPs' surface [45], Figure 4. On the other hand, conjugation on metallic NPs can be effectively carried out via the thiol moiety. The interaction occurs by reaction of sulfhydryl (RSH) groups on the metallic nanoparticles.

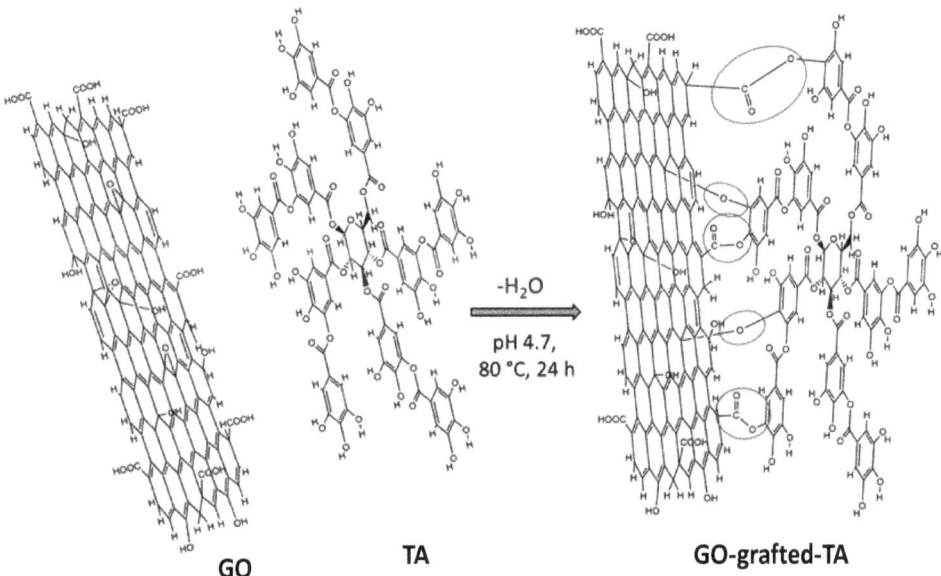

Figure 4. Representation of the covalent functionalization of graphene oxide (GO) with a biological macromolecule, tannic acid (TA) via formation of ether and ester linkages. Reproduced from Ref. [45], copyright 2022, with permission from Elsevier.

2.2. Functionalization by Noncovalent Conjugation

The noncovalent bonding comprises the attachment of molecules on the surface of nanomaterials without chemical bonding via physical adsorption and/or wrapping of molecules by weak interactions such as hydrophobic (Van der Waals), H-bonding, cation−π, anion−π, π−π, and H−π, that preserve the intrinsic properties of the nanomaterial [43]. This approach has some benefits over the covalent way: (i) it takes place under moderate conditions (water solution at room temperature), thus avoiding structural damage of the nanomaterial; and (ii) enables control of the amount of adsorbed/wrapped molecule. The versatility of this route enables a large number of substances to be coupled to the nanomaterials including polymers, solvents, surfactants, aromatic compounds, etc. In order to offer steric stabilization, nanomaterials have been anchored to biocompatible polymers such as polyethylene glycol (PEG) [46].

2.3. Functionalization by Biomolecules

Biomolecules are outstanding candidates to apply in the surface engineering of nanoparticles. Biomolecule-coated nanoparticles have features that are troublesome or inconceivable to attain with synthetic materials, such as excellent bio-macromolecule distribution with little cytotoxicity. Biomolecules such as proteins, peptides, antibodies, and oligonucleotides can be very valuable for targeting NPs to cancer cells where particular receptors are overexpressed. The synthesis of gold–thiol bonds to create oligonucleotide–AuNP conjugates was one of the first bio-nanotechnology examples reported in the literature [34]. Proteins or peptides boost the penetration of NPs into cells via receptor-mediated endocytosis. On the other hand, transferrin is a glycoprotein that can bind to specific receptors on the cell membrane. A few articles have reported the benefits of using this protein as a target for Au, MSN, and poly(lactic-co-glycolic acid) (PLGA) nanoparticles [30].

Albumins are a class of naturally occurring proteins that, besides being applied to load imaging and therapeutic agents, are valuable for modification of numerous types of NPs, as depicted in Figure 5 [47]. Surface modification of NPs with albumins, such as bovine serum albumin (BSA), provides higher water solubility, increased biocompatibility and blood circulation time, and improved stability and cellular interactions compared with uncoated nanoparticles. Different strategies for conjugation of NPs such as AuNPs with albumin have been reported [48], including: (1) Passive adsorption, so that the charged groups of the protein are anchored to the NP surface via covalent or noncovalent interactions. (2) Active adsorption, which involves the use of modified albumin in order to strengthen the albumin-NP interactions. (3) The use of this protein for NP synthesis, either as a reagent (i.e., reducing agent), foaming, stabilizer, or building block for NP synthesis, resulting in NPs with an albumin coating [49]. The use of albumin encapsulation methods provides some profits, such as the loading of agents with low solubility in order to protect them from degradation. (4) Desolvation cross-linking (coacervation process), used to produce core–shell albumin-NPs. This strategy allows chemical agents to become trapped within albumin capsules, which are very stable and protect from degradation. (5) Emulsification: an albumin solution and a nonaqueous phase are mixed, giving rise to an emulsion, and the NP is dissolved in the oil phase. This methodology is used for the encapsulation of lipophilic drugs and enhances aqueous solubility and biocompatibility. (6) Thermal gelation: an albumin water solution is heated to induce protein unfolding, which results in protein–protein interactions by disulfide and hydrogen bonding, as well as hydrophobic and electrostatic forces. Besides, unfolding induces NP–protein interactions, leading to a protein coating onto the NP surface.

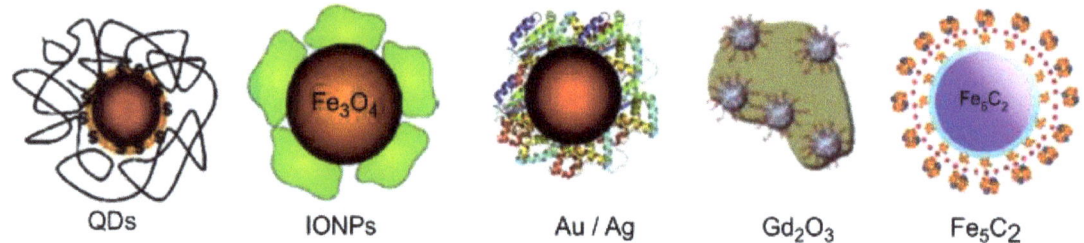

Figure 5. Representation of the surface modification of different types of NPs with albumin. Reproduced from Ref. [47], copyright 2016, with permission from John Wiley and Sons.

2.4. Functionalization by Polymers

A large number of biocompatible, commercially accessible polymers can be used for functionalization, and are typically chosen based on their specific properties such as hydrophobicity, melting point, and functional groups. Polymers frequently used as NP coatings comprise synthetic polymers (i.e., PEG [50] and PLGA [51]) and natural polymers (such as chitosan [52,53]). Polymers have been used for both covalent and noncovalent conjugation of a wide range of nanomaterials. The covalent approach involves the "grafting" (chemical anchoring) of polymeric segments to the NM surface, and can be implemented via "grafting to", "grafting from", "grafing through" and "in situ" tactics (Figure 6). The former is based on the synthesis of a modified polymer prone to react with the functional groups on the surface of the nanomaterial [46]. A shortcoming of this tactic is that the amount of polymer grafted to the nanomaterial is restricted, owed to the low reactivity and large steric barrier of the polymeric segments. In the "grafting from" path the polymer is grown from the NM surface via polymerization of monomers [43]. This approach is effective and manageable, owed to the high reactivity of monomers, allowing a high grafting level. A variation of this strategy is to carry it out via "in situ" polymerization in the presence of the inorganic precursor. Nonetheless, this method requests precise monitor of the amounts of each reagent and the polymerization conditions. In the "grafting through", a low molecular weight monomer is radically copolymerized with a polymerizable macromonomer in the presence of an initiator.

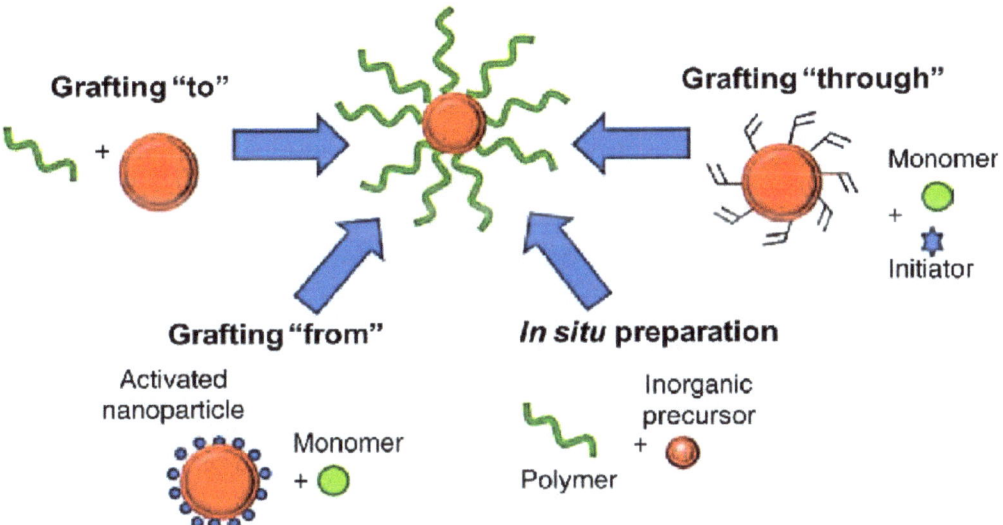

Figure 6. Schematic illustration of polymer grafting approaches: "grafting from", "grafting to", "grafting through" and "in situ" preparation in the presence of an inorganic precursor.

Polymers are suitable for functionalization because they create a physical barrier around the NPs, preventing the core of the NPs from coming into direct contact with biological receptors. Polymers can produce a physical barrier but with a reduced hydrodynamic radius. As a result, polymer coatings outperform small molecule ligands when imparting macromolecular system characteristics to the particle surface, similar to biological proteins. The use of polymers such as PEG to coat nanoparticles improves passive tumor tissue targeting, increasing permeability, and retention (EPR), as well as biocompatibility [54]. This PEG and other polymer coatings decrease blood serum protein adsorption, lengthen circulation duration, and promote particle absorption into tumor tissues [34]. Using AuNPs synthesized by stacking cationic polyallylamine and anionic poly (acrylic acid) polyelectrolyte layers, Kleinfeldt and coworkers [55] developed an excessively hydrophilic and biocompatible coating that enables colloidal stability. Makvandi et al. [56] investigated the functionalization of various polymers (glyclusters, glydendrimers, glycopolymers) and nanomaterials (Ag_2O, CuO, ZnO, Fe_3O_4, MgO, TiO_2, Se, Ni, Pd) for water purification, food containers, fabrics, and medical applications. The benefits and drawbacks of polymer functionalization were investigated and explored in that study. When natural or synthetic polymers are used to functionalize NPs, photo/thermo-responsive properties can be achieved [57]. For instance, chitosan grafted with poly-L-lactide using thiourea-functionalized, and poly-N-isopropyl acrylamide were used to synthesize photo/thermo-triggered micelles [58].

2.5. Functionalization by Small Ligands

Small ligands are a common selection for functionalizing nanomaterial since they are relatively simple to chemically bond to surfaces via functional moieties in their structure. They are an appropriate choice to adjust the nanomaterial properties such as hydrophilicity or charge with a view to improve their biological activity and interaction with other biologically essential ligands, as well as their stability, aqueous solubility, drug loading, and so forth. For instance, silica NPs can be straightforwardly tailored with organosilane molecules such as 3-(aminopropyl) triethoxysilane (APTES) through silane chemistry. It has been reported that APTES-functionalization is an effective method for adjusting drug loading and discharge from mesoporous silica nanoparticles (MSN) [59]. Besides, it is beneficial for many aims, such as the release of low soluble drugs, the targeting of drugs to a chosen position, or to make multifunctional drug delivery and imaging devices. Other ligands such as drugs have been used for tailoring NP surfaces. For example, doxorubicin (DOX, a frequently applied anticancer drug) has been conjugated to Fe_3O_4 NPs with the aim to develop dual-functional NPs [60]. These modified NPs can destroy tumor cells via the conjugated DOX, and concurrently enable magnetic resonance imaging of the tumor, which is highly valuable. Other small drugs such as methotrexate, that can target the folate receptor on cancer cells, ciprofloxacin [25,61], and so forth, have been conjugated with different nanomaterials.

Various nanomaterials functionalized by small ligands can be added as signal reporters or as carriers for loading more signal reporters in biosensors for analyte detection [62]. Mahmoudpour et al. [63] designed a method for producing aptameric functionalized materials (AFMs). Optical indicators, conducting transducers, carriers, catalysts, and other features, were combined to develop advanced AFMs. Drug delivery, bioimaging, and appropriate sensing have been highlighted as biological uses of improved AFMs. Aptamers have been identified among the most promising prospects for constructing a broad range of sensing platforms due to their unique properties, such as outstanding specificity and sensitivity, easiness of fabrication, and excellent durability in a variety of circumstances. For the manufacturing of aptamer-based nanoprobes, many signals-transduction approaches have been developed. Incorporating numerous aptasensing techniques with NPs has improved biosensor selectivity and sensitivity in recent years [64].

3. Functionalized Nanomaterials

3.1. Metallic Nanoparticles

Metal-based nanomaterials consist of nanoparticles of raw metal, such as gold (Au) and silver (Ag). AuNPs are inert in bulk, while become highly reactive in nanoparticle form [65]. The exciting surface chemistry of AuNPs opens up novel routes for the progress of unexplored multifunctional instruments for biomedical and nanotechnological applications [66]. Nanotechnology applications have drawn a great deal of interest since the late 1980s [67]. The exceptional electrical and optical properties of Au boost their use in biosensing and bioimaging. The use of organic molecules to functionalize Au NPs aids the conjugation of drugs for delivery systems. Thus, AuNPs can be used as photothermal therapeutic agents [56]. For instance, surface-modified AuNPs have been prepared via a layer-by-layer procedure with alternating polyelectrolyte layers of cationic polyallylamine and anionic poly(acrylic acid). Subsequently, papain was covalently immobilized on the modified AuNPs via amide bond between the NH_2 groups of papain and the terminal COOH groups of the modified NPs, using EDC and sulfonated NHS as coupling agents, as depicted in Figure 7, to produce a heterogeneous biocatalyst that has been applied in bioanalysis and biopharmaceutical analysis [68].

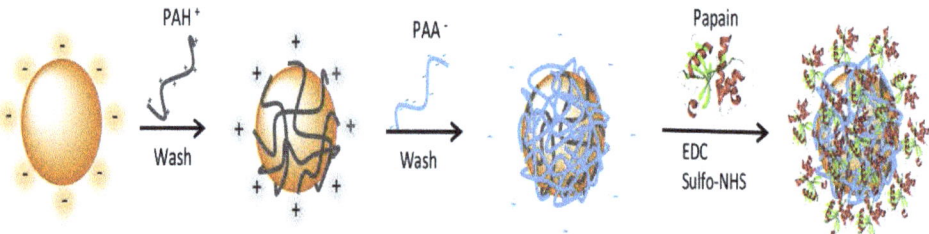

Figure 7. Schematic representation of the functionalization process of AuNPs with papain. The surface modification was achieved by a layer-by-layer (LbL) approach via activation of COOH groups of the modified AuNPs with EDC and sulfonated NHS as coupling agents, followed by amide bonding with the NH_2 groups of papain. (PAH+, polyallylamine hydrochloride; PAA−, polyacrylic acid sodium). Adapted from Ref. [68], copyright 2017, with permission from Elsevier.

The conjugation of gold nanorods (AuNRs) onto micelles, via gold-thiolate complex formation, brings photosensitivity to the nanoassembly. The size and surface morphology characterization via TEM (Figure 8) indicated that the mean micellar size was around 15 nm, and the thickness and length of the AuNRs was about 20 and 65 nm, respectively. The percentage of conjugated AuNRs to the micelles was roughly 12%. The attachment of chitosan transfers the photosensitivity of functionalized AuNRs to micelles, and the micelle thermal shrinkage induces the release of paclitaxel, a drug widely used to treat breast cancer [69].

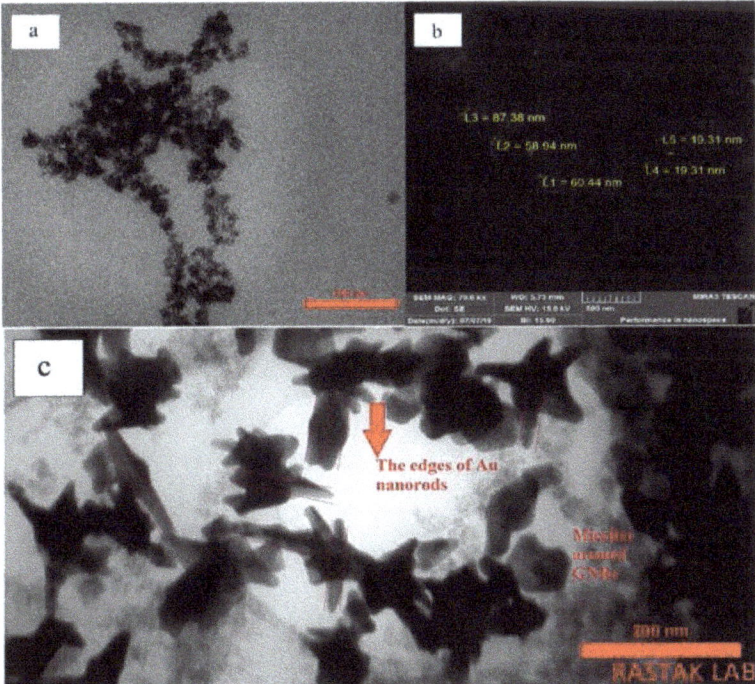

Figure 8. (a) TEM image of polymeric micelles, (b) SEM image of AuNRs, and (c) TEM image of AuNRs coated by polymeric micelles. Adapted from Ref. [69], copyright 2020, with permission from Elsevier.

Nejati et al. [19] examined functionalized AuNPs in biomedical applications. To attain this goal, their structure, production, and functionalization were extensively explored and discussed. Gold NPs have been utilized in biological applications, electrochemical technology, and radiation oncology. Multifunctionalization, that is, functionalization that allows for the provision of more than one attribute at a time, provides added value to these NPs due to synergistic effects. Multifunctionalized gold NPs have been discovered to be a viable choice in biomedicine for delivering anticancer drugs and antibiotics for combined photothermal and chemical therapy [70]. AuNPs are suitable for the delivery of the drugs to cellular destinations due to their ease of synthesis, functionalization, and biocompatibility. Figure 9 depicts functionalization of AuNPs for gene and drug delivery. AuNPs functionalized with targeted particular biomolecules can successfully kill tumor cells or bacteria (Figure 9). Large surface-to-volume ratio of AuNPs can carry a huge amount of drug molecules. AuNPs have been applied for the codispensation of protein drugs owed to their skill in penetrating cell membranes, probably because they can interact with the lipids present on the cell surface [70].

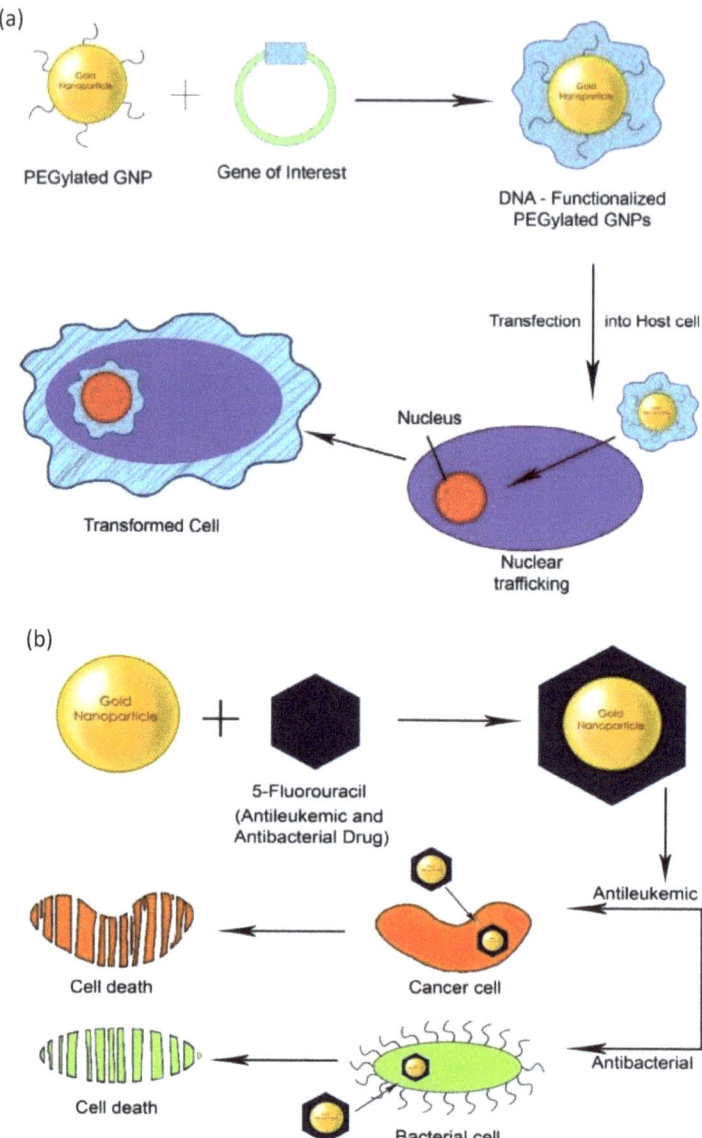

Figure 9. (a) PEGylated gold nanoparticles for gene delivery. (b) Functionalized gold nanoparticles for drug delivery. Adapted from Ref. [70], copyright 2011, with permission from MDPI.

Despite the efforts carried out, more studies into intelligent drug delivery based on nanoparticles, particularly gold NPs, is required. Despite numerous publications, only a few clinically authorized drug delivery nano systems are currently accessible in the industry. As a result, an immediate need is found to incorporate animal-model research into clinical practice [70]. Donoso–Gonzalez et al. [71] used cationic cyclodextrin-based polymer (CCD/P) to load phenylethylamine (PhEA) and piperine (PIP) onto gold nanostars (AuNSs). They evaluated the product potential for simultaneous drug loading and SERS-based detection. In addition to PhEA and PIP, the polymer contained AuNSs that had been functionalized with PhEA and PIP, resulting in a unique AuNS-CCD/P-PHEA-PIP nanosystem with an optimum size and Z potential for biomedical applications. Hybrid materials incorporating carbon nanomaterials and AuNPs have also been synthesized. For instance, Shon et al. [72] reported the synthesis of soluble fullerene-linked AuNPs using a modified Brust reaction and subsequent ligand exchange reaction of hexanethiolate-protected Au NPs with 4-aminothiophenol. Amination of C60 with 4-aminothiophenoxide ligands produced the C60-linked AuNPs. This approach enables the control of the optical and photochemical properties of the nanoparticles. Sudeep and coworkers [73] developed a self-assembled photoactive system comprising AuNPs as the central core and fullerene moieties as the photoreceptive hydrophobic shell via functionalization of the NPs with a thiolated fullerene derivative. Yaseen et al. [74] used C60-terminated alkanethiol to synthesize novel fullerenethiol-functionalized gold nanoparticles (C60−AuNPs) of 2 nm diameter with an extremely narrow size distribution. The fullerene-thiol moiety was inserted into the fullerene by the ligand exchange method. Liz–Marzán et al. [75] developed Au core/SiO_2 shell nanocomposites with tailorable thickness and good dimensional stability. Citrate-capped AuNPs were first synthesized and then reacted with aminopropyl trimethoxysilane, a widely used coupling agent, which anchored onto the NPs via silanol groups. Active silica was subsequently added, leading to the formation of a fine, dense, and fairly homogeneous silica layer wrapping the NPs.

On the other hand, AgNPs are antibacterial and anti-inflammatory, and possess excellent biocompatibility [76]. The AgNPs can be straightforwardly synthesized via simple, fast, nontoxic, and environmentally friendly means so that they can produce NPs with perfectly defined morphology and size. They were applied as a coating for cardiovascular implants to improve their biocompatibility. Additionally, their antimicrobial, antifungal, antiviral antiangiogenic, and anticancer properties make them suitable in a large number of biomedical and health care areas including device coatings, drug delivery systems, wound dressings, the textile industry, photothermal therapy, and so forth. The biological activity of AgNPs is influenced by many parameters such as the NP shape, size, morphology, state of dispersion, solution rate, reactivity, and ion discharge efficiency, amongst others, which condition their cytotoxicity. The design of AgNPs with uniform functionality, size, and morphology is crucial from a practical viewpoint. Other metallic NPs such as ruthenium and selenium have been applied in nanomedicine [77], in particular for drug delivery.

3.2. Metal Oxide-Based Nanomaterials

A wide number of variations of metal oxide NPs have been used in nanomedicine, such as iron oxide (Fe_2O_3, Fe_3O_4), CeO_2, titania (TiO_2), ZnO, NiO, silica (SiO_2), and so forth [78–80]. Iron oxide NPs are a fascinating family of nanostructures that have attracted much interest in the medical area because of their negligible toxicity, high biocompatibility, and inherent magnetic properties, which make them perfect candidates for therapeutic and diagnostic goals, particle imaging, and as contrast agents in magnetic resonance imaging (MRI) and ultrasonic techniques [81]. The incorporation of Fe_3O_4 NPs also enhances the antimicrobial properties [82].

The five most popular strategies to generate hollow iron oxide NPs are the Kirkendall effect, galvanic substitution, chemical etching, nano-template-mediated, and hydrothermal/solvothermal routes [83]. Cheah et al. [84] synthesized iron oxide NPs in diethylene glycol (DEG) by thermal decomposition of iron (III) acetylacetonate ($Fe(acac)_3$), and subsequently changed the surface of the NPs by adding surface ligands (Figure 10). Using this easy production process, surface modification of iron oxide NPs with various covering substances such as dopamine (DOPA), polyethylene glycol with thiol end group (thiol-PEG), and poly(acrylic acid) (PAA) is achievable. The size of these NPs can be precisely controlled at the nanometer scale by continuous growth. TEM images confirmed that the morphology did not change upon functionalization (Figure 10). Besides, NPs with PAA coating can be used as contrast agents. The surface change of oleic-acid-coated iron oxide NPs (Fe_3O_4-OA) (made by coprecipitation method) with tetraethylorthosilicate was studied by Nayeem et al. [85] using an inverse microemulsion approach (TEOS). To obtain thermally sensitive magnetic nanocomposites (MNCs), Fe$_3$O$_4$/SiO$_2$/P(NIPAm-co-AMPTMA), the surface of iron oxide nanoparticles was tailored using a multistep approach with poly [N-isopropylacrylamide-co-(3-acrylamidopropyl) trimethylammonium chloride], P(NIPAm-co-AMPTMA). Magnetic nanoparticles (MNPs) have been extensively studied as MRI contrast agents to aid in the detection, diagnosis, and treatment of solid cancers. The absorption of superparamagnetic iron oxide NPs (SPIONs) in the endothelial reticulum system (RES) can be used in medical imaging to detect liver neoplasms and metastases. It can also currently differentiate tiny lesions of 2–3 mm. Furthermore, ultrasmall superparamagnetic iron oxide NPs (USPIONs) show promising utility in MRI exams for the identification of lymph node metastases that are 5–10 mm wide [86]. By utilizing the distinct molecular fingerprints of these disorders, the future iteration of active targeting MNPs, which has recently been explored, has the capacity to enhance tumor detection and characterization [86].

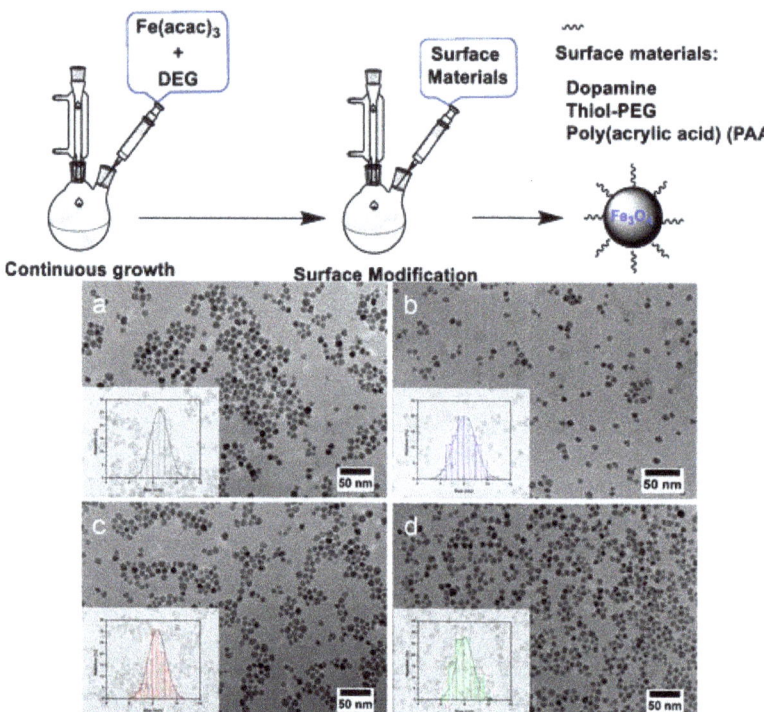

Figure 10. (**Top**) Synthesis of Fe$_3$O$_4$ NPs in diethylene glycol (DEG) by thermal decomposition of acetylacetonate (Fe(acac)$_3$), and surface modification by adding surface ligands. (**Bottom**) TEM images of Fe$_3$O$_4$ NPs (**a**), Fe$_3$O$_4$ NPs functionalized with dopamine (**b**), Fe$_3$O$_4$ NPs surface modified with polyethylene glycol with thiol end group (thiol-PEG) (**c**), and Fe$_3$O$_4$ NPs modified with poly(acrylic acid) (PAA) (**d**). Adapted from Ref. [84], copyright 2021, with permission from the American Chemical Society.

Cerium oxide (CeO$_2$) NPs, named as nanoceria, have the unique property of anti-inflammation. They have better redox as well as potential antioxidant properties with therapeutic characteristics. TiO$_2$ has the unique properties of high chemical stability, cytocompatibility, and optical properties [87]. The biocompatible properties of TiO$_2$ NPs have increased their usage in drug delivery, bone substitute materials, bone regeneration, cell and tissue behavior modulation, vascular stents, scaffolds, bioimaging, and biosensors [78]. MSN also have great potential for nanomedicine. In fact, upon functionalization, they can be efficiently targeted to cancer cells [59] and be used for encapsulation and controlled release of drugs [27]. For biomedical applications, ZnO possesses the properties of low toxicity and biodegradability. It can be used for the purpose of drug delivery, gene delivery, biosensing, bioimaging, etc. [88]. CuO NPs have also been used for targeted drug delivery in breast cancer therapy [35].

3.3. Ceramic-Based Nanomaterials

A wide range of ceramics, including $Ca_3(PO_4)_2$, bioactive glass, Al_2O_3, ZrO_2, $CaCO_3$ and so forth, are getting countless interest in the biomedical field, particularly in the tissue engineering arena. Thus, their outstanding osteoconductivity, resorbability, biocompatibility, biodegradability, and hydrophilicity make them appropriate for numerous hard tissue applications [89,90]. They can be divided into three types: bioinert, bioactive, and resorbable. Resorbable ceramics are progressively adsorbed and substituted by endogenous tissue. They can be synthesized in the forms of nanocrystals, NPs, nanopowders, or nanocoatings. The most popular is $Ca_3(PO_4)_2$, which is widely applied in the form of NPs and nanocements for orthopaedic and dental uses. The optimal surface charge density, functionality, and solution characteristics of this ceramic account for its fittingness in drug delivery and growth factor uses. Bioactive ceramics such as hydroxyapatite (HDA) NPs are a type of calcium phosphates that have been comprehensively investigated in bone regeneration and antibacterial applications [91,92]. They are osteoconductive and can link to bone tissues via chemical bonding, following the rule of bonding osteogenesis. Furthermore, for bone tissue engineering, bioactive glass is crucial, owed to its outstanding osteoconductivity, osteoinductivity, and biocompatibility [93]. Bioinert bioceramics such as ZrO_2 have great chemical stability and in vivo mechanical strength. This oxide is regarded as a nontoxic material and has strong resistance to acids; hence, it is widely used in coatings for metallic load-bearing implants and dentistry. Another widely used oxide is Al_2O_3, which possesses high hardness and superior heat resistance, and has been applied in arthroplasty, dentistry, and as an antimicrobial coating [94].

3.4. Carbon-Based Nanomaterials

Within carbon-based nanomaterials, carbon nanotubes (CNTs), graphene oxide (GO), and graphene quantum dots (GQDs) have been broadly explored in biomedical applications [13,95,96]. Purification, separation, dispersion, stability, alignment, functionalization, and arrangement of CNTs are critical parameters to be controlled prior to their applications [97]. Since the discovery of CNTs, numerous physical and chemical techniques have been developed to attain these goals [98,99]. Polysaccharides with a broad range of characteristics, large-scale production, and low prices have shown to be highly suitable for CNT functionalization. The use of chitosan for CNT purification and functionalization has been proven to be a strategy to make drug release easier and more effective. Dou et al. [100] described a one-pot tactic for the development of chitosan-coated CNTs via a combination of Diels–Alder reaction and mercaptoacetic acid locking imine (MALI) reaction (Figure 11). Taking into account the broad use of Diels–Alder chemistry and MALI reaction, several carbon nanomaterials with different functional groups might be synthesized and applied to biomedicine.

Graphene and graphene oxide (GO) are 2D carbon-based nanostructures, in the form of nanosheets, that show an optimal combination of biocompatibility, strength, flexibility, and optical transparency, which made them suitable for the design of selective and sensitive sensors of biomolecules, which is crucial for medical sciences and the healthcare industry in order to assess physiological and metabolic parameters [101]. Besides, they show antibacterial and antiviral properties [96,102,103]. Graphene-based systems have proven to be effective via direct interaction with viruses and through photo-induced mechanisms, as well as platforms for other particles or molecules with antiviral properties. GO inactivates the virus by physical disruption: it can adhere to the structure of virus spikes and destroy them with the sharp edges of the GO layers. Its antiviral activity is effective on both

DNA and RNA viruses, and depends on the concentration and incubation time. Reduced graphene oxide (rGO) and GO show similar antiviral activity, pointing towards a minor influence of the surface functional groups. The physical interaction of the viruses with their sharp edges seems to be the leading cause for the antiviral activity. Besides, they are negatively charged, which enables electrostatic interaction with the positively charged viruses. The higher interactions result in the destruction and inactivation of the virus. The viruses captured by GO have shown a loss of structural integrity: an RNA is released. The virus can then be identified using the recovered RNA [104]. Another method to inhibit the virus activity is using the GO photocatalytic activity. This approach has been developed by Hu et al. [105] to synthesize GO-aptamer nanosheets that were used to capture MS2 bacteriophage viruses, a small icosahedral nonenveloped RNA virus, which infects E. coli bacteria. This was used as a model for testing the antiviral properties of GO upon illumination with UV light. In this case, the leakage of the virus protein capsid predominates over the physical disruption produced by the sharp edges of the GO sheets.

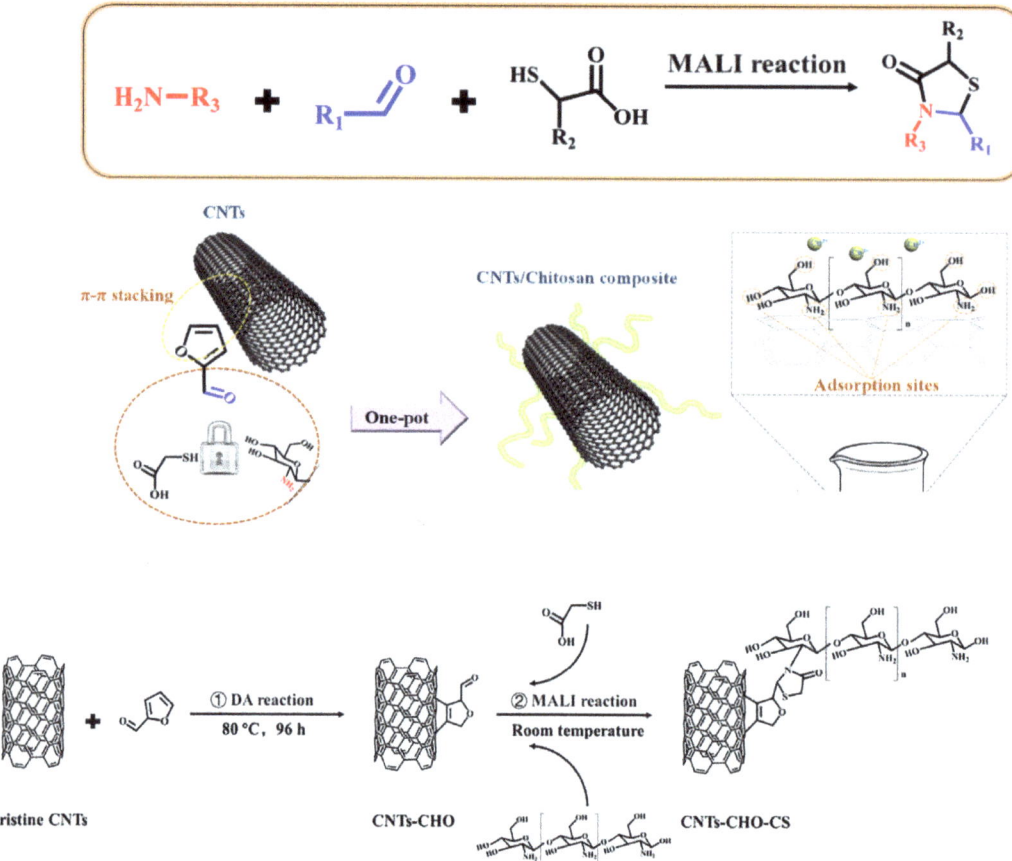

Figure 11. Functionalization of carbon nanotubes with chitosan based on MALI reaction. Adapted from Ref. [100], copyright 2019, with permission from Elsevier.

Carbon quantum dots (CQDs), 0D carbon-based nanomaterials with fluorescence characteristics, also exhibit antimicrobial and antiviral properties [106]. These include amorphous carbon nanoparticles, graphene quantum dots (GQDs), partially graphitized core–shell carbon NPs, and amorphous fluorescent polymeric NPs. Their activity is attributed to the functional groups on their surface. CQDs functionalized with boronic acid demonstrated antiviral efficacy against HCoV-229E Human Coronavirus. HCoV-229E is an enveloped, single-stranded RNA coronavirus. It is one of the viruses that produce the common cold (Coronaviridae family, Human coronavirus 229E species), with a diameter in the range of 120–160 nm. Figure 12 shows two pathways for antiviral activity: (1) the attachment of CQDs (with a mean diameter of about 7 nm to the S-protein of viruses) to prevent infectious contacts between host cells and viruses; and (2) the capacity of CQDs to disrupt RNA genomic replication. Boronic acid functions were crucial in determining antiviral efficacy [107].

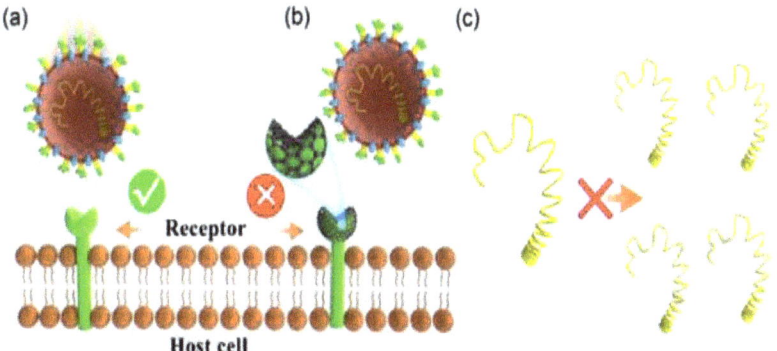

Figure 12. Scheme of the antiviral action of functionalized graphene quantum dots (GQDs). (a) Viral illnesses are caused by binding between the coronavirus (HCoV-229E) S-protein and the host cell receptor. (b) The presence of GQDs can prevent such binding. (c) This mechanism can inhibit the viral genome replication. Adapted from Ref. [107], copyright 2021, with permission from Elsevier.

Bai et al. [108] developed a molecularly imprinted fluorescent sensor for selective identification of a model drug: paclitaxel. A molecularly imprinted polymer (MIP) shell was grafted on the surface of silane-functionalized Mn:ZnS QDs using a free radical polymerization procedure (Figure 13). Methacryl polyhedral oligomeric silsesquioxane (M-POSS) was utilized to provide a porous structure.

Van Tam et al. [106] used microwave-assisted pyrolysis of fructose to synthesize aniline-functionalized graphene quantum dots (a-GQDs). Then, phenyl boric acid (PBA) was used to modify the a-GQDs, leading to a fluorescence-quenching effect. The a-GQDs/PBA nanomaterial was tested as a fluorescence turn-on sensor for glucose detection, based on the specific interaction between PBA and glucose.

QDs also have great potential for cancer treatment. The selective attachment of FR-positive tumor cells with folic acid/folate (FA) was reported as a fast and easy technique for determining folate receptor (FR) expression in cancer cells. MKN 45, HT 29, and MCF 7 cancer cells were selectively marked using graphene quantum dots with folate coating and nitrogen doping (N-GQDs) [109]. DNA-functionalized QDs have drawn considerable attention in sensing and imaging, as well as cancer therapy [110]. Covalent conjugation, electrostatic interaction, direct dative interactions, and other ways for conjugating DNA to QDs have been documented in the literature [111]. In vitro photothermal imaging was described by Wang et al. [112] as AuNPs-QD complexes combined with DNA as a template. Horo et al. [52] developed DOX-loaded chitosan-AuNPs and beads, both of which were implanted with functionalized silk fibroin. Chitosan was used as a reduction and stabilizing agent to synthesize NPs with dimensions in the range of 3-8 nm. Compared to uncoated materials, coated materials demonstrated a delay in drug release. As a result, drug delivery strategies based on functionalized silk-coated substances may be useful for producing localized and protracted drug release. Figure 13 depicts the synthesis process and the potential detection mechanism of the drug.

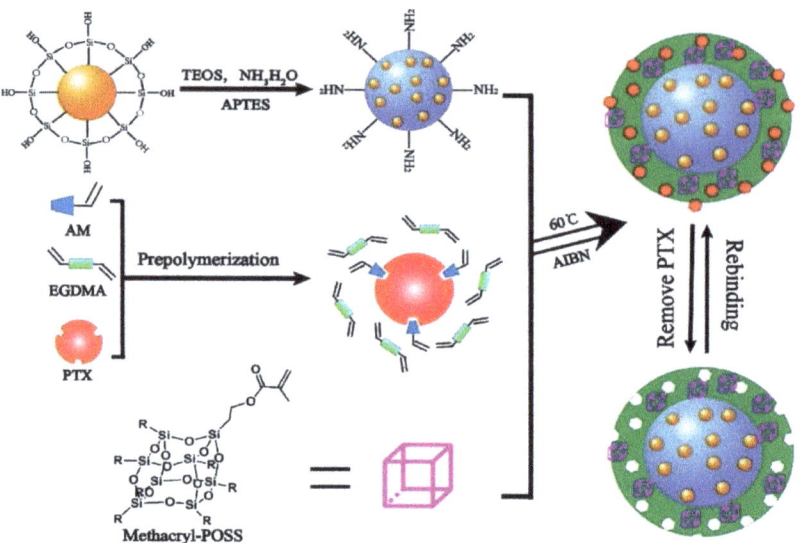

Figure 13. Schematic illustration for the preparation process and possible detection principle of the POSS-MIP/QDs. Adapted from Ref. [108], copyright 2021, with permission from Elsevier.

3.5. Polymeric Nanomaterials

Polymeric NPs are colloidal particles in the range of 10 nm–1 μm made up of polymers that can be straightforwardly synthesized through chemical reactions in order to tailor the loading and release of drugs and genes. The benefits of these NPs are their easiness to synthesize, high stability, biodegradability, nontoxicity, lengthy blood circulation time, and sustained and targeted delivery. Furthermore, they can be tailored according to their shape, size, surface functional groups, degree of porosity, as well as their mechanical characteristics [113]. They are divided into three main groups: natural, biosynthesized, and chemically synthesized. They can be fabricated into different shapes, including liposomes, dendrimers, nanospheres, nanocapsules, nanogels, and micelles (Figure 2). They are used in wound dressings, pharmaceutical excipients, medical devices, dental materials, and scaffolds [114]. Biodegradable polymers frequently used for the development of polymeric NPs are poly(lactide) (PLA), poly(ε-caprolactone) (PCL), PLGA and polycarbohydrates such as alginate, chitosan, and gelatin.

Overall, because of their excellent chemical, physical, and mechanical properties and their versatility of synthesis, functionalized nanomaterials can be employed in a variety of ways. Although functionalized nanoparticles are hardly used in the industrial field up to date, they can aid in developing novel concepts in a variety of industries. Functionalized nanomaterials promise to produce better and cost-effective consumer products and industrial operations. An inappropriate use can have a detrimental effect on surroundings, public health, and safety in various ways [115–117].

4. Cytotoxicity: The Role of Functionalization

Chemical composition, crystalline structure, size, and density are parameters that strongly influence nanomaterial toxicity and cytotoxicity [28,118]. Nanomaterial absorption and intracellular localization can be linked to some health hazards due to the nature of nanomaterials and their chemical interactions with cells. Chemical composition, for example, might cause oxidative stress in cells [119]. CNTs are believed to be more poisonous than carbon black or silica nanoparticles and can induce severe lung damage [120]. Asbestos is less hazardous than TiO_2, Fe_3O_4, and ZrO [121]. Another indicator of cytotoxicity based on membrane integrity damage is lactate dehydrogenase (LDH) leakage. Additionally, DNA damage in primary mouse embryofibroblasts (PMEF) treated in vitro with different amounts (5, 10, 20, 50, and 100 µg mL^{-1}) of manufactured nanoparticles (Figure 14) revealed that CNTs and ZnO caused more DNA damage than carbon black (CB) and SiO_2 NPs [121].

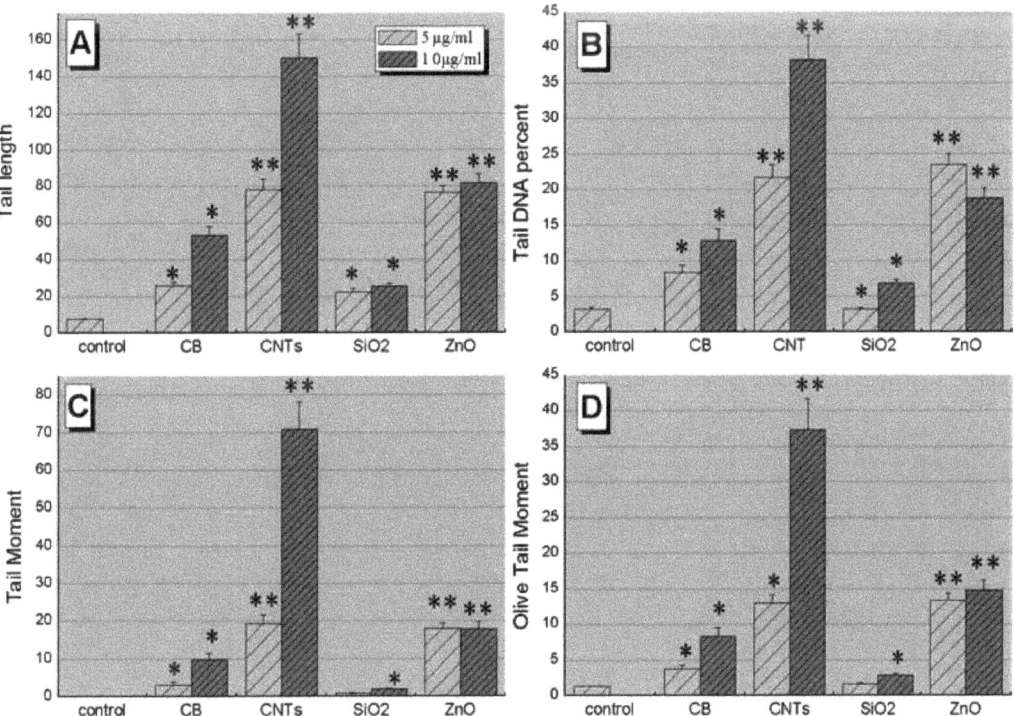

Figure 14. DNA damage determined by comet assay in PMEF cells exposed to NPs. Cells were respectively treated with 5 µg mL^{-1} of CB, CNT, SiO_2, and ZnO for one day. Damage was evaluated by (**A**) tail length, (**B**) tail DNA, (**C**) tail moment, (**D**) Olive tail moment. Values shown are the mean from 50 images. * $p < 0.05$; ** $p < 0.01$ in comparison to blank. Adapted from Ref. [121], copyright 2009, with permission from Wiley & Sons, Inc.

The crystalline structure also has a strong effect on NPs' toxicity [122]. For instance, TiO_2 NPs, which can naturally appear in three different crystalline forms, i.e., anatase, rutile, and brookite, are reported to have cytotoxic and genotoxic effects. Rutile titania is slightly more lethal than anatase TiO_2 NPs. This might be elucidated considering the different reactivity of the two forms: rutile TiO_2 NPs are more photocatalytic than anatase and therefore, are capable of producing larger quantities of oxygenated free radicals on their surface. On the other hand, other allotropes have a significant influence on cell viability and, as a result, on human health. Sato et al. [123] discovered that TiO_2 allotrope toxicity is affected by the NPs' environment's ambient conditions. In the absence of UV radiation, rutile TiO_2 NPs (200 nm) caused oxidative DNA injury, whereas TiO_2 NPs (10–20 nm) caused oxygen-reactive species (ROS) generation.

Another key component in minimizing nanomaterial toxicity is particle size [124]. Nanoparticles with a smaller size are more prone to pass through biological barriers. Phagocytosis or other pathways can facilitate the entrance of small NPs to cells. NPs can discriminate between adhesive connections because of their ability to infiltrate cells. It can produce forces such as Van der Waals, steric interactions, and electrostatic charges [125]. Furthermore, unlike big nanoparticles, NPs in the size range of 1 to 100 nm are not phagocytized but instead taken up via RME routes. In the lack of specific cell surface receptors, NPs can be absorbed. Most cells can effectively assimilate NPs with size of 50 nm or smaller (causing cytotoxicity). NPs smaller than 20 nm can easily pass through blood arteries and concentrate in organs [126]. NPs with a large surface area, such as NiO (diameter < 25 nm), clump together in liquids, and engage and induce oxidation and DNA damage by interacting with molecules including proteins and DNA [127]. The mechanisms of cell damage by NPs are depicted in Figure 15 [124].

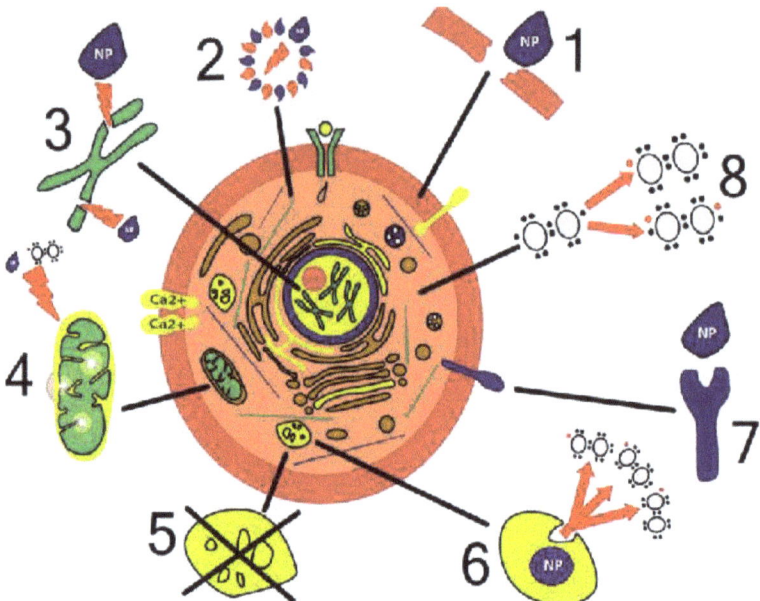

Figure 15. Mechanisms of cell damage by nanoparticles. (1) Physical damage of membranes. (2) Structural changes in cytoskeleton components. (3) Disturbance of transcription and oxidative damage of DNA. (4) Damage of mitochondria. (5) Disturbance of lysosome functioning. (6) Generation of reactive oxygen species. (7) Disturbance of membrane protein functions. (8) Synthesis of inflammatory factors and mediators. Adapted from Ref. [124], copyright 2018, with permission from Springer Nature.

5. Cost-Effective Functionalization

The functionalization of AuNPs using a mixture of DNA and PEG polymers is the most cost-effective and satisfactory method available for nanomaterial cofunctionalization [128,129]. To obtain a comparable level of gold NP binding effectiveness with DNA origami nanostructures, Wang et al. [130] used a significantly smaller amount of thiol-DNA in their technique than pure DNA functionalization. Because of the decreased DNA consumption and lower costs, the use of DNA–NP conjugates in nanotechnology can be scaled up. Figure 16 shows the functionalization process of AuNPs with DNA/PEG polymers [130].

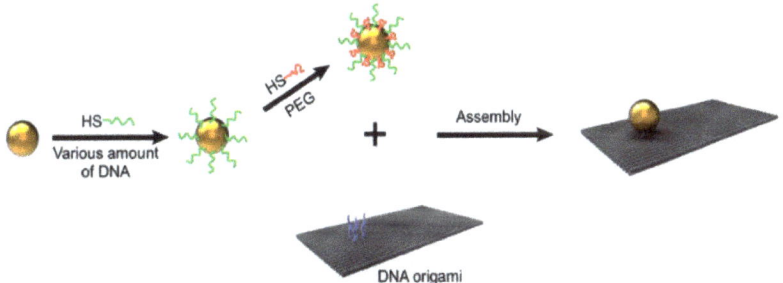

Figure 16. AuNPs are functionalized in two stages: first with DNA/PEG polymers comprising variable amounts of DNA, and then with rectangular DNA origami. Adapted from Ref. [130], copyright 2017, with permission from The Royal Society of Chemistry.

6. Applications of Functionalized Nanomaterials in Biomedicine

Medical diagnosis [131], immunization [132], treatment [133], and even healthcare services have been transformed and influenced by nanotechnology [134]. Chemical functionalization, physical functionalization, and surface synthesis link biological agents with various NPs. It is possible to classify the biomedical applications of nanotechnology into different areas, as summarized in Figure 17 [135]. Additionally, some relevant examples have been provided for each category in Table 1.

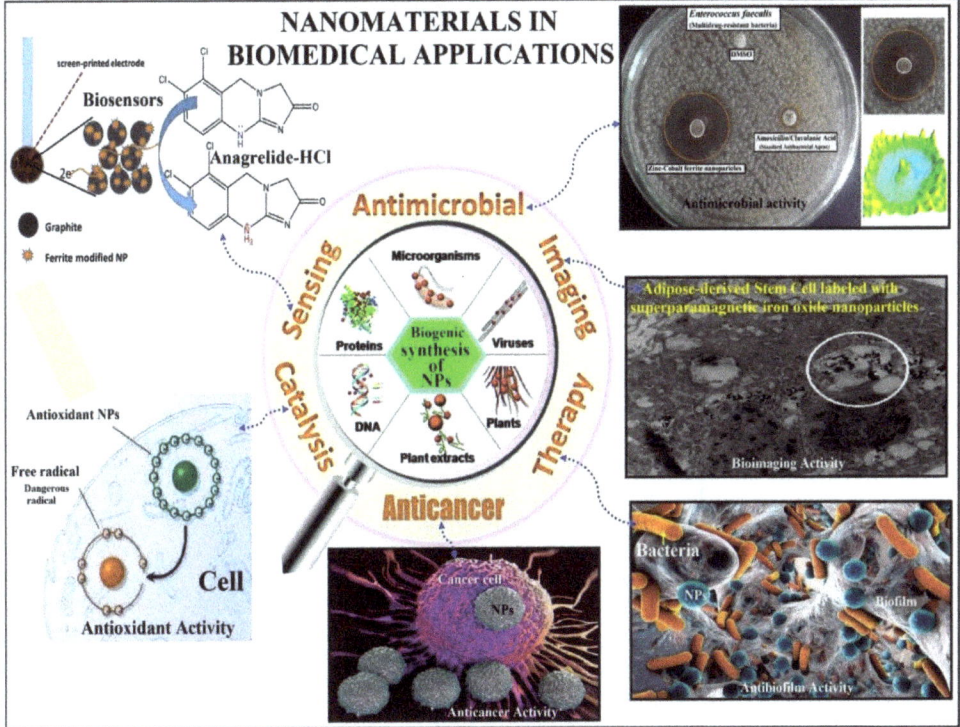

Figure 17. Summary of the applications of nanomaterials in biomedicine. Adapted from Ref. [135], copyright 2019, with permission from Elsevier.

Table 1. Applications of functionalized nanomaterials in nanomedicine.

Application	Example	Ref.
Diagnostic Imaging	X Ray	[136]
	Tomography	[137]
	Magnetic resonance imaging	[138]
	Photothermal imaging	[112]
Therapy	Drug delivery	[139]
	Gene and stem cell therapy	[140]
	Hair growth	[141]
Medical implants	Orthopaedic	[142]
	Cardiovascular	[143]
	Neurological	[144]
	Dental	[145]
Tissue Engineering	Bone	[54]
	Cartilage	[146]
Anticancer	Paclitaxel	[108]
	DOX	[60]
	Docetaxel	[30]
	Gambogic acid	[147]
Sensing	Glucose	[106]
	Insulin	[148]
	Metabolic biomarkers	[38]
Antimicrobial and Antiviral	Streptomycin, penicillin	[149]
	Coronavirus	[107]
	E. coli	[104]
	Airborne viruses	[150]

6.1. Diagnostic Implications of Functionalized Nanomaterials

Nanomaterials are extensively employed in imaging modes, such as optical coherence tomography and MRI. QDs are semiconductor nanocrystals commonly employed in optical imaging [151]. Imamura et al. [152] used PbS QDs for noninvasive scanning of septic encephalopathy in mice, suggesting that these nanomaterials can be used to image a variety of vascular systems. NIR fluorescence imaging of the mouse brain during therapy with Pbs QDs is shown in Figure 18 [152]. Before administration of QDs, only low-intensity NIR fluorescence signals were distinguished (Figure 18b, middle), due to the extremely low background fluorescence in this spectral zone. When QDs were intravascularly inserted into the mouse, the fluorescence signals arising from the mouse head augmented, and the vascular structure of cerebral blood vessels became visible (Figure 18b, right).

The development of nanoparticles with fluorescence characteristics for in vivo imaging is currently in progress. Because silicon nanocrystals are cell-safe, abundant, and more appealing than QDs [153], they do not necessitate a dense surface coating to protect the nanocrystal center from oxidation and the environment.

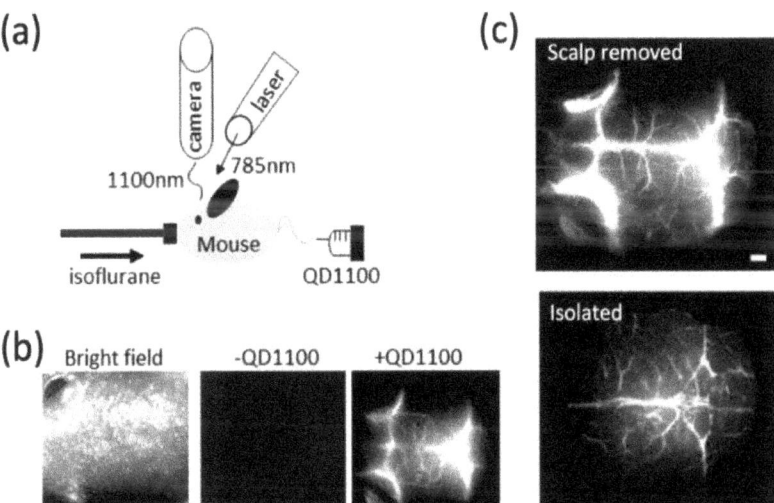

Figure 18. (a) Setup for NIR fluorescence imaging of cerebral arteries. (b) Imaging of a mouse head. Bright field micrograph (**left**), NIR fluorescence image without (**middle**), and with QDs (**right**). (c) NIR fluorescence pictures of cerebral blood vessels. The upper image shows the fluorescence after the scalp has been removed, whereas the lower micrograph shows the fluorescence after separation—with one-millimeter scale bars. Taken from Ref. [152], copyright 2016, with permission from MDPI.

6.2. Therapeutic Applications of Functionalized Nanomaterials

Magnetic nanoparticles, AuNPs, and CNTs have been utilized in the field of biomedicine. The application of NPs in postoperative treatment has attracted the attention of many researchers [153]. The use of superparamagnetic Fe_3O_4, GO, and doxorubicin-incorporated nanofibers has been claimed to reduce the localized regression of breast cancer and develop tissue regeneration [60]. A functionalized peptide that provides specific drug delivery possibilities with improved drug permeability, noteworthy aggregation in the desired target, and high therapeutic efficacy can help with the liposomal formulation in cancer treatment [20]. Docetaxel is a widely used anticancer chemotherapy drug, and transferrin is a blood–plasma glycoprotein that plays a key role in iron metabolism. Fernandes et al. [30] synthesized docetaxel-loaded liposomes functionalized with transferrin (LIP-DTX-TF), and their effects on prostate neoplasms were studied. TEM images demonstrated that the systems were spherical and nanometric in size (Figure 19a) and that the presence of DTX aided in vesicle size reduction, resulting in improved liposome stability (Figure 19b).

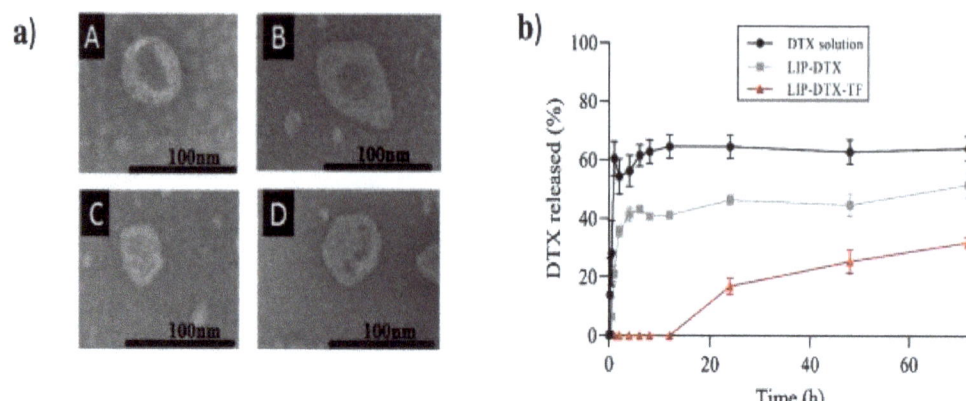

Figure 19. (a) Transmission Electron Microscopy of liposomes: (**A**) empty liposomes (LIP); (**B**) docetaxel-loaded liposomes (LIP-DTX); (**C**) empty transferrin functionalized liposomes (LIP-TF); and (**D**) docetaxel-loaded liposomes functionalized with transferrin (LIP-DTX-TF). (**b**) In vitro release profile of free DTX and encapsulated in liposomes in PBS buffer pH 7.4. Adapted from Ref. [30], copyright 2021, with permission from Elsevier.

7. Functionalized Nanomaterials: Drug/Gene Delivery

Nanomaterials can be functionalized for different purposes, including drug delivery carriers or therapeutic agents for cure and treatment, diagnostic applications in biological imaging, cell labeling, biosensors, and the use of moieties for medical devices such as stents or lenses [154–156]. Functionalization can improve biocompatibility and uptake efficiency and simultaneously minimize immune system activation, increasing the material's bioavailability inside the body. These modifications are beneficial for some drug delivery strategies to ensure that the appropriate doses of the drug are released to the correct area while limiting the detrimental effects of drug molecules on other organs [157]. Drug delivery systems are necessary to improve the efficacy of drug biodistribution. Nanomaterials have been used to carry drugs and genes in passive, active, and direct methods [49,158,159]. Due to the small size of nanoparticles, they can pass across cellular membranes and boundaries. Moreover, the increased surface-to-volume ratio of nanoparticles leads to improved drug loading [160]. Figure 20 displays biological ligands used for active targeting of NP drug carriers [161], and Table 2 summarizes different functionalized nanomaterials applied in drug/gene delivery.

Table 2. Functionalized nanomaterials used for drug/gene delivery.

Nanomaterial	Function	Size (nm)	Drug/Gene	Target Organ & Indication	Ref.
Porous $CaCO_3$	Intranasal drug carrier	2000–3200	Insulin	Postprandial hyperglycemia in diabetes	[162]
$CaCO_3$ NPs	Drug/gene delivery	116	Ciprofloxacin HCl	S. Aureus	[163]
$CaCO_3$	Drug delivery	40–200	Hydrophilic drugs and bioactive proteins (validamycin)	Inflamed region	[164]
Cationic NPs	Gene delivery	50–100	Raf gene, ATP^{μ}-Raf	Angiogenic blood vessels (tumor-bearing mice)	[165]
Fe_3O_4@GO	Drug release and antitumor therapy	200–1000	Hybrid microcapsule	Tumor cells targeting	[166]

Table 2. Cont.

Nanomaterial	Function	Size (nm)	Drug/Gene	Target Organ & Indication	Ref.
GO flakes	Drug release	1000–2000	DOX microcapsules	-	[167]
AuNPs	Drug delivery	100	–	Nasopharyngeal carcinoma cells	[168]
FA-Au-FITC [1]	Drug delivery for cancer therapy	4–7	DOX	Cytoplasm	[169]
HLA [2]-Si/Fe_3O_4 NPs	Drug delivery for cancer therapy	40–110	DOX	Tumor tissues	[170]
Fe_3O_4-SA-PVA-BSA [3]	Drug delivery	240–460	DOX	Cancer cells	[171]
CS-HYL-5-FU-PEG-G [4]	Drug delivery	300–580	COLO-205 and HT-29 colon	Cancer cells	[172]
SA/PVA/Ca [5]	Drug delivery system	500–1000	Diclofenac sodium	-	[173]
PLGA [6]-Fe_3O_4	Drug delivery system	67	5-Fluorouracil	Prostate carcinoma cell	[113]
HLA-Nanoemulsion	Drug delivery system	–	Ciprofloxacin	-	[174]
Fe_3O_4	Drug delivery system	20	Gambogic acid	Capan-1 pancreatic cancer cells	[147]
PLGA-Fe_3O_4 NPs	Intratumoral drug delivery	200–300	DOX	Murine Lewis lung carcinoma cells	[175]
Fe_3O_4 conjugate oleate/oleylamine	Drug release	12	Chromone	HeLa cells	[176]
Fe_3O_4/DPA-PEG-COOH [7]	Drug delivery	9	Dextran, PEG	Macrophage Cells	[177]
Thiolated starch-coated Fe_3O_4	Drug delivery	40–50	Isoniazid	Human body cells	[178]
Zn-doped Fe_3O_4 nano-octahedral core	Drug delivery	10–20	DOX and HSP70/HSP90 siRNAs	Tumor cells	[138]
Arginine-NCQDs [8]	Gene delivery	6–11	EGFP gene	Mammalian cells	[179]

[1] Folic acid-coated gold nanoparticles conjugated with a fluorophore; [2] Hyaluronic acid-modified mesoporous silica-coated Fe_3O_4 NPs; [3] Fe_3O_4 nanoparticles coated with a mixture of sodium alginate (SA), polyvinyl alcohol (PVA), and bovine serum albumin (BSA); [4] Polyethylene glycol-gelatin-chitosan-hyaluronidase-5-fluorouracil; [5] Sodium alginate/polyethylene glycol (vinyl alcohol); [6] Poly(lactic-co-glycolic acid); [7] Dopamine-polyethylene glycol-carboxylic acid; [8] Nitrogen-doped carbon quantum dots.

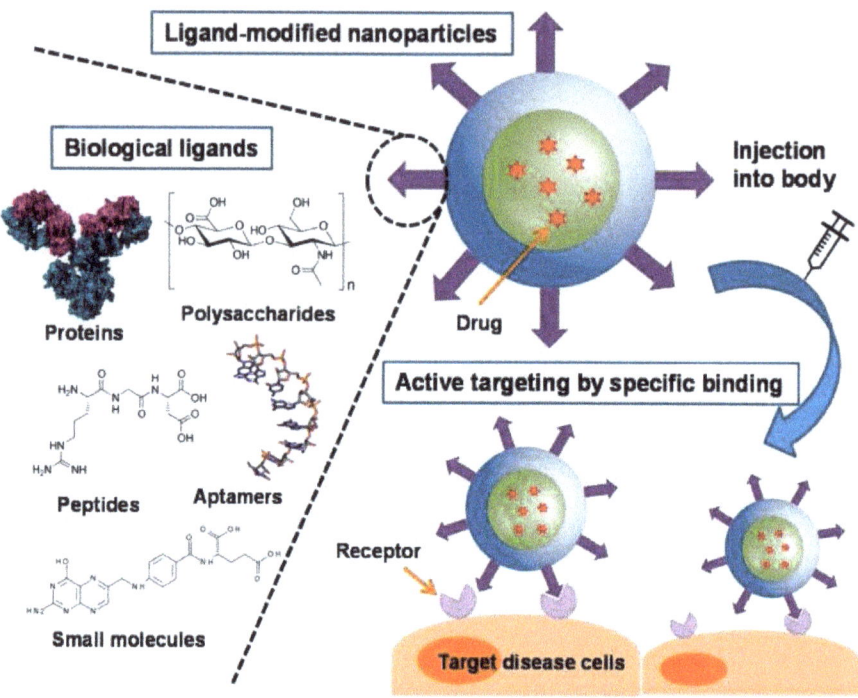

Figure 20. Illustration of biological ligands for active targeting of nanoparticle drug carriers. Taken from Ref. [161], copyright 2019, with permission from MDPI.

8. Functionalized Nanomaterials: Regenerative Medicine

Reparative and restorative medicine and nanotechnology have gained popularity in recent years, resulting in significant improvements in medical research and clinical practice [180]. Tissue engineering, cell therapy, diagnostics, medication, and gene delivery are examples of regenerative medicine applications that use various functionalized nanoscale materials [24,181,182]. Restorative medicine is a vast field of nanotechnology that strives to regenerate cells and tissues similar enough to their original design and function. Three main types of therapeutic techniques can be found in regenerative medicine: tissue-engineering treatments based on cells; biomaterials; and a combination of the two. Stem cell biology, nanotechnology, and bioengineering have progressed significantly, potentially paving the way for real regenerative medicine for various diseases [183]. Stem cells are known for their capacity to maintain their differentiating potential while intersecting to generate numerous daughter cells. Such daughter cells lack "stem-ness" and use controlled proliferation to produce adult cells of all origins throughout the body (self-renewal) [184]. Using tissue-specific or therapeutic genes, as well as primary cells that overexpress these genes, genetically engineered cell treatment can manufacture proteins with a therapeutic intent, to be used at regeneration platforms or discriminate new cells into the appropriate cellular lineage, assisting in tissue restoration [185].

When bone is formed, it comprises mostly collagen fibers and calcium phosphate, which is converted into hydroxyapatite (HDA). Bone tissue also contains several cellular structures, such as osteoblasts, osteocytes, and osteoclasts, which contribute to its calcification [186]. For bone repair, nanoscaffolds with adequate biophysical characteristics, such as stiffness and cell proliferation, have been employed. A variety of nanofiber matrices have been synthesized in recent years. The vast majority of nanoscaffolds are developed to match genuine bone's structural, compositional, and biological features [187]. Zhang et al. [188] prepared a chitosan/HDA biomimetic nanocomposite scaffold for assessing the effect of bone marrow MSC mesenchymal stem cells (BMSCs) growth, and explored the molecular mechanism both in vivo and in vitro. It was reported that this hybrid scaffold could encourage the proliferation of BMSCs and trigger the integrin-BMP/Smad signal pathway of BMSCs. In addition, HAD can also be used combined with other polymeric materials such as PEG, PCL, and PLGA, which have displayed improved effects in bone regeneration/repair (Figure 21) [189,190].

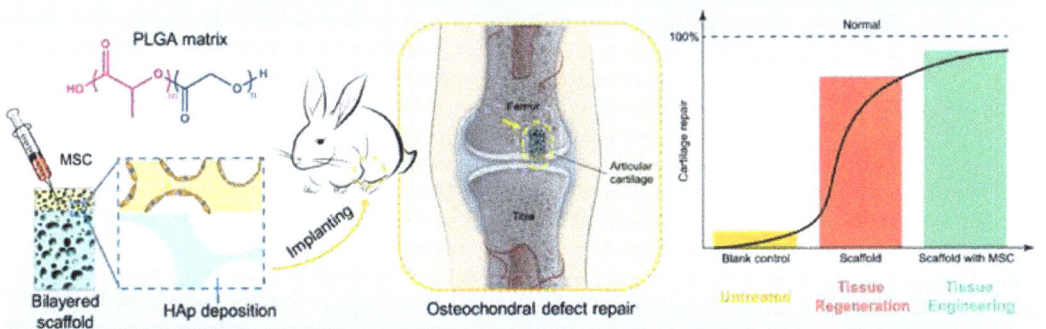

Figure 21. Illustration of hydroxyapatite-based scaffold-induced regeneration of bone. Reprinted from Ref. [189], copyright 2018, with permission from the American Chemical Society.

Besides nano-HDA, collagen, electrospun silk, anodized titanium, and nanostructured titanium surfaces are some of the primary constituents of materials that mimic the bone extracellular matrix [191]. In primary osteoblasts, nanofibers have improved osteogenesis and biomineralization. Main osteoblasts are limited in their application due to (i) restricted accessibility and intrinsic donor site malady; (ii) limited scaling capability; (iii) age-related behavior; or (iv) possibility of dedifferentiation occurring during in vitro cultivation [94].

Transient gene delivery [192], cell therapy without the need for genetic modification [193], and genetically modified cells [194] are currently three of the most exciting new procedures in the field of tissue engineering. Gene delivery is a therapeutic approach to introducing foreign genetic material directly into host cells in vivo. These genes immediately affect the host tissue, causing it to remodel [195].

Achieving better cell adhesion, motility, and differentiation through nanomedicine is possible thanks to the development of interfaces, components, and substances that mimic the cells' natural environment. Scientists have developed complex tissue/organ constructions by combining stem cells with scaffolds and stimulating factors as the basis of their tissue-engineering experiments [145,196]. Some of these are currently being utilized therapeutically as part of the standard treatment for various disorders. Scaffolds are transformed into three-dimensional structures that have the appropriate shape, size, architecture, and physical properties for different applications and environments. For this reason, tissue-engineering products are designed to look and behave like natural tissues. In addition to biocompatibility and controllable porosity and permeability, important scaffold characteristics include mechanical and degradation kinetics comparable to those of the desired tissue and support for cell adhesion and proliferation by adding nanotopographies to the biomaterial surface [197].

Biodegradability is a critical property that nanoparticles must possess to be employed safely inside the body. This is a crucial aspect to consider when building scaffolds for tissue engineering and reparative and restorative medicine [198]. Table 3 summarizes functionalized nanomaterials that have been utilized in tissue engineering.

Table 3. Functionalized nanomaterials utilized in tissue engineering.

Nanomaterial	Function	Size (nm)	Tissue	Purpose & Outcomes	Ref.
PEG-GO	Tissue engineering	50	Bone	Improved thermal stability, hydrophilicity, water absorption, biodegradation, mechanical, viscoelastic, and antibacterial properties	[54]
Oxidized alginate/gelatin hydrogel	Tissue regeneration	100–200	Cartilage regeneration for the treatment of osteoarthritis	Usefulness of the hydrogel in encouraging cellular migration and proliferation	[146]
OCMC [1]	Tissue engineering	2000–4000	BALB/c3T3 cells in rates	Biocompatibility, spinnability of hydrogel through electrospinning	[199]
Pd/PPy/rGO NC [2]	Tissue engineering	2–4	Bone	Biocompatibility, osteoproliferation, and bacterial infection prevention	[200]
3D macro-rGO/PPY	Bone tissue engineering	100–400	Backbone	Casein phosphopeptide as bioactive for bone engineering, osteoblastic performance, biological properties	[201]
Chitosan-ZnO	Soft tissue engineering	180		Improved hydrophilicity, porosity, water absorption, oxygen permeability, biodegradability, antibacterial and wound healing	[202]
Biphasic Calcium Phosphate	Bone tissue engineering	1–2	MG63 cells	Micropores and collagen coating influence cellular function, in vitro cellular behavior, scaffold–osteoblast interactions	[203]
AuNPs/glass-ceramic matrix	Bone tissue engineering	5–10	Bone	In vitro hydroxyapatite synthesis, controlled release of gold species, biocompatibility, and antibacterial activity of AuNPs	[91]
AuNPs	Tissue engineering	20	Rat brain	AuNP biochemical effects on the rat brain, biomarkers of AuNP toxicity	[204]
AuNPs	Tissue Engineering	10–50	Cardiac tissue	Effects of AuNPs on the histological deformities of rat heart tissue, toxicity, therapeutic and diagnostic potential of NPs, and their interaction with proteins and other cells	[205]
AuNPs	Tissue engineering	30 nm	Subsets of cells in human organs	NP toxicity in human blood, hemolysis, development of ROS [3], platelet condensation in cell subsets	[206]
AuNPs/polymeric coatings	Tissue engineering	18, 35, 65	Endothelial cells from human dermis	NP toxicity, uptake behavior, and uptake quantification	[207]
Bioactive glass scaffolds	Tissue engineering	50–100	Bone repair	Osteoblastic cells for bone reconstruction	[208]
$Na_2Ca_2Si_3O_9$	Bone tissue engineering	500	Bone	Bioactive and biodegradable scaffold effects, mechanical support	[209]
Bioactive glass-ceramics/apatite	Bone tissue engineering	8–20	Bone	Crystallization rate of bioactive glasses on the kinetics of HAD formation	[90]
$Ca_{10}(PO_4)_6(OH)_2$	Bone tissue engineering	1000–2000	Trabecular bone	Extent and nature of carbonate substitution on HDA	[93]

Table 3. *Cont.*

Nanomaterial	Function	Size (nm)	Tissue	Purpose & Outcomes	Ref.
GO/Chitosan Scaffold	Cardiac tissue	—	Cardiac tissue	Investigate cell survival, cell adhesion, development of intercellular networks, genes, and proteins expression	[210]
GO/Chitosan Scaffold	Cartilage repair	35–60	Cartilage tissue	Nanocomposite effect on human tissue, effects of GO	[211]
GO-coated collagen scaffolds	Tissue engineering	—	Mouse osteoblastic MC3T3-E1 cells	Influence of the GO coating on cell growth and differentiation, biocompatibility and biodegradability of collagen scaffolds, bioactivity studies	[212]
Nanocrystalline apatite/AuNPs	Tissue engineering	2–25	Bone tissue reconstruction	Toxicity of NPS in simulated physiological fluid	[66]

[1] Gelatin – oxidized carboxymethyl cellulose. [2] Nanocrystalline cellulose. [3] Reactive oxygen species.

9. Functionalized Nanomaterials: Cancer Therapy

Theranostic nanoprobes for tumors and malignancies have become a prominent focus of research since NP functionalization has been able to be used simultaneously in diagnostic and therapeutic purposes. Surface modification of NPs has been proven to generate targeted accumulation in tumor tissue due to the enhanced permeability and retention (EPR) effect [29,213]. Tumors have more permeable vasculature, a poorly defined lymphatic system, and various substances that aid in increased targeting, as contrasted to normal tissue, such as VEGF and basic fibroblast growth factor [214]. In cancer immunotherapies, NPs can keep track of critical immune cells during metastasis. Different tumor ablation therapies with magnetic NPs such as Fe_3O_4 have been reported [215] (Figure 22): (a) Magnetic hyperthermia, in which an alternating magnetic field induces NPs to produce heat, boosting tumor necrosis. (b) Photothermal ablation, in which the light absorbed by the NPs is transformed into thermal energy, producing cell death in the neighborhood. (c) Photodynamic therapy, in which photosensitizing agents anchored to NPs are activated via an external light source to make singlet oxygen species that are cytotoxic to cells. As a result, NPs have a high level of target-cell selectivity [216]. Table 4 displays functionalized nanomaterials that have been utilized for cancer treatment.

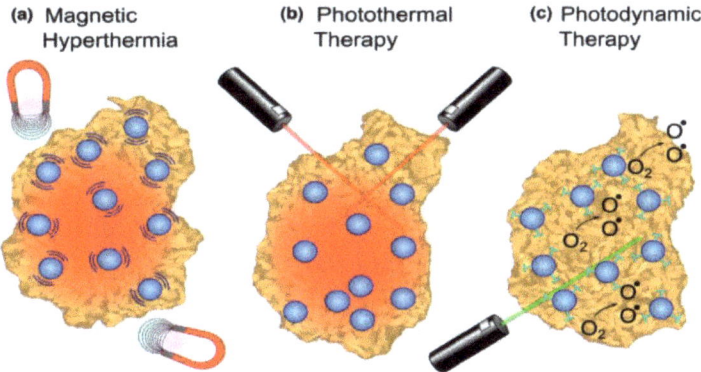

Figure 22. Schematic representation of tumor ablation therapies with iron oxide nanoparticles (NPs). Reproduced from Ref. [215], copyright 2016, with permission from Elsevier.

Table 4. Functionalized nanomaterials used for cancer therapy.

Nanomaterial	Functionalization Agent	Size (nm)	Drug	Purpose & Outcomes	Ref.
ZnO NPs	PBA	40	Curcumin	High drug-loading and release rates, in vitro and in vivo antitumor efficacy	[88]
AuNP's	Beta-cyclodextrin with PEG, biotin, PTX, rhodamine B	30–50	PTX	Cytocompatibility, stability, and biomolecule binding ease	[217]
SPION	5TR1 Aptamer	57	Epirubicin	Magnetic resonance (MR) traceability, nontoxicity, increased permeability, retention effect	[82]
Fe_3O_4 NPs	Glycerol monooleate	144	PTX, rapamycin, alone or combined	Intravenous administration of hydrophobic drugs	[218]
rGO [1]	Fe_3O_4 NPs	54.8	Camptothecin	pH-responsive drug release profile, good biocompatibility, excellent photodynamic	[219]
Fe_3O_4 MNPs+ PLGA	citric acid	130–140	DOX, verapamil	Loading hydrophilic and hydrophobic drugs	[220]
MSN [2]	β-cyclodextrin with hydroxyl, amino, and thiol groups	75.5	DOX	Higher mucoadhesive on the urothelium	[221]
rGO	HA-PEG-g-poly(dimethylaminoethyl methacrylate)	120–190	-	Biocompatibility, in vitro cellular uptake sensitive to cancer cells	[222]
MSN	Galactose	277	Camptothecin	MSN targeting to cancer cells	[59]
rPEI- Cdots [3]	FA	143	-	Biocompatible, good siRNA gene delivery carrier	[223]
PLGA NPs	bis(sulfosuccinimidyl) suberate (BS3)	184	Curcumin	Promote the loading of low-soluble drugs and aid in sustained released	[114]
ZnO NPs	PBA	414	Curcumin	Curcumin distribution to the sialic acid is much easier by PBA conjugation	[88]
Se NPs	(Arg–Gly–Asp–d-Phe–Cys [RGDfC]) cyclic peptide	18	DOX	Antitumor efficacy in vivo, effective cellular uptake A549	[224]
CuO NPs	FA, starch	108.83	Cytochrome C	Antioxidants, anticancer, antimicrobial, drug-carrier	[35]
MoS_2	FA, BSA	133	DOX	Excellent photothermal conversion ability	[225]

[1] Reduced graphene oxide. [2] Mesoporous silica nanoparticles. [3] Reducible polyethyleneimine passivated carbon dots.

10. Functionalized Nanomaterials: Medical Implants

Recently, the influence of nanotechnology on the implant field has increased strongly. Nanomaterials with biological-inspired structures are motivating scientists to investigate their potential for enhancing the performance of conventional implants [91,142]. Nanotechnology has the skill to economically substitute many traditional implants and offer numerous novel applications. It can result in more efficient and longer-lasting implants, with reduced infection rates and enhanced bone or tendon healing. In orthopedics, the goal of biomaterials is to substitute injured bone. Improved mechanical properties (e.g., strength, flexibility, hardness, elastic modulus), wear, hydrolysis and corrosion resistance, biocompatibility, osseointegration, bioinertness, and ease of surgical application are required properties to be used in orthopedics [226,227]. Nanomaterials offer an enlarged surface area, a superior stiffness, and a high roughness that can improve the adhesion and proliferation of bone-related proteins and the deposition minerals incorporating Ca [228]. Besides, FMNs can mimic the amounts of the components of natural bones and can aid in sustaining biologically active growth factors and exploit the potential of BMSCs. Numerous studies [229] have been developed to examine the optimal surface properties of FMNs that may support or assist specific protein adsorption, improved osteoblast anchoring, osteoblast differentiation, and new bone formation (Figure 23) [142].

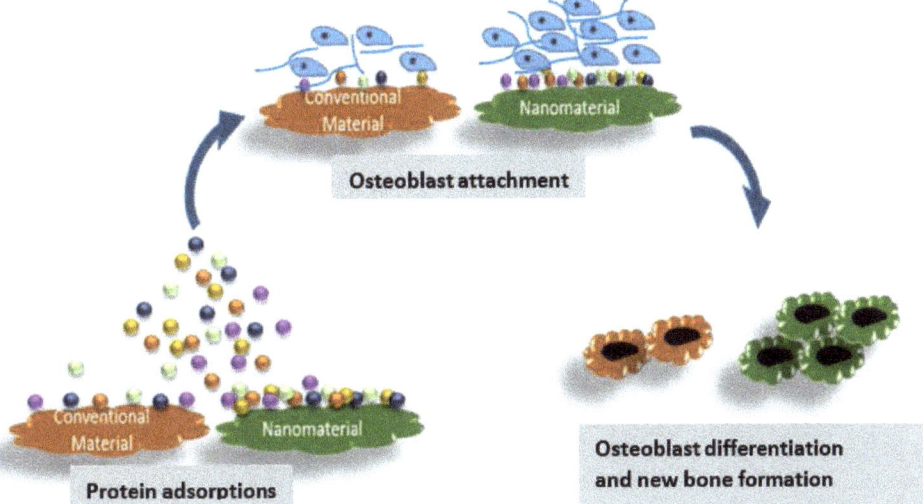

Figure 23. Comparison of bone regeneration using nanomaterials and traditional materials. Nanomaterials show improved protein adsorption, osteoblast anchoring, and differentiation compared to traditional materials. Reproduced from Ref. [142], copyright 2020, with permission from Elsevier.

Surface adjustment of nanomaterials is a prospective method to expand the performance and durability as well as to reduce the hazardous side effects that might take place during implant degradation. Surface characteristics have a key role on modulating biological interactions. Specifically, engineered nanomaterials can have a significant impact on molecular and cellular actions; this issue aids in conditioning the comprehensive biological response of an implant (i.e., protein adsorption, cell adhesion, and proliferation). Therefore, several approaches have been settled to modify nanomaterials for orthopedic implants such as anodic oxidation [230], plasma electrolytic oxidation [231], electrochemical plating [232], chemical conversion coating [233], physical vapor deposition, laser surface alloying [234], thermal spraying [235], organic coating [236], and so forth. These methods provide new implant surfaces with tailorable characteristics at the nanoscale. The particular procedure can be chosen based on different factors/goals, including to attain complex geometries and to be suitable for large-scale processing. Metal oxide NPs such as TiO_2, ZrO_2, and Al_2O_3 have been used as nanocoatings to enhance the mechanical and biochemical properties of conventional metallic implants [237].

11. Conclusions

Nanotechnology has opened up vast techniques to manipulate and transform the current medical devices or materials utilized for therapy in biomedical sciences and engineering. Numerous nanomaterials can be used in biomedical applications, both organic, such as CNTs, GO, GQDs, and polymeric NPs, and inorganic, such as metallic NPs (Au, Ag), metal oxide NPs (TiO_2, Fe_3O_4, mesoporous SiO_2), and ceramic (HAD, $CaCO_3$). Over the last years, numerous approaches have been developed to synthesize surface-engineered nanomaterials, in particular NPs, for drug/gene delivery, diagnostics, cancer therapy, tissue engineering, and medical implants, and the structure–function relationship of these functionalized nanoparticles has been widely explored. The NPs' surface modification is a potent strategy to improve biocompatibility and uptake, as corroborated by the huge quantity of scientific documents published on this subject. Investigations prove that the conjugation of polymers, biomolecules, and small ligands on the NP surface can successfully increase biocompatibility both in vivo and in vitro, due to the alteration of surface charge and to the inactivation of sensitive functional groups that can influence the stability of the cell membrane. Besides, the incorporation of certain molecules can improve NPs' passive and active uptake, reducing systemic toxicity in vivo and enabling high precision therapy and/or diagnosis. The binding of functionalization agents on the NP surface can be achieved via covalent and noncovalent tactics. The first is broadly used to link proteins, antibodies, aptamers, and peptides utilized to boost uptake and to achieve active targeting, whereas the second is frequently used for the loading of drugs and other molecules that need to be liberated into the cells. The promise of tissue and organ-specific regeneration therapy has become a reality due to major advances in regenerative medicine and nanomedicine over the previous decade. Preliminary clinical results have shown that functionalization of NPs with specific recognition surface moieties results in improved efficacy and reduced side effects, due to properties such as directed localization in tumors and active cellular uptake. Even though remarkable improvements have been attained, this research arena is still in its early stages, and significant efforts are needed in order to be able to scale up the functionalization approaches developed at the laboratory level and make them reproducible. A prerequisite for progressing in this research area is the development of novel chemical methods to conjugate chemical moieties onto NPs in a safe and consistent manner. In addition, smart and innovative nano-based technologies can offer particular physicochemical properties that could aid in fixing crucial issues associated with the treatments of viral infections such as SARS-CoV-2. Researchers may find this study valuable in analyzing past studies on the topic matter to attain commercial success.

Funding: This research received no external funding.

Institutional Review Board Statement: Not applicable.

Informed Consent Statement: Not applicable.

Data Availability Statement: This article's data sharing is not applicable as no new data were created or analyzed in this study.

Conflicts of Interest: The author declares that there is no conflict of interest regarding the publication of this article.

Abbreviations

aniline functionalized graphene quantum dots	a-GQDs
aptameric functionalized materials	AFMs
carbon nanotubes	CNTs
carbon quantum dots	CQDs
cationic β-cyclodextrin-based polymer	CCD/P
docetaxel-loaded liposomes functionalized with transferrin	LIP-DTX-TF
dopamine	DOPA
dopamine-polyethylene glycol-carboxylic acid	DPA-PEG-COOH
enhanced permeability and retention	EPR
extra-cellular matrix	ECM
folate receptor	FR
folic acid	FA

folic acid-coated gold nanoparticles conjugated with fluorophore	FA-Au-FITC
functionalized nanoparticles	FNPs
gold nanoparticles	AuNPs
graphene oxide	GO
hyaluronic acid	HLA
hydroxyapatite	HDA
lactate dehydrogenase	LDH
poly(lactic-co-glycolic acid)	PLGA
magnetic nanoparticles	MNPs
magnetic resonance imaging	MRI
mesoporous silica nanoparticles	MSN
methacryl polyhedral oligomeric silsesquioxane	M-POSS
nanoparticles	NPs
near-infrared	NIR
nitrogen-doped carbon quantum dots	NCQDs
nitrogen-doped graphene quantum dots	N-GQDs
oleic acid-coated iron oxide NPs	Fe_3O_4-OA
phenyl boronic acid	PBA
phenylethylamine	PhEA
photodynamic therapy	PDT
photothermal	PT
piperine	PIP
poly(acrylic acid)	PAA
polyethylene glycol	PEG
polyethylene glycol with thiol end group	thiol-PEG
polyethylene glycol-gelatin-chitosan-hyaluronidase-5-fluorouracil	CS-HYL-5-FU-PEG-G
positron emission tomography	PET
quantum dots	QDs
receptor-mediated endocytosis	RME
reticulum endothelial system	RES
sodium alginate (SA)–polyvinyl alcohol (PVA)–bovin serum albumin	SA-PVA-BSA
sodium alginate/polyethylene glycol (vinyl alcohol)	SA/PVA/Ca
superparamagnetic iron oxide NPs	SPIONs
tetraethylorthosilicate	TEOS
ultra-small superparamagnetic iron oxide NP	USPIONs

References

1. Patra, J.K.; Das, G.; Fraceto, L.F.; Campos, E.V.R.; del Pilar Rodriguez-Torres, M.; Acosta-Torres, L.S.; Diaz-Torres, L.A.; Grillo, R.; Swamy, M.K.; Sharma, S.; et al. Nano based drug delivery systems: Recent developments and future prospects. *J. Nanobiotechnol.* **2018**, *16*, 71. [CrossRef] [PubMed]
2. Bilal, M.; Qindeel, M.; Raza, A.; Mehmood, S.; Rahdar, A. Stimuli-responsive nanoliposomes as prospective nanocarriers for targeted drug delivery. *J. Drug Deliv. Sci. Technol.* **2021**, *66*, 102916. [CrossRef]
3. Rauf, A.; Tabish, T.A.; Ibrahim, I.M.; Hassan, M.R.U.; Tahseen, S.; Sandhu, M.A.; Shahnaz, G.; Rahdar, A.; Cucchiarini, M.; Pandey, S. Design of Mannose-Coated Rifampicin nanoparticles modulating the immune response and Rifampicin induced hepatotoxicity with improved oral drug delivery. *Arab. J. Chem.* **2021**, *14*, 103321. [CrossRef]
4. Hong, E.J.; Choi, D.G.; Shim, M.S. Targeted and effective photodynamic therapy for cancer using functionalized nanomaterials. *Acta Pharm. Sin. B* **2016**, *6*, 297–307. [CrossRef]
5. Rahdar, A.; Hajinezhad, M.R.; Hamishekar, H.; Ghamkhari, A.; Kyzas, G.Z. Copolymer/graphene oxide nanocomposites as potential anticancer agents. *Polym. Bull.* **2021**, *78*, 4877–4898. [CrossRef]
6. Zhang, G.; Khan, A.A.; Wu, H.; Chen, L.; Gu, Y.; Gu, N. The Application of Nanomaterials in Stem Cell Therapy for Some Neurological Diseases. *Curr. Drug Targets* **2018**, *19*, 279–298. [CrossRef]
7. Theus, A.S.; Ning, L.; Jin, L.; Roeder, R.K.; Zhang, J.; Serpooshan, V. Nanomaterials for bioprinting: Functionalization of tissue-specific bioinks. *Essays Biochem.* **2021**, *65*, 429–439. [CrossRef]
8. Lloyd, J.R.; Byrne, J.M.; Coker, V.S. Biotechnological synthesis of functional nanomaterials. *Curr. Opin. Biotechnol.* **2011**, *22*, 509–515. [CrossRef]
9. Díez-Pascual, A.M. Hot Topics in Macromolecular Science. *Macromol* **2021**, *1*, 173–176. [CrossRef]
10. Díez-Pascual, A.M. Nanoparticle reinforced polymers. *Polymers* **2019**, *11*, 625. [CrossRef]

11. Kobayashi, K.; Wei, J.; Iida, R.; Ijiro, K.; Niikura, K. Surface engineering of nanoparticles for therapeutic applications. *Polym. J.* 2014, *46*, 460–468. [CrossRef]
12. Sanità, G.; Carrese, B.; Lamberti, A. Nanoparticle Surface Functionalization: How to Improve Biocompatibility and Cellular Internalization. *Front. Mol. Biosci.* 2020, *7*, 587012. [CrossRef] [PubMed]
13. Díez-Pascual, A.M. Effect of Graphene Oxide on the Properties of Poly(3-Hydroxybutyrate-co-3-Hydroxyhexanoate. *Polymers* 2021, *13*, 2233. [CrossRef] [PubMed]
14. Razzaq, S.; Rauf, A.; Raza, A.; Akhtar, S.; Tabish, T.A.; Sandhu, M.A.; Zaman, M.; Ibrahim, I.M.; Shahnaz, G.; Rahdar, A.; et al. Multifunctional Polymeric Micelle for Targeted Delivery of Paclitaxel by the Inhibition of the P-Glycoprotein Transporters. *Nanomaterials* 2021, *11*, 2858. [CrossRef]
15. Rahdar, A.; Hasanein, P.; Bilal, M.; Beyzaei, H.; Kyzas, G.Z. Quercetin-loaded F127 nanomicelles: Antioxidant activity and protection against renal injury induced by gentamicin in rats. *Life Sci.* 2021, *276*, 119420. [CrossRef]
16. Er, S.; Laraib, U.; Arshad, R.; Sargazi, S.; Rahdar, A.; Pandey, S.; Thakur, V.K.; Díez-Pascual, A.M. Amino Acids, Peptides, and Proteins: Implications for Nanotechnological Applications in Biosensing and Drug/Gene Delivery. *Nanomaterials* 2021, *11*, 3002. [CrossRef]
17. Mrówczyński, R.; Grześkowiak, B.F. Biomimetic Catechol-Based Nanomaterials for Combined Anticancer Therapies. *Nanoeng. Biomater. Biomed. Appl.* 2022, *2*, 145–180. [CrossRef]
18. Jamir, M.; Islam, R.; Pandey, L.M.; Borah, J. Effect of surface functionalization on the heating efficiency of magnetite nanoclusters for hyperthermia application. *J. Alloy. Compd.* 2021, *854*, 157248. [CrossRef]
19. Nejati, K.; Dadashpour, M.; Gharibi, T.; Mellatyar, H.; Akbarzadeh, A. Biomedical Applications of Functionalized Gold Nanoparticles: A Review. *J. Clust. Sci.* 2021, *33*, 1–16. [CrossRef]
20. Sonju, J.J.; Dahal, A.; Singh, S.S.; Jois, S.D. Peptide-functionalized liposomes as therapeutic and diagnostic tools for cancer treatment. *J. Control. Release* 2021, *329*, 624–644. [CrossRef]
21. Jazayeri, M.H.; Amani, H.; Pourfatollah, A.A.; Pazoki-Toroudi, H.; Sedighimoghaddam, B. Various methods of gold nanoparticles (GNPs) conjugation to antibodies. *Sens. Bio-Sens. Res.* 2016, *9*, 17–22. [CrossRef]
22. Li, H.; Wang, Q.; Liang, G. Phase Transfer of Hydrophobic Nanoparticles Functionalized with Zwitterionic Bisphosphonate Ligands for Renal-Clearable Imaging Nanoprobes. *ACS Appl. Nano Mater.* 2021, *4*, 2621–2633. [CrossRef]
23. Karthik, V.; Selvakumar, P.; Kumar, P.S.; Vo, D.-V.N.; Gokulakrishnan, M.; Keerthana, P.; Elakkiya, V.T.; Rajeswari, R. Graphene-based materials for environmental applications: A review. *Environ. Chem. Lett.* 2021, *19*, 3631–3644. [CrossRef]
24. Díez-Pascual, A.M.; Diez-Vicente, A. L Antibacterial SnO2 nanorods as efficient fillers of poly(propylene fumarate-co-ethylene glycol) biomaterials. *Mater. Sci. Eng. C* 2017, *78*, 806–816. [CrossRef]
25. Alshamrani, M. Broad-Spectrum Theranostics and Biomedical Application of Functionalized Nanomaterials. *Polymers* 2022, *14*, 1221. [CrossRef]
26. Xia, Q.; Huang, J.; Feng, Q.; Chen, X.; Liu, X.; Li, X.; Zhang, T.; Xiao, S.; Li, H.; Zhong, Z.; et al. Size- and cell type-dependent cellular uptake, cytotoxicity and in vivo distribution of gold nanoparticles. *Int. J. Nanomed.* 2019, *14*, 6957–6970. [CrossRef]
27. Hirayama, H.; Amolegbe, S.A.; Islam, M.S.; Rahman, M.A.; Goto, N.; Sekine, Y.; Hayami, S. Encapsulation and controlled release of an antimalarial drug using surface functionalized mesoporous silica nanocarriers. *J. Mater. Chem. B* 2021, *9*, 5043–5046. [CrossRef]
28. Guadagnini, R.; Halamoda Kenzaoui, B.; Walker, L.; Pojana, G.; Magdolenova, Z.; Bilanicova, D.; Saunders, M.; Juillerat-Jeanneret, L.; Marcomini, A.; Huk, A. Toxicity screenings of nanomaterials: Challenges due to interference with assay processes and components of classic in vitro tests. *Nanotoxicology* 2015, *9*, 13–24. [CrossRef]
29. Ellah, N.A.; Abouelmagd, S. Surface functionalization of polymeric nanoparticles for tumor drug delivery: Approaches and challenges. *Expert Opin. Drug Deliv.* 2017, *14*, 201–214. [CrossRef]
30. Fernandes, M.A.; Eloy, J.O.; Luiz, M.T.; Junior, S.L.R.; Borges, J.C.; De la Fuente, L.R.; Luis, C.O.-D.S.; Marchetti, J.M.; Santos-Martinez, M.J.; Chorilli, M. Transferrin-functionalized liposomes for docetaxel delivery to prostate cancer cells. *Colloids Surf. A Physicochem. Eng. Asp.* 2021, *611*, 125806. [CrossRef]
31. Rong, L.; Qin, S.-Y.; Zhang, C.; Cheng, Y.-J.; Feng, J.; Wang, S.-B.; Zhang, X.-Z. Biomedical applications of functional peptides in nano-systems. *Mater. Today Chem.* 2018, *9*, 91–102. [CrossRef]
32. Xie, S.; Ai, L.; Cui, C.; Fu, T.; Cheng, X.; Qu, F.; Tan, W. Functional Aptamer-Embedded Nanomaterials for Diagnostics and Therapeutics. *ACS Appl. Mater. Interfaces* 2021, *13*, 9542–9560. [CrossRef] [PubMed]
33. Farahavar, G.; Abolmaali, S.S.; Gholijani, N.; Nejatollahi, F. Antibody-guided nanomedicines as novel breakthrough therapeutic, diagnostic and theranostic tools. *Biomater. Sci.* 2019, *7*, 4000–4016. [CrossRef] [PubMed]
34. Mout, R.; Moyano, D.F.; Rana, S.; Rotello, V.M. Surface functionalization of nanoparticles for nanomedicine. *Chem. Soc. Rev.* 2012, *41*, 2539–2544. [CrossRef]
35. Mariadoss, A.V.A.; Saravanakumar, K.; Sathiyaseelan, A.; Venkatachalam, K.; Wang, M.-H. Folic acid functionalized starch encapsulated green synthesized copper oxide nanoparticles for targeted drug delivery in breast cancer therapy. *Int. J. Biol. Macromol.* 2020, *164*, 2073–2084. [CrossRef]
36. Wei, W.; Zhang, X.; Zhang, S.; Wei, G.; Su, Z. Biomedical and bioactive engineered nanomaterials for targeted tumor photothermal therapy: A review. *Mater. Sci. Eng. C* 2019, *104*, 109891. [CrossRef]
37. Azevedo, C.; Macedo, M.H.; Sarmento, B. Strategies for the enhanced intracellular delivery of nanomaterials. *Drug Discov. Today* 2018, *23*, 944–959. [CrossRef]

38. Farzin, L.; Shamsipur, M.; Samandari, L.; Sheibani, S. Advances in the design of nanomaterial-based electrochemical affinity and enzymatic biosensors for metabolic biomarkers: A review. *Microchim. Acta* **2018**, *185*, 1–25. [CrossRef]
39. Gravely, M.; Safaee, M.M.; Roxbury, D. Biomolecular Functionalization of a Nanomaterial To Control Stability and Retention within Live Cells. *Nano Lett.* **2019**, *19*, 6203–6212. [CrossRef]
40. Montaseri, H.; Kruger, C.A.; Abrahamse, H. Review: Organic nanoparticle based active targeting for photodynamic therapy treatment of breast cancer cells. *Oncotarget* **2020**, *11*, 2120–2136. [CrossRef]
41. Gole, B.; Sanyal, U.; Banerjee, R.; Mukherjee, P.S. High Loading of Pd Nanoparticles by Interior Functionalization of MOFs for Heterogeneous Catalysis. *Inorg. Chem.* **2016**, *55*, 2345–2354. [CrossRef] [PubMed]
42. Bertella, S.; Luterbacher, J.S. Lignin Functionalization for the Production of Novel Materials. *Trends Chem.* **2020**, *2*, 440–453. [CrossRef]
43. Díez-Pascual, A.M. Chemical Functionalization of Carbon Nanotubes with Polymers: A Brief Overview. *Macromol* **2021**, *1*, 64–83. [CrossRef]
44. Díez-Pascual, A.M. Carbon-Based Nanomaterials. *Int. J. Mol. Sci.* **2021**, *22*, 7726. [CrossRef] [PubMed]
45. Sainz-Urruela, C.; Vera-López, S.; Andrés, M.P.S.; Díez-Pascual, A.M. Surface functionalization of graphene oxide with tannic acid: Covalent vs non-covalent approaches. *J. Mol. Liq.* **2022**, *357*, 119104. [CrossRef]
46. Díez-Pascual, A.M. Development of Graphene-Based Polymeric Nanocomposites: A Brief Overview. *Polymers* **2021**, *13*, 2978. [CrossRef]
47. Chen, Q.; Liu, Z. Albumin carriers for cancer theranostics: A conventional platform with new promise. *Adv. Mater.* **2016**, *28*, 10557–10566. [CrossRef]
48. Bolaños, K.; Kogan, M.J.; Araya, E. Capping gold nanoparticles with albumin to improve their biomedical properties. *Int. J. Nanomed.* **2019**, *14*, 6387–6406. [CrossRef]
49. Chakraborty, A.; Dhar, P. A review on potential of proteins as an excipient for developing a nano-carrier delivery system. *Crit. Rev. Ther. Drug Carr. Syst.* **2017**, *34*, 453–488. [CrossRef]
50. Díez-Pascual, A.M.; García-García, D.; Andrés, M.P.S.; Vera, S. Determination of riboflavin based on fluorescence quenching by graphene dispersions in polyethylene glycol. *RSC Adv.* **2016**, *6*, 1968. [CrossRef]
51. Díez-Pascual, A.M.; Díez-Vicente, A.L. Multifunctional poly(glycolic acid-co-propylene fumarate) electrospun fibers reinforced with graphene oxide and hydroxyapatite nanorods. *J. Mater. Chem. B* **2017**, *5*, 4084–4096. [CrossRef] [PubMed]
52. Horo, H.; Bhattacharyya, S.; Mandal, B.; Kundu, L.M. Synthesis of functionalized silk-coated chitosan-gold nanoparticles and microparticles for target-directed delivery of antitumor agents. *Carbohydr. Polym.* **2021**, *258*, 117659. [CrossRef] [PubMed]
53. Díez-Pascual, A.M.; Díez-Vicente, A.L. Electrospun fibers of chitosan-grafted polycaprolactone/poly(3-hydroxybutyrate-co-3-hydroxyhexanoate) blends. *J. Mater. Chem. B* **2016**, *4*, 600–612. [CrossRef] [PubMed]
54. Diez-Pascual, A.M.; Diez-Vicente, A.L. Poly(propylene fumarate)/Polyethylene Glycol-Modified Graphene Oxide Nanocomposites for Tissue Engineering. *ACS Appl. Mater. Interfaces* **2016**, *8*, 17902–17914. [CrossRef] [PubMed]
55. Kleinfeldt, L.; Gädke, J.; Biedendieck, R.; Krull, R.; Garnweitner, G. Spray-Dried Hierarchical Aggregates of Iron Oxide Nanoparticles and Their Functionalization for Downstream Processing in Biotechnology. *ACS Omega* **2019**, *4*, 16300–16308. [CrossRef]
56. Makvandi, P.; Wang, C.Y.; Zare, E.N.; Borzacchiello, A.; Niu, L.N.; Tay, F.R. Metal-Based Nanomaterials in Biomedical Applications: Antimicrobial Activity and Cytotoxicity Aspects. *Adv. Funct. Mater.* **2020**, *30*, 1910021. [CrossRef]
57. Mazzotta, E.; Orlando, C.; Muzzalupo, R. New Nanomaterials with Intrinsic Antioxidant Activity by Surface Functionalization of Niosomes with Natural Phenolic Acids. *Pharmaceutics* **2021**, *13*, 766. [CrossRef]
58. Delfi, M.; Ghomi, M.; Zarrabi, A.; Mohammadinejad, R.; Taraghdari, Z.; Ashrafizadeh, M.; Zare, E.; Agarwal, T.; Padil, V.; Mokhtari, B.; et al. Functionalization of Polymers and Nanomaterials for Biomedical Applications: Antimicrobial Platforms and Drug Carriers. *Prosthesis* **2020**, *2*, 117–139. [CrossRef]
59. Gary-Bobo, M.; Hocine, O.; Brevet, D.; Maynadier, M.; Raehm, L.; Richeter, S.; Charasson, V.; Loock, B.; Morère, A.; Maillard, P.; et al. Cancer therapy improvement with mesoporous silica nanoparticles combining targeting, drug delivery and PDT. *Int. J. Pharm.* **2012**, *423*, 509–515. [CrossRef]
60. Jose, J.; Kumar, R.; Harilal, S.; Mathew, G.E.; Parambi, D.G.T.; Prabhu, A.; Uddin, M.S.; Aleya, L.; Kim, H.; Mathew, B. Magnetic nanoparticles for hyperthermia in cancer treatment: An emerging tool. *Environ. Sci. Pollut. Res.* **2020**, *27*, 19214–19225. [CrossRef]
61. Sharmeen, S.; Rahman, A.M.; Lubna, M.M.; Salem, K.S.; Islam, R.; Khan, M.A. Polyethylene glycol functionalized carbon nanotubes/gelatin-chitosan nanocomposite: An approach for significant drug release. *Bioact. Mater.* **2018**, *3*, 236–244. [CrossRef]
62. Muhammad, M.; Shao, C.S.; Huang, Q. Aptamer-functionalized Au nanoparticles array as the effective SERS biosensor for label-free detection of interleukin-6 in serum. *Sens. Actuators B Chem.* **2021**, *334*, 129607. [CrossRef]
63. Mahmoudpour, M.; Ding, S.; Lyu, Z.; Ebrahimi, G.; Du, D.; Dolatabadi, J.E.N.; Torbati, M.; Lin, Y. Aptamer functionalized nanomaterials for biomedical applications: Recent advances and new horizons. *Nano Today* **2021**, *39*, 101177. [CrossRef]
64. Mahmoudpour, M.; Karimzadeh, Z.; Ebrahimi, G.; Hasanzadeh, M.; Ezzati Nazhad Dolatabadi, J. Synergizing Functional Nanomaterials with Aptamers Based on Electrochemical Strategies for Pesticide Detection: Current Status and Perspectives. *Crit. Rev. Anal. Chem.* **2021**, 1–28. [CrossRef] [PubMed]
65. Hassanisaadi, M.; Bonjar GH, S.; Rahdar, A.; Pandey, S.; Hosseinipour, A.; Abdolshahi, R. Environmentally Safe Biosynthesis of Gold Nanoparticles Using Plant Water Extracts. *Nanomaterials* **2021**, *11*, 2033. [CrossRef] [PubMed]

66. Simon, S.; Ciceo-Lucacel, R.; Radu, T.; Baia, L.; Ponta, O.; Iepure, A.; Simon, V. Gold nanoparticles developed in sol–gel derived apatite—bioactive glass composites. *J. Mater. Sci. Mater. Med.* **2012**, *23*, 1193–1201. [CrossRef]
67. Qingling, F.; Wei, J.; Aifantis, K.E.; Fan, Y.; Feng, Q.; Cui, F.-Z.; Watari, F. Current investigations into magnetic nanoparticles for biomedical applications. *J. Biomed. Mater. Res. Part A* **2016**, *104*, 1285–1296. [CrossRef]
68. Liu, S.; Höldrich, M.; Sievers-Engler, A.; Horak, J.; Lämmerhofer, M. Papain-functionalized gold nanoparticles as heterogeneous biocatalyst for bioanalysis and biopharmaceuticals analysis. *Anal. Chim. Acta* **2017**, *963*, 33–43. [CrossRef]
69. Pourjavadi, A.; Bagherifard, M.; Doroudian, M. Synthesis of micelles based on chitosan functionalized with gold nanorods as a light sensitive drug delivery vehicle. *Int. J. Biol. Macromol.* **2020**, *149*, 809–818. [CrossRef]
70. Tiwari, P.M.; Vig, K.; Dennis, V.A.; Singh, S.R. Functionalized gold nanoparticles and their biomedical applications. *Nanomaterials* **2011**, *1*, 31–63. [CrossRef]
71. Donoso-González, O.; Lodeiro, L.; Aliaga, Á.E.; Laguna-Bercero, M.A.; Bollo, S.; Kogan, M.J.; Yutronic, N.; Sierpe, R. Functionalization of gold nanostars with cationic β-cyclodextrin-based polymer for drug co-loading and SERS monitoring. *Pharmaceutics* **2021**, *13*, 261. [CrossRef] [PubMed]
72. Shon, Y.S.; Choo, H. [60]Fullerene-linked gold nanoparticles: Synthesis and layer-by-layer growth on a solid surface. *Chem. Commun.* **2002**, *21*, 2560–2561. [CrossRef]
73. Sudeep, P.K.; Ipe, B.I.; Thomas, K.G.; George, M.V.; Barazzouk, S.; Hotchandani, S.; Kamat, P.V. Fullerene-functionalized gold nanoparticles. A self-assembled photoactive antenna-metal nanocore assembly. *Nano Lett.* **2002**, *2*, 29–35. [CrossRef]
74. Yaseen, M.; Humayun, M.; Khan, A.; Usman, M.; Ullah, H.; Tahir, A.; Ullah, H. Preparation, functionalization, modification, and applications of nanostructured gold: A critical review. *Energies* **2021**, *14*, 1278. [CrossRef]
75. Liz-Marzán, L.M.; Giersig, M.; Mulvaney, P. Synthesis of nanosized gold−silica core−shell particles. *Langmuir* **1996**, *12*, 4329–4335. [CrossRef]
76. Bhargava, A.; Dev, A.; Mohanbhai, S.J.; Pareek, V.; Jain, N.; Choudhury, S.R.; Panwar, J.; Karmakar, S. Pre-coating of protein modulate patterns of corona formation, physiological stability and cytotoxicity of silver nanoparticles. *Sci. Total Environ.* **2021**, *772*, 144797. [CrossRef]
77. Matsuo, T. Functionalization of Ruthenium Olefin-Metathesis Catalysts for Interdisciplinary Studies in Chemistry and Biology. *Catalysts* **2021**, *11*, 359. [CrossRef]
78. Díez-Pascual, A.M.; Díez-Vicente, A.L. Nano-TiO$_2$ Reinforced PEEK/PEI Blends as Biomaterials for Load-Bearing Implant Applications. *ACS Appl. Mater. Interfaces* **2015**, *7*, 5561–5573. [CrossRef]
79. Díez-Pascual, A.M.; Díez-Vicente, A.L. Effect of TiO$_2$ Nanoparticles on the Performance of Polyphenylsulfone Biomaterial for Orthopaedic Implants. *J. Mater. Chem. B* **2014**, *2*, 7502–7514. [CrossRef]
80. Díez-Pascual, A.M.; Díez-Vicente, A.L. High-Performance Aminated Poly(phenylene sulfide)/ZnO Nanocomposites for Medical Applications. *ACS Appl. Mater. Interfaces* **2014**, *6*, 10132–101045. [CrossRef]
81. Díez-Pascual, A.M.; Díez-Vicente, A.L. Antimicrobial and sustainable food packaging based on poly(butylene adipate-co-terephthalate) and electrospun chitosan nanofibers. *RSC Adv.* **2015**, *5*, 93095. [CrossRef]
82. Jalalian, S.H.; Taghdisi, S.M.; Hamedani, N.S.; Kalat, S.A.M.; Lavaee, P.; ZandKarimi, M.; Ghows, N.; Jaafari, M.R.; Naghibi, S.; Danesh, N.M.; et al. Epirubicin loaded super paramagnetic iron oxide nanoparticle-aptamer bioconjugate for combined colon cancer therapy and imaging in vivo. *Eur. J. Pharm. Sci.* **2013**, *50*, 191–197. [CrossRef] [PubMed]
83. Wei, R.; Xu, Y.; Xue, M. Hollow iron oxide nanomaterials: Synthesis, functionalization, and biomedical applications. *J. Mater. Chem. B* **2021**, *9*, 1965–1979. [CrossRef] [PubMed]
84. Cheah, P.; Brown, P.; Qu, J.; Tian, B.; Patton, D.L.; Zhao, Y. Versatile Surface Functionalization of Water-Dispersible Iron Oxide Nanoparticles with Precisely Controlled Sizes. *Langmuir* **2021**, *37*, 1279–1287. [CrossRef] [PubMed]
85. Nayeem, J.; Al-Bari, A.A.; Mahiuddin; Rahman, A.; Mefford, O.T.; Ahmad, H.; Rahman, M. Silica coating of iron oxide magnetic nanoparticles by reverse microemulsion method and their functionalization with cationic polymer P(NIPAm-co-AMPTMA) for antibacterial vancomycin immobilization. *Colloids Surf. A Physicochem. Eng. Asp.* **2021**, *611*, 125857. [CrossRef]
86. Sun, C.; Lee, J.S.H.; Zhang, M. Magnetic nanoparticles in MR imaging and drug delivery. *Adv. Drug Deliv. Rev.* **2008**, *60*, 1252–1265. [CrossRef]
87. Díez-Pascual, A.M.; Díez-Vicente, A.L. Development of Linseed Oil/TiO$_2$ Green Nanocomposites as Antimicrobial Coatings. *J. Mater. Chem. B* **2015**, *3*, 4458–4471. [CrossRef]
88. Kundu, M.; Sadhukhan, P.; Ghosh, N.; Chatterjee, S.; Manna, P.; Das, J.; Sil, P.C. pH-responsive and targeted delivery of curcumin via phenylboronic acid-functionalized ZnO nanoparticles for breast cancer therapy. *J. Adv. Res.* **2019**, *18*, 161–172. [CrossRef]
89. Al-Harbi, N.; Mohammed, H.; Al-Hadeethi, Y.; Bakry, A.S.; Umar, A.; Hussein, M.A.; Abbassy, M.A.; Vaidya, K.G.; Berakdar, G.A.; Mkawi, E.M.; et al. Silica-Based Bioactive Glasses and Their Applications in Hard Tissue Regeneration: A Review. *Pharmaceuticals* **2021**, *20*, 75. [CrossRef]
90. Filho, O.P.; La Torre, G.P.; Hench, L.L. Effect of crystallization on apatite-layer formation of bioactive glass 45S5. *J. Biomed. Mater. Res. Off. J. Soc. Biomater. Jpn. Soc. Biomater.* **1996**, *30*, 509–514. [CrossRef]
91. Aina, V.; Cerrato, G.; Martra, G.; Bergandi, L.; Costamagna, C.; Ghigo, D.; Malavasi, G.; Lusvardi, G.; Menabue, L. Gold-containing bioactive glasses: A solid-state synthesis to produce alternative biomaterials for bone implantations. *J. R. Soc. Interface* **2013**, *10*, 20121040. [CrossRef] [PubMed]

92. Naffakh, M.; Diez-Pascual, A.M. Thermoplastic Polymer Nanocomposites Based on Inorganic Fullerene-like Nanoparticles and Inorganic Nanotubes. *Inorganics* **2014**, *2*, 291–312. [CrossRef]
93. Gibson, I.R.; Bonfield, W. Novel synthesis and characterization of an AB-type carbonate-substituted hydroxyapatite. *J. Biomed. Mater. Res. Off. J. Soc. Biomater. Jpn. Soc. Biomater. Aust. Soc. Biomater. Korean Soc. Biomater.* **2002**, *59*, 697–708. [CrossRef] [PubMed]
94. Ursino, H.L.; James, B.D.; Ludtka, C.M.; Allen, J.B. Bone tissue engineering. In *Tissue Engineering Using Ceramics and Polymers*; Elsevier: Amsterdam, The Netherlands, 2022; pp. 587–644.
95. Díez-Pascual, A.M.; Rahdar, A. LbL Nano-Assemblies: A Versatile Tool for Biomedical and Healthcare Applications. *Nanomaterials* **2022**, *12*, 949. [CrossRef] [PubMed]
96. Díez-Pascual, A.M. State of the Art in the Antibacterial and Antiviral Applications of Carbon-Based Polymeric Nanocomposites. *Int. J. Mol. Sci.* **2021**, *22*, 10511. [CrossRef]
97. Schnorr, J.M.; Swager, T.M. Emerging applications of carbon nanotubes. *Chem. Mater.* **2011**, *23*, 646–657. [CrossRef]
98. Naffakh, M.; Díez-Pascual, A.M.; Gómez-Fatou, M.A. New hybrid nanocomposites containing carbon nanotubes, inorganic fullerene-like WS2 nanoparticles and poly(ether ether ketone) (PEEK). *J. Mater. Chem.* **2011**, *21*, 7425. [CrossRef]
99. Díez-Pascual, A.M.; Martínez, G.; González-Domínguez, J.M.; Ansón, A.; Martínez, M.T.; Gómez, M.A. Grafting of a hydroxylated poly(ether ether ketone) to the surface of single-walled carbon nanotubes. *J. Mater. Chem.* **2020**, *20*, 8285. [CrossRef]
100. Dou, J.; Gan, D.; Huang, Q.; Liu, M.; Chen, J.; Deng, F.; Zhu, X.; Wen, Y.; Zhang, X.; Wei, Y. Functionalization of carbon nanotubes with chitosan based on MALI multicomponent reaction for Cu2+ removal. *Int. J. Biol. Macromol.* **2019**, *136*, 476–485. [CrossRef]
101. Sainz-Urruela, C.; Vera-López, S.; San Andrés, M.P.; Díez-Pascual, A.M. Graphene-Based Sensors for the Detection of Bioactive Compounds: A Review. *Int. J. Mol. Sci.* **2021**, *22*, 3316. [CrossRef]
102. Díez-Pascual, A.M. Antibacterial Action of Nanoparticle Loaded Nanocomposites Based on Graphene and Its Derivatives: A Mini-Review. *Int. J. Mol. Sci.* **2020**, *21*, 3563. [CrossRef] [PubMed]
103. Innocenzi, P.; Stagi, L. Carbon-based antiviral nanomaterials: Graphene, C-dots, and fullerenes. A perspective. *Chem. Sci.* **2020**, *11*, 6606–6622. [CrossRef] [PubMed]
104. Xin, Q.; Shah, H.; Nawaz, A.; Xie, W.; Akram, M.Z.; Batool, A.; Tian, L.; Jan, S.U.; Boddula, R.; Guo, B.; et al. Antibacterial carbon-based nanomaterials. *Adv. Mater.* **2019**, *31*, e1804838. [CrossRef] [PubMed]
105. Hu, X.; Mu, L.; Wen, J.; Zhou, Q. Covalently synthesized graphene oxide-aptamer nanosheets for efficient visible-light photocatalysis of nucleic acids and proteins of viruses. *Carbon* **2012**, *50*, 2772–2781. [CrossRef]
106. Van Tam, T.; Hur, S.H.; Chung, J.S.; Choi, W.M. Novel paper- and fiber optic-based fluorescent sensor for glucose detection using aniline-functionalized graphene quantum dots. *Sens. Actuators B Chem.* **2021**, *329*, 129250. [CrossRef]
107. Seifi, T.; Kamali, A.R. Antiviral performance of graphene-based materials with emphasis on COVID-19: A review. *Med. Drug Discov.* **2021**, *11*, 100099. [CrossRef]
108. Bai, J.; Chen, L.; Zhu, Y.; Wang, X.; Wu, X.; Fu, Y. A novel luminescence sensor based on porous molecularly imprinted polymer-ZnS quantum dots for selective recognition of paclitaxel. *Colloids Surf. A Physicochem. Eng. Asp.* **2020**, *610*, 125696. [CrossRef]
109. Soleymani, J.; Hasanzadeh, M.; Somi, M.H.; Ozkan, S.A.; Jouyban, A. Targeting and sensing of some cancer cells using folate bioreceptor functionalized nitrogen-doped graphene quantum dots. *Int. J. Biol. Macromol.* **2018**, *118*, 1021–1034. [CrossRef]
110. Banerjee, A.; Pons, T.; Lequeux, N.; Dubertret, B. Quantum dots–DNA bioconjugates: Synthesis to applications. *Interface Focus* **2016**, *6*, 20160064. [CrossRef]
111. Sun, D.; Gang, O. DNA-Functionalized Quantum Dots: Fabrication, Structural, and Physicochemical Properties. *Langmuir* **2013**, *29*, 7038–7046. [CrossRef]
112. Wang, G.; Li, Z.; Luo, X.; Yue, R.; Shen, Y.; Ma, N. DNA-templated nanoparticle complexes for photothermal imaging and labeling of cancer cells. *Nanoscale* **2018**, *10*, 16508–16520. [CrossRef] [PubMed]
113. Hajikarimi, Z.; Khoei, S.; Khoee, S.; Mahdavi, S.R. Evaluation of the cytotoxic effects of PLGA coated iron oxide nanoparticles as a carrier of 5-fluorouracil and mega-voltage X-ray radiation in DU145 prostate cancer cell line. *IEEE Trans. Nanobioscience* **2014**, *13*, 403–408. [CrossRef] [PubMed]
114. Thamake, S.I.; Raut, S.2.; Ranjan, A.P.; Gryczynski, Z.; Vishwanatha, J.K. Surface functionalization of PLGA nanoparticles by non-covalent insertion of a homo-bifunctional spacer for active targeting in cancer therapy. *Nanotechnology* **2010**, *22*, 035101. [CrossRef] [PubMed]
115. Du, H.; Parit, M.; Liu, K.; Zhang, M.; Jiang, Z.; Huang, T.-S.; Zhang, X.; Si, C. Multifunctional Cellulose Nanopaper with Superior Water-Resistant, Conductive, and Antibacterial Properties Functionalized with Chitosan and Polypyrrole. *ACS Appl. Mater. Interfaces* **2021**, *13*, 32115–32125. [CrossRef] [PubMed]
116. Sofla, R.L.M.; Rezaei, M.; Babaie, A. Investigation of the effect of graphene oxide functionalization on the physical, mechanical and shape memory properties of polyurethane/reduced graphene oxide nanocomposites. *Diam. Relat. Mater.* **2019**, *95*, 195–205. [CrossRef]
117. Yan, S.; Wang, W.; Li, X.; Ren, J.; Yun, W.; Zhang, K.; Li, G.; Yin, J. Preparation of mussel-inspired injectable hydrogels based on dual-functionalized alginate with improved adhesive, self-healing, and mechanical properties. *J. Mater. Chem. B* **2018**, *6*, 6377–6390. [CrossRef] [PubMed]
118. Saifi, M.A.; Khan, W.; Godugu, C. Cytotoxicity of Nanomaterials: Using Nanotoxicology to Address the Safety Concerns of Nanoparticles. *Pharm. Nanotechnol.* **2018**, *6*, 3–16. [CrossRef]

119. Srivastava, V.; Gusain, D.; Sharma, Y.C. Critical Review on the Toxicity of Some Widely Used Engineered Nanoparticles. *Ind. Eng. Chem. Res.* **2015**, *54*, 6209–6233. [CrossRef]
120. Madani, S.Y.; Mandel, A.; Seifalian, A.M. A concise review of carbon nanotube's toxicology. *Nano Rev.* **2013**, *4*, 21521. [CrossRef]
121. Yang, H.; Liu, C.; Yang, D.; Zhang, H.; Xi, Z. Comparative study of cytotoxicity, oxidative stress and genotoxicity induced by four typical nanomaterials: The role of particle size, shape and composition. *J. Appl. Toxicol.* **2009**, *29*, 69–78. [CrossRef]
122. Katsumiti, A.; Berhanu, D.; Howard, K.T.; Arostegui, I.; Oron, M.; Reip, P.; Valsami-Jones, E.; Cajaraville, M. Cytotoxicity of TiO_2 nanoparticles to mussel hemocytes and gill cells in vitro: Influence of synthesis method, crystalline structure, size and additive. *Nanotoxicology* **2015**, *9*, 543–553. [CrossRef] [PubMed]
123. Sato, S.; Nakamura, R.; Abe, S. Visible-light sensitization of TiO2 photocatalysts by wet-method N doping. *Appl. Catal. A Gen.* **2005**, *284*, 131–137. [CrossRef]
124. Sukhanova, A.; Bozrova, S.; Sokolov, P.; Berestovoy, M.; Karaulov, A.; Nabiev, I. Dependence of Nanoparticle Toxicity on Their Physical and Chemical Properties. *Nanoscale Res. Lett.* **2018**, *13*, 1–21. [CrossRef] [PubMed]
125. Wang, J.; Yao, H.; Shi, X. Cooperative entry of nanoparticles into the cell. *J. Mech. Phys. Solids* **2014**, *73*, 151–165. [CrossRef]
126. Elrahman, A.A.; Mansour, F. Targeted magnetic iron oxide nanoparticles: Preparation, functionalization and biomedical application. *J. Drug Deliv. Sci. Technol.* **2019**, *52*, 702–712. [CrossRef]
127. Kheirallah, D.A.M.; El-Samad, L.M.; Abdel-Moneim, A.M. DNA damage and ovarian ultrastructural lesions induced by nickel oxide nano-particles in Blaps polycresta (Coleoptera: Tenebrionidae). *Sci. Total Environ.* **2021**, *753*, 141743. [CrossRef] [PubMed]
128. Du, Y.; Jin, J.; Liang, H.; Jiang, W. Structural and physicochemical properties and biocompatibility of linear and looped polymer-capped gold nanoparticles. *Langmuir* **2019**, *35*, 8316–8324. [CrossRef]
129. Sen, G.T.; Ozkemahli, G.; Shahbazi, R.; Erkekoglu, P.; Ulubayram, K.; Kocer-Gumusel, B. The effects of polymer coating of gold nanoparticles on oxidative stress and DNA damage. *Int. J. Toxicol.* **2020**, *39*, 328–340. [CrossRef]
130. Wang, R.; Bowling, I.; Liu, W. Cost effective surface functionalization of gold nanoparticles with a mixed DNA and PEG monolayer for nanotechnology applications. *RSC Adv.* **2017**, *7*, 3676–3679. [CrossRef]
131. Singh, A.; Amiji, M.M. Application of nanotechnology in medical diagnosis and imaging. *Curr. Opin. Biotechnol.* **2022**, *74*, 241–246. [CrossRef]
132. El-Sayed, A.; Kamel, M. Advances in nanomedical applications: Diagnostic, therapeutic, immunization, and vaccine production. *Environ. Sci. Pollut. Res.* **2020**, *27*, 19200–19213. [CrossRef] [PubMed]
133. Mi, Y.; Shao, Z.; Vang, J.; Kaidar-Person, O.; Wang, A.Z. Application of nanotechnology to cancer radiotherapy. *Cancer Nanotechnol.* **2016**, *7*, 1–16. [CrossRef] [PubMed]
134. Aithal, P.S. Nanotechnology Innovations & Business Opportunities: A Review. *Int. J. Manag. IT Eng.* **2016**, *6*, 182–204.
135. Elkodous, M.A.; El-Sayyad, G.S.; Abdelrahman, I.Y.; El-Bastawisy, H.S.; Mohamed, A.E.; Mosallam, F.M.; Nasser, H.; Gobara, M.; Baraka, A.; Elsayed, M.; et al. Therapeutic and diagnostic potential of nanomaterials for enhanced biomedical applications. *Colloids Surf. B Biointerfaces* **2019**, *180*, 411–428. [CrossRef] [PubMed]
136. Chen, X.; Song, J.; Chen, X.; Yang, H. X-ray-activated nanosystems for theranostic applications. *Chem. Soc. Rev.* **2019**, *48*, 3073–3101. [CrossRef] [PubMed]
137. Gharpure, K.; Wu, S.; Li, C.; Lopez-Berestein, G.; Sood, A.K. Nanotechnology: Future of oncotherapy. *Clin. Cancer Res.* **2015**, *21*, 3121–3130. [CrossRef]
138. Chen, Z.; Peng, Y.; Li, Y.; Xie, X.; Wei, X.; Yang, G.; Zhang, H.; Li, N.; Li, T.; Qin, X. Aptamer-Dendrimer Functionalized Magnetic Nano-Octahedrons: Theranostic Drug/Gene Delivery Platform for Near-Infrared/Magnetic Resonance Imaging-Guided Magnetochemotherapy. *ACS Nano* **2021**, *15*, 16683–16696. [CrossRef]
139. Zeytunluoglu, A.; Arslan, I. Current perspectives on nanoemulsions in targeted drug delivery: An overview. In *Handbook of Research on Nanoemulsion Applications in Agriculture, Food, Health, and Biomedical Sciences*; IGI Global: Hershey, PA, USA, 2022; pp. 118–140.
140. Dong, Y.; Wu, X.; Chen, X.; Zhou, P.; Xu, F.; Liang, W. Nanotechnology shaping stem cell therapy: Recent advances, application, challenges, and future outlook. *Biomed. Pharmacother.* **2021**, *137*, 111236. [CrossRef]
141. Pereira, M.N.; Ushirobira, C.Y.; Cunha-Filho, M.S.; Gelfuso, G.M.; Gratieri, T. Nanotechnology advances for hair loss. *Ther. Deliv.* **2018**, *9*, 593–603. [CrossRef]
142. Kumar, S.; Nehra, M.; Kedia, D.; Dilbaghi, N.; Tankeshwar, K.; Kim, K.-H. Nanotechnology-based biomaterials for orthopaedic applications: Recent advances and future prospects. *Mater. Sci. Eng. C* **2020**, *106*, 110154. [CrossRef]
143. Zheng, X.; Zhang, P.; Fu, Z.; Meng, S.; Dai, L.; Yang, H. Applications of nanomaterials in tissue engineering. *RSC Adv.* **2021**, *11*, 19041–19058. [CrossRef] [PubMed]
144. Kumar, R.; Aadil, K.R.; Ranjan, S.; Kumar, V.B. Advances in nanotechnology and nanomaterials based strategies for neural tissue engineering. *J. Drug Deliv. Sci. Technol.* **2020**, *57*, 101617. [CrossRef]
145. Bakopoulou, A.; Papachristou, E.; Bousnaki, M.; Hadjichristou, C.; Kontonasaki, E.; Theocharidou, A.; Papadopoulou, L.; Kantiranis, N.; Zachariadis, G.; Leyhausen, G. Human treated dentin matrices combined with Zn-doped, Mg-based bioceramic scaffolds and human dental pulp stem cells towards targeted dentin regeneration. *Dent. Mater.* **2016**, *32*, e159–e175. [CrossRef] [PubMed]
146. Balakrishnan, B.; Joshi, N.; Jayakrishnan, A.; Banerjee, R. Self-crosslinked oxidized alginate/gelatin hydrogel as injectable, adhesive biomimetic scaffolds for cartilage regeneration. *Acta Biomater.* **2014**, *10*, 3650–3663. [CrossRef]

147. Wang, C.; Zhang, H.; Chen, B.; Yin, H.; Wang, W. Study of the enhanced anticancer efficacy of gambogic acid on Capan-1 pancreatic cancer cells when mediated via magnetic Fe_3O_4 nanoparticles. *Int. J. Nanomed.* **2011**, *6*, 1929.
148. Haruta, S.; Hanafusa, T.; Fukase, H.; Miyajima, H.; Oki, T. An effective absorption behavior of insulin for diabetic treatment following intranasal delivery using porous spherical calcium carbonate in monkeys and healthy human volunteers. *Diabetes Technol. Ther.* **2003**, *5*, 1–9. [CrossRef]
149. Zazo, H.; Colino, C.I.; Lanao, J.M. Current applications of nanoparticles in infectious diseases. *J. Control. Release* **2016**, *224*, 86–102. [CrossRef]
150. Rudramurthy, G.R.; Swamy, M.K.; Sinniah, U.R.; Ghasemzadeh, A. Nanoparticles: Alternatives against drug-resistant pathogenic microbes. *Molecules* **2016**, *21*, 836. [CrossRef]
151. Smith, B.R.; Gambhir, S.S. Nanomaterials for in vivo imaging. *Chem. Rev.* **2017**, *117*, 901–986. [CrossRef]
152. Imamura, Y.; Yamada, S.; Tsuboi, S.; Nakane, Y.; Tsukasaki, Y.; Komatsuzaki, A.; Jin, T. Near-infrared emitting PbS quantum dots for in vivo fluorescence imaging of the thrombotic state in septic mouse brain. *Molecules* **2016**, *21*, 1080. [CrossRef]
153. Wang, Y.; Wu, H.; Lin, D.; Zhang, R.; Li, H.; Zhang, W.; Liu, W.; Huang, S.; Yao, L.; Cheng, J.; et al. One-dimensional electrospun ceramic nanomaterials and their sensing applications. *J. Am. Ceram. Soc.* **2022**, *105*, 765–785. [CrossRef]
154. Liu, J.; Wang, Z.; Zhao, S.; Ding, B. Multifunctional nucleic acid nanostructures for gene therapies. *Nano Res.* **2018**, *11*, 5017–5027. [CrossRef]
155. Patil-Sen, Y. Advances in nano-biomaterials and their applications in biomedicine. *Emerg. Top. Life Sci.* **2021**, *5*, 169–176. [CrossRef] [PubMed]
156. Wu, G.; Li, P.; Feng, H.; Zhang, X.; Chu, P.K. Engineering and functionalization of biomaterials via surface modification. *J. Mater. Chem. B* **2015**, *3*, 2024–2042. [CrossRef] [PubMed]
157. Singh, T.V.; Shagolsem, L.S. Biopolymer based nano-structured materials and their applications. In *Nanostructured Materials and Their Applications*; Springer: Singapore, 2021; pp. 337–366.
158. Sargazi, S.; Mukhtar, M.; Rahdar, A.; Barani, M.; Pandey, S.; Díez-Pascual, A.M. Active Targeted Nanoparticles for Delivery of Poly(ADP-ribose) Polymerase (PARP) Inhibitors: A Preliminary Review. *Int. J. Mol. Sci.* **2021**, *22*, 10319. [CrossRef] [PubMed]
159. Sivasankarapillai, V.S.; Das, S.S.; Sabir, F.; Sundaramahalingam, M.A.; Colmenares, J.C.; Prasannakumar, S.; Rajan, M.; Rahdar, A.; Kyzas, G.Z. Progress in natural polymer engineered biomaterials for transdermal drug delivery systems. *Mater. Today Chem.* **2021**, *19*, 100382. [CrossRef]
160. Bouchoucha, M.; Gaudreault, R.C.; Fortin, M.A.; Kleitz, F. Mesoporous silica nanoparticles: Selective surface functionalization for optimal relaxometric and drug loading performances. *Adv. Funct. Mater.* **2014**, *24*, 5911–5923. [CrossRef]
161. Yoo, J.; Park, C.; Yi, G.; Lee, D.; Koo, H. Active Targeting Strategies Using Biological Ligands for Nanoparticle Drug Delivery Systems. *Cancers* **2019**, *11*, 640. [CrossRef]
162. Huang, Y.; Cao, L.; Parakhonskiy, B.V.; Skirtach, A.G. Hard, Soft, and Hard-and-Soft Drug Delivery Carriers Based on $CaCO_3$ and Alginate Biomaterials: Synthesis, Properties, Pharmaceutical Applications. *Pharmaceutics* **2022**, *14*, 909. [CrossRef]
163. Dizaj, S.M.; Lotfipour, F.; Barzegar-Jalali, M.; Zarrintan, M.-H.; Adibkia, K. Ciprofloxacin HCl-loaded calcium carbonate nanoparticles: Preparation, solid state characterization, and evaluation of antimicrobial effect against *Staphylococcus aureus*. *Artif. Cells Nanomed. Biotechnol.* **2016**, *45*, 535–543. [CrossRef]
164. Ueno, Y.; Futagawa, H.; Takagi, Y.; Ueno, A.; Mizushima, Y. Drug-incorporating calcium carbonate nanoparticles for a new delivery system. *J. Control. Release* **2005**, *103*, 93–98. [CrossRef] [PubMed]
165. Hood, J.D.; Bednarski, M.; Frausto, R.; Guccione, S.; Reisfeld, R.A.; Xiang, R.; Cheresh, D.A. Tumor regression by targeted gene delivery to the neovasculature. *Science* **2002**, *296*, 2404–2407. [CrossRef]
166. Deng, L.; Li, Q.; Al-Rehili, S.; Omar, H.; Almalik, A.; Alshamsan, A.; Zhang, J.; Khashab, N.M. Hybrid iron oxide–graphene oxide–polysaccharides microcapsule: A micro-matryoshka for on-demand drug release and antitumor therapy in vivo. *ACS Appl. Mater. Interfaces* **2016**, *8*, 6859–6868. [CrossRef]
167. Ye, C.; Combs, Z.A.; Calabrese, R.; Dai, H.; Kaplan, D.L.; Tsukruk, V.V. Robust microcapsules with controlled permeability from silk fibroin reinforced with graphene oxide. *Small* **2014**, *10*, 5087–5097. [CrossRef] [PubMed]
168. Yang, P.-H.; Sun, X.; Chiu, J.-F.; Sun, H.; He, Q.-Y. Transferrin-mediated gold nanoparticle cellular uptake. *Bioconjugate Chem.* **2005**, *16*, 494–496. [CrossRef] [PubMed]
169. Cheng, J.; Gu, Y.-J.; Cheng, S.H.; Wong, W.-T. Surface functionalized gold nanoparticles for drug delivery. *J. Biomed. Nanotechnol.* **2013**, *9*, 1362–1369. [CrossRef]
170. Fang, Z.; Li, X.; Xu, Z.; Du, F.; Wang, W.; Shi, R.; Gao, D. Hyaluronic acid-modified mesoporous silica-coated superparamagnetic Fe_3O_4 nanoparticles for targeted drug delivery. *Int. J. Nanomed.* **2019**, *14*, 5785. [CrossRef]
171. Prabha, G.; Raj, V. Sodium alginate–polyvinyl alcohol–bovin serum albumin coated Fe_3O_4 nanoparticles as anticancer drug delivery vehicle: Doxorubicin loading and in vitro release study and cytotoxicity to HepG2 and L02 cells. *Mater. Sci. Eng. C* **2017**, *79*, 410–422. [CrossRef]
172. Rajan, M.; Raj, V.; Al-Arfaj, A.A.; Murugan, A.M. Hyaluronidase enzyme core-5-fluorouracil-loaded chitosan-PEG-gelatin polymer nanocomposites as targeted and controlled drug delivery vehicles. *Int. J. Pharm.* **2013**, *453*, 514–522. [CrossRef]
173. Hua, S.; Ma, H.; Li, X.; Yang, H.; Wang, A. pH-sensitive sodium alginate/poly (vinyl alcohol) hydrogel beads prepared by combined Ca^{2+} crosslinking and freeze-thawing cycles for controlled release of diclofenac sodium. *Int. J. Biol. Macromol.* **2010**, *46*, 517–523. [CrossRef]

174. Arshad, R.; Tabish, T.A.; Kiani, M.H.; Ibrahim, I.M.; Shahnaz, G.; Rahdar, A.; Kan, M.; Pandey, S. Hyaluronic Acid Functionalized Self-Nano-Emulsifying Drug Delivery System (SNEDDS) for Enhancement in Ciprofloxacin Targeted Delivery against Intracellular Infection. *Nanomaterials* **2021**, *11*, 1086. [CrossRef] [PubMed]
175. Guo, Q.; Jia, Y.; Yuan, M.; Huang, X.; Sui, X.; Tang, F.; Peng, J.; Chen, J.; Lu, S.; Cui, X.; et al. Co-encapsulation of magnetic Fe_3O_4 nanoparticles and doxorubicin into biodegradable PLGA nanocarriers for intratumoral drug delivery. *Int. J. Nanomed.* **2012**, *7*, 1697–1708. [CrossRef] [PubMed]
176. Wang, B.; Xu, C.; Xie, J.; Yang, Z.; Sun, S. pH controlled release of chromone from chromone-Fe_3O_4 nanoparticles. *J. Am. Chem. Soc.* **2008**, *130*, 14436–14437. [CrossRef] [PubMed]
177. Xie, J.; Xu, C.; Kohler, N.; Hou, Y.; Sun, S. Controlled PEGylation of monodisperse Fe3O4 nanoparticles for reduced non-specific uptake by macrophage cells. *Adv. Mater.* **2007**, *19*, 3163–3166. [CrossRef]
178. Saikia, C.; Hussain, A.; Ramteke, A.; Sharma, H.K.; Maji, T.K. Crosslinked thiolated starch coated Fe3O4 magnetic nanoparticles: Effect of montmorillonite and crosslinking density on drug delivery properties. *Starch-Stärke* **2014**, *66*, 760–771. [CrossRef]
179. Rezaei, A.; Hashemi, E. A pseudohomogeneous nanocarrier based on carbon quantum dots decorated with arginine as an efficient gene delivery vehicle. *Sci. Rep.* **2021**, *11*, 1–10. [CrossRef]
180. Di Marzio, N.; Eglin, D.; Serra, T.; Moroni, L. Bio-Fabrication: Convergence of 3D Bioprinting and Nano-Biomaterials in Tissue Engineering and Regenerative Medicine. *Front. Bioeng. Biotechnol.* **2020**, *8*, 326. [CrossRef]
181. Kapat, K.; Shubhra, Q.T.H.; Zhou, M.; Leeuwenburgh, S. Piezoelectric nano-biomaterials for biomedicine and tissue regeneration. *Adv. Funct. Mater.* **2020**, *30*, 1909045. [CrossRef]
182. Díez-Pascual, A.M.; Diez-Vicente, A.L. Epoxidized Soybean Oil/ZnO Biocomposites for Soft Tissue Applications: Preparation and Characterization. *ACS Appl. Mater. Interfaces* **2014**, *6*, 17277–17288. [CrossRef]
183. Giubilato, E.; Cazzagon, V.; Amorim, M.J.B.; Blosi, M.; Bouillard, J.; Bouwmeester, H.; Costa, A.L.; Fadeel, B.; Fernandes, T.F.; Fito, C.; et al. Risk management framework for nano-biomaterials used in medical devices and advanced therapy medicinal products. *Materials* **2020**, *13*, 4532. [CrossRef]
184. Rana, D.; Ramasamy, K.; Leena, M.; Jiménez, C.; Campos, J.; Ibarra, P.; Haidar, Z.S.; Ramalingam, M. Surface functionalization of nanobiomaterials for application in stem cell culture, tissue engineering, and regenerative medicine. *Biotechnol. Prog.* **2016**, *32*, 554–567. [CrossRef] [PubMed]
185. Labusca, L.; Herea, D.-D.; Mashayekhi, K. Stem cells as delivery vehicles for regenerative medicine-challenges and perspectives. *World J. Stem Cells* **2018**, *10*, 43–56. [CrossRef] [PubMed]
186. Lyons, J.G.; Plantz, M.A.; Hsu, W.K.; Hsu, E.L.; Minardi, S. Nanostructured biomaterials for bone regeneration. *Front. Bioeng. Biotechnol.* **2020**, *8*, 922. [CrossRef]
187. Zhu, L.; Luo, D.; Liu, Y. Effect of the nano/microscale structure of biomaterial scaffolds on bone regeneration. *Int. J. Oral Sci.* **2020**, *12*, 1–15. [CrossRef] [PubMed]
188. Zhang, Y.; Venugopal, J.R.; El-Turki, A.; Ramakrishna, S.; Su, B.; Lim, C.T. Electrospun biomimetic nanocomposite nanofiber of hydroxiapatite/chitosan for bone tissue engineering. *Biomaterials* **2008**, *29*, 4314–4322. [CrossRef] [PubMed]
189. Liang, X.Y.; Duan, P.G.; Gao, J.M.; Guo, R.S.; Qu, Z.H.; Li, X.F.; He, Y.; Yao, H.Q.; Ding, J.D. Bilayered PLGA/PLGA-HAp Composite Scaffold for Osteochondral Tissue Engineering and Tissue Regeneration. *ACS Biomater. Sci. Eng.* **2018**, *4*, 3506–3521. [CrossRef]
190. Naffakh, M.; Diez-Pascual, A.M. WS2 inorganic nanotubes reinforced poly(L-lacticacid)/hydroxyapatite hybrid composite biomaterials. *RSC Adv.* **2015**, *5*, 65514. [CrossRef]
191. Cui, Y.; Li, H.; Li, Y.; Mao, L. Novel insights into nanomaterials for immunomodulatory bone regeneration. *Nanoscale Adv.* **2022**, *4*, 334–352. [CrossRef]
192. McMillan, A.; Nguyen, M.K.; Gonzalez-Fernandez, T.; Ge, P.; Yu, X.; Murphy, W.L.; Kelly, D.; Alsberg, E. Dual non-viral gene delivery from microparticles within 3D high-density stem cell constructs for enhanced bone tissue engineering. *Biomaterials* **2018**, *161*, 240–255. [CrossRef]
193. Perez, J.R.; Kouroupis, D.; Li, D.J.; Best, T.M.; Kaplan, L.; Correa, D. Tissue engineering and cell-based therapies for fractures and bone defects. *Front. Bioeng. Biotechnol.* **2018**, *6*, 105. [CrossRef]
194. Veatch, J.R.; Singhi, N.; Srivastava, S.; Szeto, J.L.; Jesernig, B.; Stull, S.M.; Fitzgibbon, M.; Sarvothama, M.; Yechan-Gunja, S.; James, S.E.; et al. A therapeutic cancer vaccine delivers antigens and adjuvants to lymphoid tissues using genetically modified T cells. *J. Clin. Investig.* **2021**, *131*, e144195. [CrossRef] [PubMed]
195. Acri, T.M.; Laird, N.Z.; Jaidev, L.R.; Meyerholz, D.K.; Salem, A.K.; Shin, K. Nonviral Gene Delivery Embedded in Biomimetically Mineralized Matrices for Bone Tissue Engineering. *Tissue Eng. Part A* **2021**, *27*, 1074–1083. [CrossRef] [PubMed]
196. Ghandforoushan, P.; Hanaee, J.; Aghazadeh, Z.; Samiei, M.; Navali, A.M.; Khatibi, A.; Davaran, S. Novel nanocomposite scaffold based on gelatin/PLGA-PEG-PLGA hydrogels embedded with TGF-β1 for chondrogenic differentiation of human dental pulp stem cells in vitro. *Int. J. Biol. Macromol.* **2022**, *201*, 270–287. [CrossRef] [PubMed]
197. Sirkkunan, D.; Pingguan-Murphy, B.; Muhamad, F. Directing Axonal Growth: A Review on the Fabrication of Fibrous Scaffolds That Promotes the Orientation of Axons. *Gels* **2022**, *8*, 25. [CrossRef]
198. Abdal-Hay, A.; Sheikh, F.A.; Gómez-Cerezo, N.; Alneairi, A.; Luqman, M.; Pant, H.R.; Ivanovski, S. A review of protein adsorption and bioactivity characteristics of poly ε-caprolactone scaffolds in regenerative medicine. *Eur. Polym. J.* **2022**, *162*, 110892. [CrossRef]

199. Joy, J.; Pereira, J.; Aid-Launais, R.; Pavon-Djavid, G.; Ray, A.R.; Letourneur, D.; Meddahi-Pellé, A.; Gupta, B. Gelatin—Oxidized carboxymethyl cellulose blend based tubular electrospun scaffold for vascular tissue engineering. *Int. J. Biol. Macromol.* **2018**, *107*, 1922–1935. [CrossRef]
200. Murugesan, B.; Pandiyan, N.; Arumugam, G.; Sonamuthu, J.; Samayanan, S.; Yurong, C.; Juming, Y.; Mahalingam, S. Fabrication of palladium nanoparticles anchored polypyrrole functionalized reduced graphene oxide nanocomposite for antibiofilm associated orthopedic tissue engineering. *Appl. Surf. Sci.* **2020**, *510*, 145403. [CrossRef]
201. Jie, W.; Song, F.; Li, X.; Li, W.; Wang, R.; Jiang, Y.; Zhao, L.; Fan, Z.; Wang, J.; Liu, B. Enhancing the proliferation of MC3T3-E1 cells on casein phosphopeptide-biofunctionalized 3D reduced-graphene oxide/polypyrrole scaffolds. *RSC Adv.* **2017**, *7*, 34415–34424. [CrossRef]
202. Díez-Pascual, A.M.; Díez-Vicente, A.L. Wound Healing Bionanocomposites Based on Castor Oil Polymeric Films Reinforced with Chitosan-Modified ZnO Nanoparticles. *Biomacromolecules* **2015**, *16*, 2631–2644. [CrossRef]
203. Lee, M.-H.; You, C.; Kim, K.-H. Combined effect of a microporous layer and type I collagen coating on a biphasic calcium phosphate scaffold for bone tissue engineering. *Materials* **2015**, *8*, 1150–1161. [CrossRef]
204. Siddiqi, N.J.; Abdelhalim, M.A.K.; El-Ansary, A.K.; Alhomida, A.S.; Ong, W.Y. Identification of potential biomarkers of gold nanoparticle toxicity in rat brains. *J. Neuroinflammation* **2012**, *9*, 1–7. [CrossRef] [PubMed]
205. Abdelhalim, M.A.K. Exposure to gold nanoparticles produces cardiac tissue damage that depends on the size and duration of exposure. *Lipids Health Dis.* **2011**, *10*, 1–9. [CrossRef] [PubMed]
206. Love, S.A.; Thompson, J.W.; Haynes, C.L. Development of screening assays for nanoparticle toxicity assessment in human blood: Preliminary studies with charged Au nanoparticles. *Nanomedicine* **2012**, *7*, 1355–1364. [CrossRef] [PubMed]
207. Freese, C.; Gibson, M.I.; Klok, H.A.; Unger, R.E.; Kirkpatrick, C.J. Size-and coating-dependent uptake of polymer-coated gold nanoparticles in primary human dermal microvascular endothelial cells. *Biomacromolecules* **2012**, *13*, 1533–1543. [CrossRef] [PubMed]
208. Fu, Q.; Rahaman, M.N.; Bal, B.S.; Brown, R.F.; Day, D.E. Mechanical and in vitro performance of bioactive glass scaffolds prepared by a polymer foam replication technique. *Acta Biomater.* **2008**, *4*, 1854–1864. [CrossRef]
209. Chen, Q.Z.; Thompson, I.D.; Boccaccini, A.R. 45S5 Bioglass®-derived glass–ceramic scaffolds for bone tissue engineering. *Biomaterials* **2006**, *27*, 2414–2425. [CrossRef]
210. Jiang, L.; Chen, D.; Wang, Z.; Zhang, Z.; Xia, Y.; Xue, H.; Liu, Y. Preparation of an electrically conductive graphene oxide/chitosan scaffold for cardiac tissue engineering. *Appl. Biochem. Biotechnol.* **2019**, *188*, 952–964. [CrossRef]
211. Shamekhi, M.A.; Mirzadeh, H.; Mahdavi, H.; Rabiee, A.; Mohebbi-Kalhori, D.; Eslaminejad, M.B. Graphene oxide containing chitosan scaffolds for cartilage tissue engineering. *Int. J. Biol. Macromol.* **2019**, *127*, 396–405. [CrossRef]
212. Nishida, E.; Miyaji, H.; Takita, H.; Kanayama, I.; Tsuji, M.; Akasaka, T.; Sugaya, T.; Sakagami, R.; Kawanami, M. Graphene oxide coating facilitates the bioactivity of scaffold material for tissue engineering. *Jpn. J. Appl. Phys.* **2014**, *53*, 06JD04. [CrossRef]
213. Arshad, R.; Fatima, I.; Sargazi, S.; Rahdar, A.; Karamzadeh-Jahromi, M.; Pandey, S.; Díez-Pascual, A.M.; Bilal, M. Novel Perspectives towards RNA-Based Nano-Theranostic Approaches for Cancer Management. *Nanomaterials* **2021**, *11*, 3330. [CrossRef]
214. Feng, X.; Jiang, D.; Kang, T.; Yao, J.; Jing, Y.; Jiang, T.; Feng, J.; Zhu, Q.; Song, Q.; Dong, N.; et al. Tumor-homing and penetrating peptide-functionalized photosensitizer-conjugated PEG-PLA nanoparticles for chemo-photodynamic combination therapy of drug-resistant cancer. *ACS Appl. Mater. Interfaces* **2016**, *8*, 17817–17832. [CrossRef] [PubMed]
215. Revia, R.A.; Zhang, M. Magnetite nanoparticles for cancer diagnosis, treatment, and treatment monitoring: Recent advances. *Mater. Today* **2016**, *19*, 157–168. [CrossRef] [PubMed]
216. Naidoo, C.; Kruger, C.A.; Abrahamse, H. Photodynamic therapy for metastatic melanoma treatment: A review. *Technol. Cancer Res. Treat.* **2018**, *17*. [CrossRef] [PubMed]
217. Heo, D.N.; Yang, D.H.; Moon, H.-J.; Lee, J.B.; Bae, M.S.; Lee, S.C.; Lee, W.J.; Sun, I.-C.; Kwon, I.K. Gold nanoparticles surface-functionalized with paclitaxel drug and biotin receptor as theranostic agents for cancer therapy. *Biomaterials* **2012**, *33*, 856–866. [CrossRef]
218. Dilnawaz, F.; Singh, A.; Mohanty, C.; Sahoo, S.K. Dual drug loaded superparamagnetic iron oxide nanoparticles for targeted cancer therapy. *Biomaterials* **2010**, *31*, 3694–3706. [CrossRef]
219. Dash, B.; Jose, G.; Lu, Y.-J.; Chen, J.-P. Functionalized reduced graphene oxide as a versatile tool for cancer therapy. *Int. J. Mol. Sci.* **2021**, *22*, 2989. [CrossRef]
220. Shen, J.M.; Gao, F.Y.; Yin, T.; Zhang, H.X.; Ma, M.; Yang, Y.J.; Yue, F. cRGD-functionalized polymeric magnetic nanoparticles as a dual-drug delivery system for safe targeted cancer therapy. *Pharmacol. Res.* **2013**, *70*, 102–115. [CrossRef]
221. Zhang, Q.; Neoh, K.G.; Xu, L.; Lu, S.; Kang, E.T.; Mahendran, R.; Chiong, E. Functionalized mesoporous silica nanoparticles with mucoadhesive and sustained drug release properties for potential bladder cancer therapy. *Langmuir* **2014**, *30*, 6151–6161. [CrossRef]
222. Kim, S.H.; Lee, J.E.; Sharker, S.M.; Jeong, J.H.; In, I.; Park, S.Y. In vitro and in vivo tumor targeted photothermal cancer therapy using functionalized graphene nanoparticles. *Biomacromolecules* **2015**, *16*, 3519–3529. [CrossRef]
223. Wu, Y.-F.; Wu, H.-C.; Kuan, C.-H.; Lin, C.-J.; Wang, L.-W.; Chang, C.-W.; Wang, T.-W. Multi-functionalized carbon dots as theranostic nanoagent for gene delivery in lung cancer therapy. *Sci. Rep.* **2016**, *6*, 1–12. [CrossRef]
224. Xia, Y.; Chen, Y.; Hua, L.; Zhao, M.; Xu, T.; Wang, C.; Li, Y.; Zhu, B. Functionalized selenium nanoparticles for targeted delivery of doxorubicin to improve non-small-cell lung cancer therapy. *Int. J. Nanomed.* **2018**, *13*, 6929–6939. [CrossRef] [PubMed]

225. Zhang, X.; Wu, J.; Williams, G.R.; Niu, S.; Qian, Q.; Zhu, L.-M. Functionalized MoS2-nanosheets for targeted drug delivery and chemo-photothermal therapy. *Colloids Surf. B Biointerfaces* **2019**, *173*, 101–108. [CrossRef] [PubMed]
226. Ammarullah, M.I.; Afif, I.Y.; Maula, M.I.; Winarni, T.I.; Tauviqirrahman, M.; Akbar, I.; Basri, H.; Van der Heide, E.; Jamari, J. Tresca Stress Simulation of Metal-on-Metal Total Hip Arthroplasty during Normal Walking Activity. *Materials* **2021**, *14*, 7554. [CrossRef] [PubMed]
227. Basri, H.; Syahrom, A.; Prakoso, A.T.; Wicaksono, D.; Amarullah, M.I.; Ramadhoni, T.S.; Nugraha, R.D. The Analysis of Dimple Geometry on Artificial Hip Joint to the Performance of Lubrication. *J. Phys. Conf. Ser.* **2019**, *1198*, 042012. [CrossRef]
228. Webster, T.J.; Siegel, R.W.; Bizios, R. Osteoblast adhesion on nanophase ceramics. *Biomaterials* **1999**, *20*, 1221–1227. [CrossRef]
229. Tran, P.A.; Sarin, L.; Hurt, R.H.; Webster, T.J. Opportunities for nanotechnologyenabled bioactive bone implants. *J. Mater. Chem.* **2009**, *19*, 2653–2659. [CrossRef]
230. Zinger, O.; Anselme, K.; Denzer, A.; Habersetzer, P.; Wieland, M.; Jeanfils, J.; Hardouin, P.; Landolt, D. Time-dependent morphology and adhesion of osteoblastic cells on titanium model surfaces featuring scale-resolved topography. *Biomaterials* **2004**, *25*, 2695–2711. [CrossRef]
231. Chu, P.K.; Chen, J.Y.; Wang, L.P.; Huang, N. Plasma-surface modification of biomaterials. *Mater. Sci. Eng. R Rep.* **2002**, *36*, 143–206. [CrossRef]
232. Yu, G.; Hu, L.; Vosgueritchian, M.; Wang, H.; Xie, X.; McDonough, J.R.; Cui, X.; Cui, Y.; Bao, Z. Solution-processed graphene/MnO$_2$ nanostructured textiles for high-performance electrochemical capacitors. *Nano Lett.* **2011**, *11*, 2905–2911. [CrossRef]
233. Rojaee, R.; Fathi, M.; Raeissi, K. Electrophoretic deposition of nanostructured hydroxyapatite coating on AZ91 magnesium alloy implants with different surface treatments. *Appl. Surf. Sci.* **2013**, *285*, 664–673. [CrossRef]
234. Vorobyev, A.Y.; Guo, C. Direct femtosecond laser surface nano/microstructuring and its applications. *Laser Photonics Rev.* **2013**, *7*, 385–407. [CrossRef]
235. Bolelli, G.; Bellucci, D.; Cannillo, V.; Lusvarghi, L.; Sola, A.; Stiegler, N.; Müller, P.; Killinger, A.; Gadow, R.; Altomare, L.; et al. Suspension thermal spraying of hydroxyapatite: Microstructure and in vitro behaviour. *Mater. Sci. Eng. C* **2014**, *34*, 287–303. [CrossRef] [PubMed]
236. Sima, F.; Davidson, P.M.; Dentzer, J.; Gadiou, R.; Pauthe, E.; Gallet, O.; Mihailescu, I.N.; Anselme, K. Inorganic-organic thin implant coatings deposited by lasers. *ACS Appl. Mater. Interfaces* **2014**, *7*, 911–920. [CrossRef] [PubMed]
237. McEntire, B.; Bal, B.S.; Rahaman, M.; Chevalier, J.; Pezzotti, G. Ceramics and ceramic coatings in orthopaedics. *J. Eur. Ceram. Soc.* **2015**, *35*, 4327–4369. [CrossRef]

Review

Latent Potential of Multifunctional Selenium Nanoparticles in Neurological Diseases and Altered Gut Microbiota

Hajra Ashraf [1,*], Davide Cossu [1], Stefano Ruberto [1], Marta Noli [1], Seyedesomaye Jasemi [1], Elena Rita Simula [1] and Leonardo A. Sechi [1,2,*]

1. Department of Biomedical Sciences, University of Sassari, 07100 Sassari, Italy
2. Complex Structure of Microbiology and Virology, AOU Sassari, 07100 Sassari, Italy
* Correspondence: hajraashraf67@gmail.com (H.A.); sechila@uniss.it (L.A.S.)

Abstract: Neurological diseases remain a major concern due to the high world mortality rate and the absence of appropriate therapies to cross the blood–brain barrier (BBB). Therefore, the major focus is on the development of such strategies that not only enhance the efficacy of drugs but also increase their permeability in the BBB. Currently, nano-scale materials seem to be an appropriate approach to treating neurological diseases based on their drug-loading capacity, reduced toxicity, targeted delivery, and enhanced therapeutic effect. Selenium (Se) is an essential micronutrient and has been of remarkable interest owing to its essential role in the physiological activity of the nervous system, i.e., signal transmission, memory, coordination, and locomotor activity. A deficiency of Se leads to various neurological diseases such as Parkinson's disease, epilepsy, and Alzheimer's disease. Therefore, owing to the neuroprotective role of Se (selenium) nanoparticles (SeNPs) are of particular interest to treat neurological diseases. To date, many studies investigate the role of altered microbiota with neurological diseases; thus, the current review focused not only on the recent advancement in the field of nanotechnology, considering SeNPs to cure neurological diseases, but also on investigating the potential role of SeNPs in altered microbiota.

Keywords: selenium; nanoparticles; neurological diseases; gut microbiota

1. Introduction

Neurological diseases are regarded as the world's leading cause of disability and mortality, and they account for 12% of global deaths. The most common neurological diseases include Alzheimer's disease, Parkinson's disease, and multiple sclerosis [1].

The central nervous system comprises the brain and spinal cord, which play an important role in neurological diseases. According to the body's function and regulation, the CNS has three predominant barriers, i.e., the blood–brain barrier (BBB), the cerebrospinal fluid–blood barrier (the avascular arachnoid epithelium), and the blood–cerebrospinal fluid barrier (the choroid plexus epithelium). Owing to these naturally existing barriers, particularly the BBB, the treatment of neurological diseases through drug delivery into the CNS is challenging [2]. However, there are some FDA-approved drugs that are currently used for neurological disease treatment (Table 1).

Currently, there is no effective therapy for many neurological diseases. Scientists and technologists from multidisciplinary fields, i.e., from behavior to the molecular level, have carried out research in multiple directions, but a truly interdisciplinary way of treatment has not yet been explored. The ultimate consequence of this is that many pathological disorders involving the central nervous system (CNS) remain untreated.

Nanoparticles (NPs) represent a promising approach in the treatment of neurodegenerative diseases, specifically Parkinson's and Alzheimer's disease (REF) [3,4]. Drug delivery through nanosized particles not only crosses the blood–brain barrier but also makes for target-specific delivery. Moreover, numerous benefits are associated with NPs to treat the CNS, i.e., high biological and chemical stability, ability to be administered by various routes, large surface-to-volume ratio, and feasibility to incorporate both hydrophobic and hydrophilic drugs [5,6].

Table 1. FDA approved drugs for neurological diseases.

FDA-Approved Drugs for Neurological Diseases			
Drug Name	Approval	Disease	Indications
Briumvi	28 December 2022	Multiple sclerosis (MS)	BRIUMVI is a CD20-directed cytolytic antibody indicated for the treatment of relapsing forms of multiple sclerosis (MS)
Relyvrio	29 September 2022	Amyotrophic lateral sclerosis (ALS)	RELYVRIO is indicated for the treatment of amyotrophic lateral sclerosis (ALS) in adults.
Aduhelm	7 June 2021	Alzheimer's disease	To treat Alzheimer's disease
Suvorexant	29 January 2020	Mild-to-moderate Alzheimer's disease (AD)	Insomnia characterized by difficulties with sleep onset and/or sleep maintenance
18F-Fluortaucipir	28 May 2020	Alzheimer's disease (AD)	Evaluation of tau neurofibrillary tangle (NFT) density and distribution with positron-emission tomography
Ozanimod	25 March 2020	Multiple sclerosis (MS)	Relapsing multiple sclerosis (MS), including clinically isolated syndrome (CIS) and active secondary progressive MS (aSPMS) in adults
Inebulizumab	12 June 2020	neuromyelitis optica spectrum disorder (NMOSD)	Antiaquaporin-4 positive (AQP4)$^+$ neuromyelitis optica spectrum disorder (NMOSD)
Satralizumab	16 August 2020	neuromyelitis optica spectrum disorder (NMOSD)	Antiaquaporin-4 positive (AQP4)$^+$ neuromyelitis optica spectrum disorder (NMOSD)
Ofatumumab	20 August 2020	Multiple sclerosis (MS)	Relapsing forms of multiple sclerosis (MS), including clinically isolated syndrome (CIS) and active secondary progressive MS (aSPMS) in adults

Selenium (Se), being an important trace element in the body, showed remarkable health benefits, i.e., improving the immune system [7], securing the nervous system's physiological activity [8], and combating oxidative damage caused by free radical species [9]. As an integral component of selenoproteins, Se has an essential role in the fundamental functioning of the CNS [10]. Therefore, a deficiency of Se contributes to the pathogenesis of various neuropathological and neurodegenerative diseases. Se supplementation has numerous beneficial impacts on neurological diseases. However, Se has a narrow range between toxic and beneficial doses. The Expert Group on Vitamins and Minerals (EVM) recommended that the daily dose of Se should be 60 μg for women and 70 μg for men [11,12], a dose above 400 μg is considered toxic and leads to a disorder known as selenosis. Owing to their incredible health benefits, Se (selenium) nanoparticles (SeNPs) gained worldwide attention due to their wide application in the field of therapeutics. SeNPs have lower toxicity, higher efficiency to resist free radical species, and acceptable bioavailability in comparison to inorganic selenium. Moreover, based on the experimental data, the toxicity of SeNPs is classified as lower than that of other organic and inorganic compounds such as selenate, selenite, and selenomethionine. SeNPs are involved in numerous physiological and metabolic processes, such as the regulation of the immune system and the antioxidant defense system [13–15]. Additionally, SeNPs have a strong capability to penetrate biological cells and tissues, suggesting their potential efficiency to inhibit oxidative stress and inflammation [16,17]. Owing to these unique advantages, recently, SeNPs have snatched a lot of attention from scientists for their use in the treatment of neurological diseases.

In light of the above-mentioned discussion, the current review summarizes the potential benefits of SeNPs to treat neurological diseases. Since recent studies have investigated the role of the altered microbiota in neurological diseases, this review also provides insight into how SeNPs can regulate the altered microbiota, a crucial step in opening new perspectives on the use of SeNPs as potential pharmacotherapy.

2. Materials and Methods

This review is based on Google Scholar and PubMed searches using the following keywords: neurological diseases, selenium nanoparticles, and microbiota. The final search was performed in December 2022, and recent papers with high relevance were selected for the review.

3. Results and Discussions

3.1. Selenium Compounds and Their Physiological Effects

Selenium is an essential trace element that plays an important role in various physiological functions, including reactive oxygen species (ROS) control and modulation in the immune system [18]. According to the European Food Safety Authority, the recommended daily allowance (RDA) of Se is 70 µg day^{-1} for men, 60 µg/day for women [19], 65 µg day^{-1} for pregnant women, and 75 µg day^{-1} for lactating women [20,21].

The main form of Se is the Se analog of the amino acid methionine known as selenomethionine (SeMet), which is absorbed and makes an entry to the methionine pool in the body after digestion [22,23]. Selenium in the form of inorganic selenate and selenite are mostly used as supplementation. Se often plays a major role in the generation of selenoproteins that is essential for the body due to their multiplex roles, i.e., protein folding, control over thyroid hormone metabolism, redox signaling, etc. [20].

Se is also known to have antibacterial, antiviral, antifungal, and antitumor properties. In addition, various studies confirmed Se's role in thyroid, cardiovascular, and neurological diseases [19,20,24]. Se adequate amount supports the immune system by enhancing the activity of natural killer (NK) cells and the proliferation of T cells against pathogens and cancer cells, also enhancing the efficacy of vaccines [25,26]. It also contributed to the reduction of risk associated with various inflammation-related diseases like rheumatoid arthritis [27]. Se maintains ROS production and enhances DNA stability while decreasing the renal and hepatic side effects of chemotherapeutic drugs [18].

3.1.1. Selenium Bioavailability, Metabolism, and Physiological Functions

The bioavailability of Se depends upon the food consumed [28], being more prevalent in animal products than in vegetables. The content of Se is more influenced by the source of the animal and also its species, i.e., fish have elevated levels of Se. SeMet is abundant in both animals and plants, whereas selenocysteine is mostly present in animals. The principal form of selenium in the body is SeMet, as it enters the Se pool directly [23,29].

Under physiological conditions, all forms of Se have an absorption rate of 70–90%, except selenite, which has a lower absorption rate of 60%. In addition, food processing also influences bioavailability, as proteins are more easily digestible at higher temperatures and Se release and bioavailability become more efficient. Due to synergistic, additive, and antagonistic interactions, total carbohydrate, fat, protein, and fiber contents also influenced Se bioavailability [30,31].

Se metabolism occurs mainly in the liver as it is responsible for selonoprotein synthesis and excretion via various selenometabolites. Mostly Se is excreted through urine, while some is significantly excreted through feces [30].

3.1.2. Se Potential Therapeutic Impact

Various studies confirmed the immunomodulatory and anti-inflammatory role of Se and its supplementation has been demonstrated to cure various anti-inflammatory diseases, i.e., chronic lymphedema, Crohn's disease, asthma, and chronic lymphedema [24,32]. Se is also known to be effective against cancer as it decreases ROS production and prevents gene dysfunction and DNA damage generated by oxidative stress in the body. It is also used as a chemotherapeutic and radiotherapy adjuvant, as its pro-oxidant effects are more effective on malignant cells than on healthy cells. [24,33].

Deficiency of Se correlates with various bacterial, parasitic, and viral infections, which show the influence of Se on the function of the immune system [25,34] as HIV, H1N1, influenza, West Nile virus infection, etc. [29]. Se supplementation has also been proven to be favorable for the treatment of numerous bacterial infections such as *Mycobacterium tuberculosis, Helicobacter pylori, Escherichia coli*, etc. Instead of provoking an immune response against the poliovirus and influenza A vaccinations, Se supplementation has also corresponded to antiparasitic properties against *Heligmosomoides bakeri* and *Trypanosoma cruzi* [25,34].

The potential role of Se in cardiovascular diseases has been revealed by numerous studies due to its protection against excessive platelet aggregation and oxidative damage, which ultimately stop the pathologies of cardiovascular diseases, i.e., heart hypertrophy, atherosclerosis, congestive failure, and hypertension [24,29,35]. Due to the regulatory effect of selenoproteins on the insulin signaling cascade, Se is also associated with the prevention of type 2 diabetes. Reduction in insulin resistance is shown to be due to selenoproteins, as they diminish pancreatic insulin production and indirectly thioredoxin reductases (TR) lower insulin resistance. However, some studies depict a higher association of Se supplementations with a greater risk of type 2 diabetes, so the role of Se in diabetes is not yet clear [36,37].

Selenium is present in glandular and gray matter regions of the brain and contributes to various dopaminergic and neurotransmission pathways; hence, it is also used as a potential biomarker in various neurological diseases, i.e., Alzheimer's, epilepsy, and Parkinson's diseases [24,38]. The antioxidant neuroprotective function of Se creates a strong impact on the hyperphosphorylation of the tau protein, cytoskeleton assembly regulation, Aβ deposition attenuation, and the tendency to bind with neurotoxic metals, which constitutes its ability to have a potential role in the development of Alzheimer's disease. Various selenoproteins were also studied to protect dopaminergic neurons, strengthening the neuroprotective role of Se against Parkinson's disease [39,40]. Considering the low level of Se in the brain, the applications of Se will only be beneficial for patients who have severe Se deficiency, lower selenoprotein production, or mutations in genes associated with the delivery of Se [39]. The main Se potential in neurological diseases is described in Figure 1.

3.2. Preparation and Characterization Methods of Se Nanoparticles (SeNPs)

SeNP Production Methods

Se can be synthesized from three common methods of obtaining nanoparticles: physical, chemical, and biological methods (Figure 2). Since chemical methods involved the use of high temperatures, dangerous chemicals, and an acidic pH for the catalytic reduction of ionic selenium, this represented a less safe method for the synthesis of Se nanoparticles (SeNPs) [41,42]. Physical methods such as electrodeposition techniques, phyto-thermal-associated synthesis, and microwave synthesis are less common than chemical methods. The third and most effective method used nowadays is the biological method, which uses algae, yeast, fungi, and plants as biological catalysts for the production of nanoparticles. The biological method is advantageous over the other two methods due to its lower cost, fast growth rate of microorganisms and plants, lower toxicity, common procedures for culturing, the nonexistence of severe extreme conditions, and eco-friendly production of nanoparticles (Table 2) [43–46].

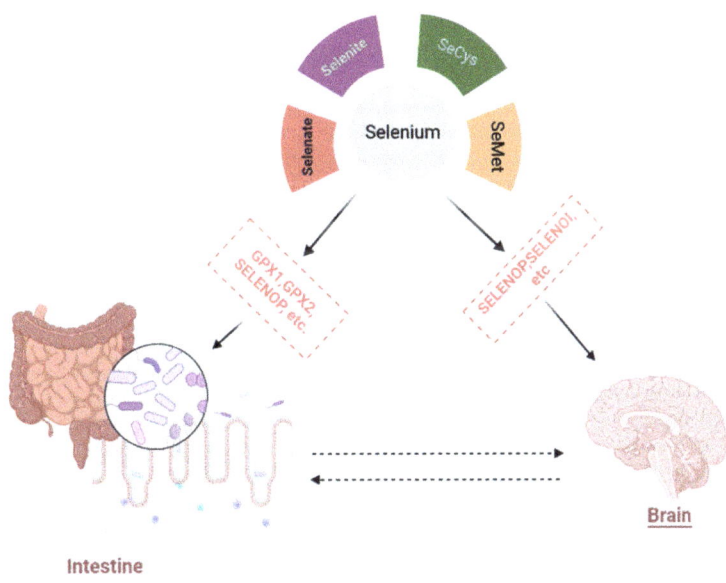

Figure 1. Different selenoproteins regulate different organs in the body. Se is mainly absorbed in the form of selenoproteins. GPX1 and GPX2 maintain the body's health by regulating the production of reactive oxygen species (ROS). SELENOP normally acts as a plasma transporter in numerous organs, while DIO1 affects thyroid hormone activity. SELENOI, on the other hand, is involved in managing the nervous system, and a deficiency of SELENOI results in the emergence of neurodegenerative diseases.

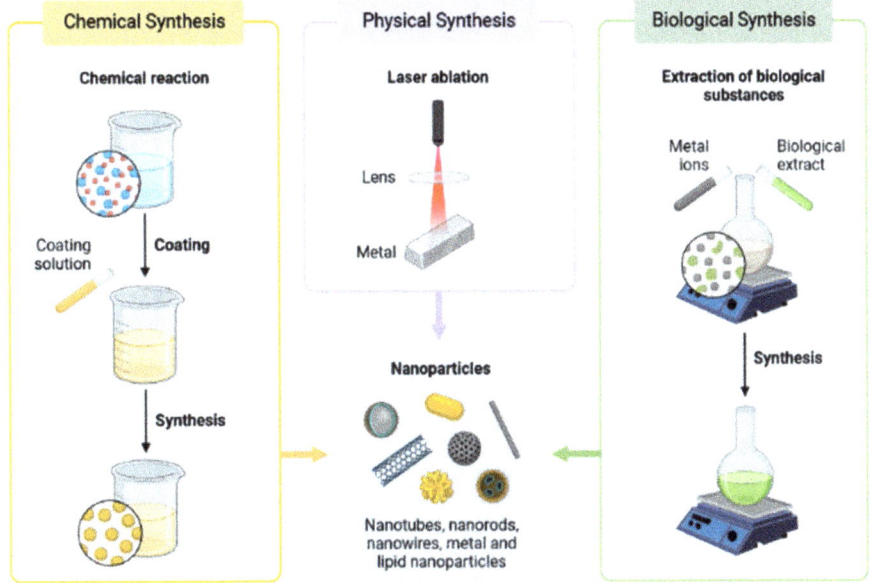

Figure 2. Nanoparticles Methods of Production.

Table 2. Comparative evaluation of SeNPs production methods.

Method of Production	Materials	Characteristics	Advantages	Disadvantages
Chemical Method	• Inorganic Se (i.e., selenate or selenite) reduction by a reducing agent. • Use of a capping agent for stabilization of nanoparticles.	• Characteristics of NPs depend upon stabilizing agents.	• Simple method without the need for technological instruments and biological incubation.	• Use of harmful chemicals that make it a less environment-friendly method.
Physical Method	• Usage of physically based methods, i.e., laser ablation, heating, etc., to induce changes in inorganic Se in the presence of the stabilizing agent.	• Characteristics of NPs depend upon stabilizing agents. • Small-sized nanoparticles production.	• Environment-friendly process. • Rapid reaction. • Less energy spent.	• Specific instrument necessities.
Biological Method	• Use of biological agent as a stabilizing and reducing agent for inorganic selenium.	• The characteristics of NPs depend upon biological organisms, i.e., plants, fungi, and yeast.	• Environment-friendly process. • No need of extra stabilizing agent as biological organisms itself acts as both reducing and stabilizing agent.	• Need for optimization of several steps and processes in order to obtain NPs.

The biosynthesis of SeNPs has been conducted using various plant extracts, i.e., *Cinnamomum zeylanicum* bark, fresh citrus and lemon fruits [47], Aloe vera leaf extracts, *Dillenia indica* [48], *Vitis vinifera* [49], *Prunus amygdalus* leaf [50], *Allium sativum* [51], etc. The main benefit of using plant extracts is that plant's secondary metabolites themselves act as natural reductant and stabilizer agents in an eco-friendly approach.

Due to the biological activities of selenium, SeNPs are widely used for various biomedicinal applications, for example, in the treatment of neurological diseases and diabetes as an antiviral, antibacterial, anti-apoptotic, and anti-inflammatory drug, and for the effective delivery of selective drugs into the tissues.

It is important to determine the physical characteristics of nanoparticles because the shape and size of nanoparticles affect their activity on cells and tissues. For example, Se nanowires have higher photoconductivity, while spherical-shaped SeNPs have been proven to have higher biological activities [21]. The antioxidant properties of nanoparticles also depend on their size: SeNPs have been shown to scavenge free radicals in a size-dependent manner (5–200 nm) [52]. The functionalization of NPs with other substances also depends upon the shape and size of NPs, i.e., the effectiveness of chitosan as an antioxidant and antitumor agent firmly depends upon SeNPS characteristics [16,53]. The synthesis method influences the shape and size of NPs and, consequently, their medicinal properties. There are different forms of SeNPSs, such as rod-like, hexagonally flowered, nanowires, nanotubes, nanoneedles, and nanorods. The spherically shaped SeNPs are more commonly used for pharmacological and biological purposes [54].

3.3. Role of SeNP in Neurodegenerative Diseases

Se, being a principal trace element in humans and animals, plays a remarkable role in regulating the standard physiological functions of the brain. It also has a neuroprotective role, and some selenoproteins also participate in the protection against neurodegenerative diseases. Studies proved that the metabolism of Se in the brain is different from that in other body organs, as Se remains preserved in the brain in the case of Se deficiency [55–57]. Currently, SeNPs' role in brain diseases has been studied because neurons are more prone to be damaged by oxidative stress due to several reasons, such as a low level of antioxidant enzymes, a high consumption of oxygen, and occupancy of the high level of polyunsaturated fats [58–60].

3.3.1. Alzheimer's Disease and SeNPs

One of the main factors in the pathogenesis of neurodegenerative diseases is oxidative stress. Numerous natural antioxidants are used as treatments, but the hurdle is their limited accuracy [56,57]. Therefore, the focus is now on the synthesis of nanoparticles that have greater antioxidant potential. Various studies showed that nanoparticles more often act as an oxidizing agent and may cause damage to neurons, decreasing the cognitive functions of Alzheimer patients' brains [61–65]. Despite that, several studies pointed out that SeNPs in Alzheimer's disease prevent the aggregation of amyloid-β (Aβ) protein and also can cross the BBB [66,67]. It has been demonstrated that SeNPs coated with epigallocatechin-3-gallate and peptide B6 had a similar effect [68]. Xianbo Zhoub et al. determined that cysteine enantiomer modified SeNPs (abbreviated as D/LSeNPs) demonstrated a strong impact on the aggregation of Aβ in the presence of metal ions, i.e., Cu^{2+} and Zn^{2+}. These SeNPs modified by the chelating agent can prevent Aβ fibril formation by blocking metal ion binding sites and by binding with Aβ. Modified SeNPs are more effective in protecting the cell because of their effective absorption by PC12 cells, protection from oxidative stress, and potential to maintain cellular redox potential [69].

A considerable therapeutic promise in Alzheimer's disease is the inhibition of amyloid β (Aβ) aggregation. Although, the non-selective disposition of drugs and BBB put a major hurdle in achieving this. A study conducted by Licong Yang et al. demonstrated that the conjugation of the targeted peptide with SeNPs acts as dual-functional NPs that not only cross the BBB but also inhibit the aggregation of Aβ [70].

Similarly, in another study conducted by Dongdong Sun et al., it was found that SeNPs coated with the chelating agent were effective in preventing Aβ aggregation, memory impairment, and ameliorating cognition [71].

Nevertheless, the current focus is on the synthesis of nanoparticles based on natural resources for the cure of AD [72], as resveratrol (Res)-polyphenol, which is mainly found in plants, has an antioxidant and especially a neuroprotective effect [73,74]. Thus, the synthesis of SeNPs with Res coating enhanced the antiaggregatory and antioxidant potency of reversatol, which was demonstrated on PC12 cells of the adrenal medulla of rats [75]. The potency of ResSeNPs to bind with Aβ42 and block the Cu^{2+} binding that leads to cell death by damaging the cell membrane has been demonstrated [75].

3.3.2. SeNPs and Parkinson's Disease

The second most progressive neurodegenerative disease is Parkinson's disease, which has the main characteristics of muscle rigidity, dyskinesia with tremors, postural instability, and bradykinesia [76–79]. Although the pathophysiology of Parkinson's disease is not yet clear, oxidative stress is regarded as one of the prime pathological markers of PD as it results in neuronal damage and ultimately death [80,81]. Yue Dong et al. evaluated the antioxidant and therapeutic potential of glycine-SeNPs. For the study of Parkinson's disease MPTP (1-methyl-4-phenyl-1,2,3,6-tetrahydropyridine) is considered a potential neurotoxin. Two animal group models were designed with and without MPTP to check the neuroprotective effect of glycine-SeNPs. Results depicted that glycine-SeNPs decreased the MDA level and increased GSH-PX activity and SOD activity, thus influencing a neuroprotective effect in comparison to MPTP-induced PD rats [82].

3.4. Selenium Nanoparticles and Gut–Brain Axis

About 2500 years ago, Hippocrates stated that the gut was responsible for the beginning of all diseases. With time, this statement gains a lot of support from the ongoing research on animal models and humans. The gut is regarded as the home of a diverse and complex ecosystem of trillions of microorganisms that include yeasts, bacteria, viruses, protozoa, and archaea [83]. The human gut microbiota is considered a unique entity that is shaped by lifestyle and diet, and as a result, the physiology of the host is shaped by microorganisms [84,85]. Host and gut microbiome symbiotic relationships start when embryonic development is shaped by maternal microbiota and initiate gut microbiota colonization during birth and development [86–88]. The microbiota influenced the maturation of the neural, immune, and endocrine systems and played a remarkable role in cognitive and postnatal brain development [89–91].

Methods in the Study of the Microbiota

High-throughput DNA sequencing technologies have made possible the detailed study of the microbiome. The two techniques that have largely been used to study the microbiome are based on whole metagenome sequencing and 16S ribosomal RNA gene sequencing (Figure 3). The initial steps for both methodologies involved the isolation of microbial cells from host cells, DNA extraction, and amplification using a random primer (for metagenomics) or gene-specific primers (16s rRNA).

The gene that encodes 16s rRNA is a unique identifier of closely related and individual species because it contains both highly conserved and hypervariable regions. The 16sRNA gene identified the bacterial species in the sample either by comparing it with the reference genome or by clustered de novo. This approach uses quantitative measures to describe species' evenness, diversity, and relative abundance of specific groups of closely related species. In the metagenomic approach, unbiased sequencing of DNA is conducted for all the microbial species present in the sample [92,93].

3.5. Gut Microbiota and Neurodegenerative Diseases

Microorganisms living in the gastrointestinal tract (GI) have gained prime interest in studies of their role in neurological diseases. The GI tract is extremely vascularized, having an enriched lymphatic system tract, and is animated by a multiplex enteric nervous system, which is renowned as "the second brain". Thus, there are numerous access points through which luminal microbes can gain access and influence the host immune response either directly or indirectly. The diverse population of microorganisms, i.e., Firmicutes and Bacteroidetes, largely participate in the colonization of the GI tract [93,94]. The gut commensal microbes enhanced the digestion and absorption of nutrients and yielded enhanced enzymatic activity by expressing unique genes [95]. Gut microbes use compounds derived from these nutrients as a source of metabolic intermediates and energy [96]. Thus, it becomes clear that the gut microbiome has a considerable role in human physiology, and dysbiosis results in a wide range of neurological and other diseases, including diabetes and obesity.

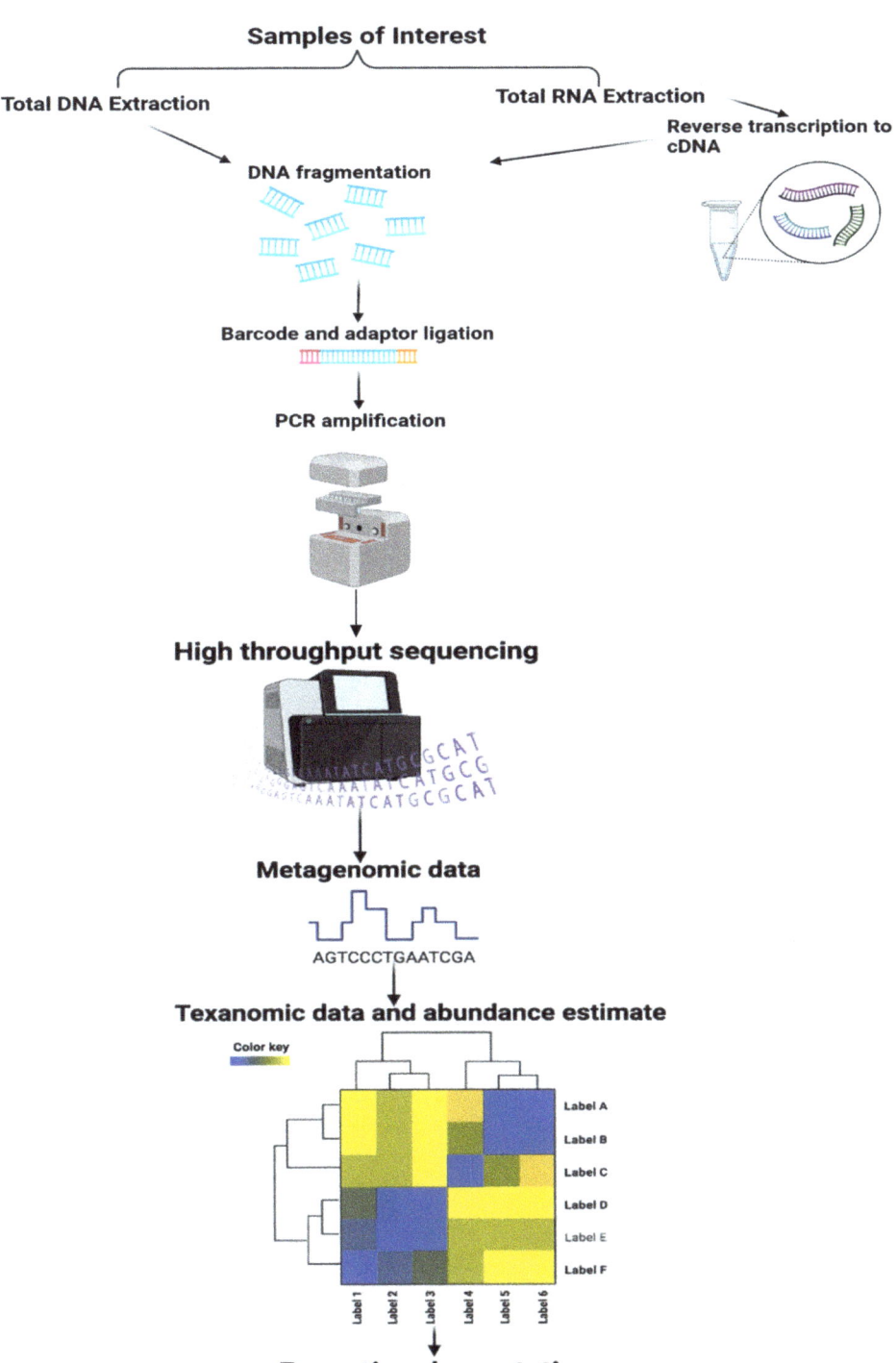

Figure 3. Overview of key steps involved in metagenome study.

3.5.1. Parkinson's Diseases (PD)

The pathology of the gut is a well-known marker of Parkinson's disease. About 60–80% of patients suffered from constipation up to 20 years before the clinical onset of PD, and it is regarded as one of the earliest symptoms [97–99]. It is noteworthy that at the earliest stage of the disease, deposition of α-synuclein is observed even before motor pathology onset [100–102]. Considering these findings, it is suspected that in the gut, the pathology of PD occurs before expanding into the brain. Chandra et al. [103] conducted a study to gain insight into the role of the gut microbiome in PD. A germ-free gnotobiotic animal model is used for the study. It was observed that ASO mice growing in germ-free conditions overexpress α-synuclein as compared to colonized ASO mice. The germ-free ASO mice were then inoculated with microbial metabolites derived from carbohydrates and short-chain fatty acids, which, as a result, promoted the pathology of PD. Additionally, antibody treatment enhanced the PD burden. Appealingly recolonization of ASO mice with the microbiota of healthy donors results in improved cognitive behavior in PD mice in comparison to ASO mice recolonized with the microbiota of PD patients. Gut microbiome dysbiosis is also revealed in human PD. Compared to control, microbial species, i.e., *Ralstonia, Coprococcu, Blautia*, and *Roseburia*, are increased in PD patients, while microbial communities belonging to the *Prevotellaceae and Faecalibacterium* families are decreased in the observed stool samples. It is also observed that *Enterobacteriaceae* family abundance is also significantly associated with gait dysfunction and postural instability [104,105].

3.5.2. Alzheimer's Disease (AD)

The relationship between gut microbiota and AD pathogenesis is well understood in the animal model. Minter et al. [106] first reported the relationship of AD with microbiota. It was observed that the murine model of AD was influenced by antibiotic-induced perturbations in the gut microbiota diversity, and as a result, amyloidosis and neuroinflammation occurred. In another study, the sequencing of 16s rRNA was performed by analyzing the fecal samples of APP transgenic mice with the control, which revealed a significant gut microbiome difference between them. In germ-free transgenic APP mice, cerebral Aβ was also reduced. However, the recolonization of germ-free transgenic APP mice with the microbiota of transgenic APP mice results enhanced the level of cerebral Aβ, and this effect was less when the microbiota of wild-type mice was used [107].

3.5.3. Multiple Sclerosis (MS)

The gnotobiotic mouse also has been effective in studying MS pathology's relationship with microbiota [108]. Transgenic EAE mice grown in sterile environments experienced no diseases or markedly attenuated disease; however, colonization with MS patients' microbiota restored the phenotype of EAE [109,110]. Further studies supported this linkage of microbiota with MS pathology, i.e., Berer k. et al. [111] observed in their study that oral administration of *Bifidobacterium animalis* and *Bacteroides fragilis* reduced the development of MS disease. The role of human gut microbiota in MS directly comes from the comparison of the microbiota of healthy controls and MS patients. One large study reported that microbial populations, i.e., *Akkermansia, Butyricimonas*, and *Methanobrevibacter*, are different between both MS patients and healthy controls [112]. Vicente Navarro et al. researched the linkage of gut microbiota with MS patients having active relapsing-remitting multiple sclerosis (RRMS). The results showed a difference in microbial species at *Clostridium, Hungatella, Lachnospiraceae, Shuttleworthia, Bilophila, Poephyromonas*, and *Ruminococcaceae* between healthy control and RRMS patients [113]. In another study, Sherein G.Elgendy et al. found that alterations in microbiota are directly linked with the exacerbation of MS. Disruption in intestinal microbiota results in the enrichment or depletion of certain bacteria that leads to MS predisposition. *Desulfovibrio, Firmicutes, Actinobacteria*, and lactic acid bacteria were higher in MS patients in comparison to healthy controls, while *Clostridium cluster IV* is comparatively lower in MS patients [114]. A new perspective on how microbiota influenced MS patients was explained by Atsushi Kadowaki et al. A study found that gut microbiota-dependent CCR9 CD4 T cells were altered in secondary progressive multiple sclerosis (SMPS), which leads to the development of SMPS [115].

3.6. Selenium Nanoparticles, Microbiota, and Neurodegenerative Diseases

The synergetic communication between the central nervous system and gut, mediated by gut microbiota, plays a significant role in the development of neurological diseases such as Alzheimer's disease [116] (Table 3). Vogt et al. [117] conducted an extensive sequencing of stool samples, showing the difference between microbiome diversity in healthy controls and AD patients. At the phylum level, actinobacteria have a lower prevalence, while Firmicutes are present in abundance. Similarly, at the genus level, *Gemella*, *Blautia*, *Alistipes*, and *Phascolarctobacterium* are at a higher level in comparison to *Clostridium* and *Bifidobacterium*, which are less abundant. The difference in the microbiome between AD and healthy controls strongly suggested that altered gut microbiota are directly linked with alternations in AD neuropathology. Another study conducted by Mancuso et al. found excessive *Shigella* abundance in comparison to *Eubacterium rectale* in amyloid-positive patients [118]. Probiotics have a significant effect on modulating the gut–brain axis.

Additionally, microbiota dysbiosis also leads to the secretion of inflammatory-related molecules, such as lipopolysaccharide and amyloids, and causes damage to the intestinal mucosal barrier, ultimately stimulating neuroinflammation and microglia activation, which are possibly involved in the progression of neurodegeneration [119]. Enhanced permeability of the intestine causes enhanced metabolite accumulation and translocation, resulting in microbial community imbalance [120]. One of the important pattern recognition receptors that are involved in brain inflammation through the activation and release of microglia and other inflammatory factors is Toll-like receptor 4 (TLR4). TLR4 is majorly activated by lipopolysaccharide (LPS), resulting in the activation of inflammation-related signaling pathways. [121]. Hou et al. [122] demonstrated that high plasma LPS levels and intestinal permeability directly correspond with inflammatory cytokine expression in mouse brains. Therefore, high levels of LPS may cause microglia activation because of intestinal barrier dysfunction. Hence, the microbiota–gut–brain axis concept was based on the communication between the brain and the gut microbiota achieved by the enteric nervous system, the vagus nerve, the immune system, and microbial metabolites, i.e., tryptophan, proteins, and short-chain fatty acids (SCFAs) (Figure 4). Current studies investigated whether the administration of probiotics enhanced the pathophysiology of autoimmune neurological diseases, i.e., AD. Akbari et al. [123] illustrated in their study that the administration of probiotics containing *Lactobacillus casei*, *Lactobacillus fermentum*, *Lactobacillus acidophilus*, and *Bifidobacterium bifidum* had a positive effect on AD patients.

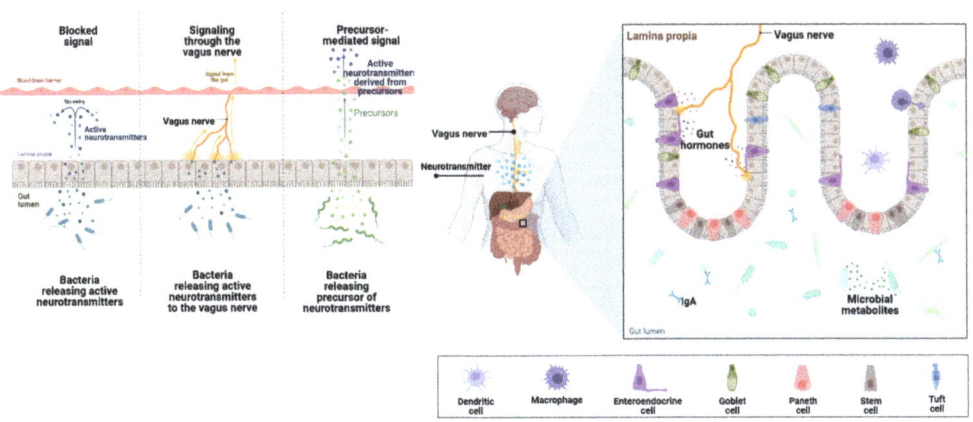

Figure 4. Gut-brain axis.

Similarly, a meta-analysis suggested that probiotics influenced the cognitive behavior of AD patients by decreasing oxidative stress and neuroinflammation levels [124]. Thus, the results strongly convinced us of the potential efficacy of probiotics in AD patients by improving cognitive dysfunction [125]. Selenium, a micronutrient, plays an important role in redox regulation because of its integration into selenoproteins. Koc E.R. et al. [126] documented a direct relationship between Se deficiency and cognitive impairment in AD patients. In another study conducted by Tamtaji et al. [127], it was demonstrated that the administration of Se in combination with multiple probiotics enhanced the metabolic profile and mini-mental state examination (MMSE) score of AD patients. Additionally, the supplementation of sodium selenite at high or super nutritional levels results in high Se uptake by the central nervous system, which significantly improves MMSE scores [128]. However, several concerns are associated with sodium selenates, which limit their implementation in the food and medicine industries, i.e., low biological activity, high toxicity, not easy absorption and utilization by the human body, and a narrow range of safe supplementation [129]. Currently, SeNPs have gained a lot of attention due to their high bioactivity, low toxicity, and high bioavailability. Moreover, based on experimentation data, Se species toxicity is ranked as selenate > selenite > selenomethionine > SeNPs.

A recent study conducted by Lei Qiao et al. [130] showed that administration of SeNPs enriched with *Lactobacillus casei* ATCC 393 averted cognitive dysfunction in AD mice through the modulation of the microbiota-gut-brain axis. ATCC 393 SeNPs minimize aggregation of amyloid beta (Aβ) protein and modulate brain-derived neurotrophic factor (BDNF) or Akt/cAMP-response element binding protein (CREB) pathways that prevent neuronal death. Additionally, SeNPs caused TAU protein hyperphosphorylation, improved cognitive dysfunction, restored gut microbiota balance, regulated immune response, and enhanced production of SCFAs, which ultimately inhibit microglia activation and protect the neuronal cells from neurotoxicity, i.e., neuroinflammation, and oxidative stress. Thus, *L. casei* ATCC 393-SeNPs may act as a safe and promising nutritional supplement to avert neurological diseases.

Licong Yang et al. [131] studied the effect of surface-modified SeNPs in Alzheimer's disease mice. SeNPs were coated with dihydromyricetin (DMY), as it was unstable under physiological conditions, so it was further coated with chitosan (CS). To cross the blood–brain barrier, CS/DMY SeNPs were further coated with the BBB-targeted peptide Tg; thus, the resultant Tg-CS/DMY@SeNPs that easily cross the BBB inhibit the aggregation of Aβ protein and reduce the secretion of inflammatory cytokines through the NF-κB pathway. Moreover, it repairs the gut barrier and regulates the gut microbiota species, i.e., *Dubosiella*, *Bifidobacterium*, and *Desulfovibri*. Moreover, Tg-CS/DMY@SeNPs enhanced the relative abundance of *Gordonibacter*, which downregulates the NLRP3 inflammasome protein expression and decreases the serum inflammatory factor concentration. Through this, it is suggested that Tg-CS/DMY@SeNPs reduce neuroinflammation in the gut microbiota-NLRP3 inflammasome brain axis.

Moreover, Tg-CS/DMY@SeNPs enhanced the relative abundance of *Gordonibacter*, which downregulates the NLRP3 inflammasome protein expression and decreases the serum inflammatory factor concentration. Through this, it is suggested that Tg-CS/DMY@SeNPs reduce neuroinflammation in the gut microbiota-NLRP3 inflammasome brain axis.

Resveratrol (Res) has a neuroprotective effect, but it has lower bioavailability. Changjiang Li et al. [132] illustrated for the first time that oral administration of resveratrol selenium peptide nanocomposites regulated gut microbiota and reduced Aβ aggregation by diminishing Alzheimer's disease-like pathogenesis. The mechanism of action involved binding with Aβ and decreasing aggregation, lowering ROS, and increasing antioxidant enzyme activity, activating the Akt signaling pathway that results in the downregulation of neuroinflammation, averting inflammatory-related gut bacteria and oxidative stress, and helping to overcome gut microbiota dysbiosis (Figure 5).

Table 3. Effect of SeNPs on neurological diseases and microbiota.

Nanomaterials	Average Size	Experimental Model	Dose	Exposure Time	Administration Way	Gut Microbiota Alteration	Effects to Host	References
TGN-Res@SeNPs	14 nm	AD model mice	50 mg/kg b.w.	16 weeks	Oral gavage	1. Decrease of Desulfovibrio, Candidatus_Saccharimonas, Ruminococcaceae_UCG-014, Lachnoclostridium, Enterorhabdus, and Faecalibaculum; 2. Increase of Lachnospiraceae_NK4A136_group, Alistipes, Odoribacter, Helicobacter and Rikenella	Alleviation of Alzheimer's disease-like pathogenesis	[132]
Biogenic SeNPs	170.5 to 182.5 nm	SD rats	0.5, 1.0 or 2.0 mg/kg	-	Administered by gavage	1. Protected the integrity of the spinal cord 2. Decreased the expression of several inflammatory factors 3. Enhanced the production of M2-type macrophages by regulating their polarization, indicating a suppressed inflammatory response	Improve the disturbed microenvironment and promote nerve regeneration	[133]
DMY@SeNPs	46.30 nm	APP/PS1 mice	50 mg/kg body weight	16 weeks	Oral gravage	Regulate the population of inflammatory-related gut microbiota such as Bifidobacterium, Dubosiella, and Desulfovibrio	Ameliorate neuroinflammation through the gut microbiota-NLRP3 inflammasome-brain axis	[131]

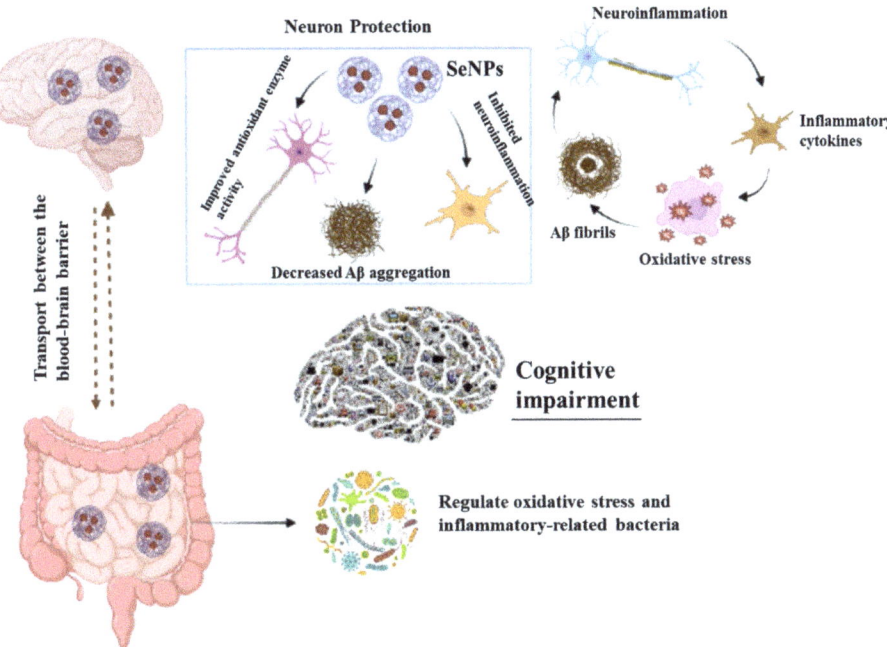

Figure 5. SeNPs' mechanism of action to diminish Alzheimer's disease-like pathogenesis.

Thus, the abovementioned studies illustrated that functionalized SeNPs are potential drug candidates for treating neurological diseases, particularly Alzheimer's disease.

4. Conclusions

In conclusion, the main aim of this review was to organize the latest data on the pharmacotherapeutic potential of SeNPs to treat neurodegenerative diseases. In addition, the well-studied role of microbiota in neurological diseases was also presented. To the best of our knowledge, this is the first-ever study that mentioned the role of SeNPs in treating both neurodegenerative diseases and altered microbiota at the same time. Though this study has illustrated that SeNPs could be a potential hallmark in neurological disease treatment. Moreover, the data presented in this study will help the researchers to quickly navigate the current research on SeNPs and their therapeutic potential in treating neurological diseases that are linked with altered microbiota. This review will also open new doors of research for scientists to find the potential of SeNPs to treat microbiota-related diseases and to overcome some major challenges associated with nanomaterial synthesis, i.e., the difficulty of assessing safety and effectiveness, the lack of specialized equipment for efficient and high-quality nanomaterial synthesis. Nevertheless, the treatment of neurological diseases, which is regarded as an uphill battle, could be easily overcome if multimodal agents are actively practiced with the help of nanotechnology.

Author Contributions: Conceptualization, H.A. and L.A.S.; Writing and compiling H.A.; Final editing E.R.S.; Advice on the interpretation of data, revised manuscript, and final approval, D.C., M.N., S.R. and S.J.; Supervision, Project funding and administration, L.A.S. All authors have read and agreed to the published version of the manuscript.

Funding: This research received was funded by Human Endogenous Retroviruses in neurological diseases (Multiple Sclerosis, Parkinson, Autism Spectrum, Disease) and Type 1 Diabetes to Leonardo A Sechi from Regione Autonoma della Sardegna legge regionale n. 22, 12 December 2022.

Institutional Review Board Statement: Not applicable.

Informed Consent Statement: Not applicable.

Data Availability Statement: Not applicable.

Acknowledgments: All the images are created with BioRender.com.

Conflicts of Interest: The authors declare no conflict of interest.

References

1. Tilleux, S.; Hermans, E. Neuroinflammation and regulation of glial glutamate uptake in neurological disorders. *J. Neurosci. Res.* **2007**, *85*, 2059–2070. [CrossRef] [PubMed]
2. Zhang, W.; Mehta, A.; Tong, Z.; Esser, L.; Voelcker, N.H. Development of polymeric nanoparticles forblood–brain barrier trans-fer—Strategies and challenges. *Adv. Sci.* **2021**, *8*, 2003937. [CrossRef] [PubMed]
3. Kassem, L.M.; Ibrahim, N.A.; Farhana, S.A. Nanoparticle Therapy Is a Promising Approach in the Management and Prevention of Many Diseases: Does It Help in Curing Alzheimer Disease? *J. Nanotechnol.* **2020**, *2020*, 8147080. [CrossRef]
4. Pichla, M.; Bartosz, G.; Sadowska-Bartosz, I. The Antiaggregative and Antiamyloidogenic Properties of Nanoparticles: A Promising Tool for the Treatment and Diagnostics of Neurodegenerative Diseases. *Oxid. Med. Cell. Longev.* **2020**, *2020*, 3534570. [CrossRef]
5. De Jong, W.H.; Borm, P.J. Drug delivery and nanoparticles: Applications and hazards. *Int. J. Nanomed.* **2008**, *3*, 133. [CrossRef] [PubMed]
6. Masserini, M. Nanoparticles for brain drug delivery. *Int. Sch. Res. Not.* **2013**, *2013*, 238428. [CrossRef] [PubMed]
7. Arthur, J.R.; McKenzie, R.C.; Beckett, G.J. Selenium in the immune system. *J. Nutr.* **2003**, *133*, 1457S–1459S. [CrossRef]
8. Steinbrenner, H.; Sies, H. Selenium homeostasis and antioxidant selenoproteins in brain: Implications for disorders in the central nervous system. *Arch. Biochem. Biophys.* **2013**, *536*, 152–157. [CrossRef]
9. Ojeda, L.; Nogales, F.; Murillo, L.; Carreras, O.; Murillo, M.L.O.; Bueno, F.N.; Taravillo, M.L.M.; Sánchez, O.C. The role of folic acid and selenium against oxidative damage from ethanol in early life programming: A review. *Biochem. Cell Biol.* **2018**, *96*, 178–188. [CrossRef]
10. Pitts, M.W.; Byrns, C.N.; Ogawa-Wong, A.N.; Kremer, P.; Berry, M.J. Selenoproteins in Nervous System Development and Function. *Biol. Trace Elem. Res.* **2014**, *161*, 231–245. [CrossRef]
11. Bisht, N.; Phalswal, P.; Khanna, P.K. Selenium nanoparticles: A review on synthesis and biomedical applications. *Mater. Adv.* **2021**, *3*, 1415–1431. [CrossRef]

12. Moreno, F.; García-Barrera, T.; Gómez-Ariza, J.L. Simultaneous speciation and preconcentration of ultra trace concentra-tions of mercury and selenium species in environmental and biological samples by hollow fiber liquid phase microextraction prior to high performance liquid chromatography coupled to inductively coupled plasma mass spectrometry. *J. Chromatogr. A* **2013**, *1300*, 43–50. [PubMed]
13. Kameswari, S.; Narayanan, A.L.; Rajeshkumar, S. Free radical scavenging and anti-inflammatory potential of Acalypha indica mediated selenium nanoparticles. *Drug Invent. Today* **2020**, *13*, 348–351.
14. Zhai, X.; Zhang, C.; Zhao, G.; Stoll, S.; Ren, F.; Leng, X. Antioxidant capacities of the selenium nanoparticles stabilized by chitosan. *J. Nanobiotechnol.* **2017**, *15*, 4. [CrossRef] [PubMed]
15. Zhao, G.; Wu, X.; Chen, P.; Zhang, L.; Yang, C.S.; Zhang, J. Selenium nanoparticles are more efficient than sodium selenite in pro-ducing reactive oxygen species and hyper-accumulation of selenium nanoparticles in cancer cells generates potent therapeutic effects. *Free Radic. Biol. Med.* **2018**, *126*, 55–66. [CrossRef]
16. Rehman, A.; John, P.; Bhatti, A. Biogenic Selenium Nanoparticles: Potential Solution to Oxidative Stress Mediated Inflammation in Rheumatoid Arthritis and Associated Complications. *Nanomaterials* **2021**, *11*, 2005. [CrossRef] [PubMed]
17. Khurana, A.; Tekula, S.; Saifi, M.A.; Venkatesh, P.; Godugu, C. Therapeutic applications of selenium nanoparticles. *Biomed. Pharmacother.* **2019**, *111*, 802–812. [CrossRef]
18. Kiełczykowska, M.; Kocot, J.; Paździor, M.; Musik, I. Selenium—A fascinating antioxidant of protective properties. *Adv. Clin. Exp. Med.* **2018**, *27*, 245–255. [CrossRef]
19. Hariharan, S.; Dharmaraj, S. Selenium and selenoproteins: It's role in regulation of inflammation. *Inflammopharmacology* **2020**, *28*, 667–695. [CrossRef]
20. Constantinescu-Aruxandei, D.; Frîncu, R.M.; Capră, L.; Oancea, F. Selenium Analysis and Speciation in Dietary Supplements Based on Next-Generation Selenium Ingredients. *Nutrients* **2018**, *10*, 1466. [CrossRef] [PubMed]
21. Ferro, C.; Florindo, H.F.; Santos, H.A. Selenium Nanoparticles for Biomedical Applications: From Development and Characteriza-tion to Therapeutics. *Adv. Health Mater.* **2021**, *10*, e2100598. [CrossRef] [PubMed]
22. Winther, K.H.; Rayman, M.P.; Bonnema, S.J.; Hegedüs, L. Selenium in thyroid disorders—Essential knowledge for clinicians. *Nat. Rev. Endocrinol.* **2020**, *16*, 165–176. [CrossRef] [PubMed]
23. Burk, R.F.; Hill, K.E. Regulation of Selenium Metabolism and Transport. *Annu. Rev. Nutr.* **2015**, *35*, 109–134. [CrossRef]
24. Roman, M.; Jitaru, P.; Barbante, C. Selenium biochemistry and its role for human health. *Metallomics* **2013**, *6*, 25–54. [CrossRef] [PubMed]
25. Avery, J.C.; Hoffmann, P.R. Selenium, selenoproteins, and immunity. *Nutrients* **2018**, *10*, 1203. [CrossRef] [PubMed]
26. Kiremidjian-Schumacher, L.; Roy, M.; Wishe, H.I.; Cohen, M.W.; Stotzky, G. Supplementation with selenium augments the functions of natural killer and lymphokine-activated killer cells. *Biol. Trace Elem. Res.* **1996**, *52*, 227–239. [CrossRef] [PubMed]
27. Liao, C.; Carlson, B.A.; Paulson, R.F.; Prabhu, K.S. The intricate role of selenium and selenoproteins in erythropoiesis. *Free. Radic. Biol. Med.* **2018**, *127*, 165–171. [CrossRef]
28. Bodnar, M.; Szczyglowska, M.; Konieczka, P.; Namiesnik, J. Methods of selenium supplementation: Bioavailability and determi-nation of selenium compounds. *Crit. Rev. Food Sci. Nutr.* **2016**, *56*, 36–55. [CrossRef] [PubMed]
29. Rayman, M.P. The importance of selenium to human health. *Lancet* **2000**, *356*, 233–241. [CrossRef]
30. Navarro-Alarcon, M.; Cabrera-Vique, C. Selenium in food and the human body: A review. *Sci. Total. Environ.* **2008**, *400*, 115–141. [CrossRef] [PubMed]
31. Moreda-Piñeiro, J.; Moreda-Piñeiro, A.; Bermejo-Barrera, P. In vivo and in vitro testing for selenium and selenium compounds bioavailability assessment in foodstuff. *Crit. Rev. Food Sci. Nutr.* **2015**, *57*, 805–833. [CrossRef] [PubMed]
32. Micke, O.; Schomburg, L.; Buentzel, J.; Kisters, K.; Muecke, R. Selenium in Oncology: From Chemistry to Clinics. *Molecules* **2009**, *14*, 3975–3988. [CrossRef] [PubMed]
33. Kipp, A.P. Selenium in colorectal and differentiated thyroid cancer. *Hormones* **2020**, *19*, 41–46. [CrossRef]
34. Steinbrenner, H.; Al-Quraishy, S.; Dkhil, M.; Wunderlich, F.; Sies, H. Dietary Selenium in Adjuvant Therapy of Viral and Bacterial Infections. *Adv. Nutr. Int. Rev. J.* **2015**, *6*, 73–82. [CrossRef] [PubMed]
35. Ingles, D.P.; Rodriguez, J.B.C.; Garcia, H. Supplemental Vitamins and Minerals for Cardiovascular Disease Prevention and Treatment. *Curr. Cardiol. Rep.* **2020**, *22*, 22. [CrossRef] [PubMed]
36. Kohler, L.N.; Foote, J.; Kelley, C.P.; Florea, A.; Shelly, C.; Chow, H.-H.S.; Hsu, P.; Batai, K.; Ellis, N.; Saboda, K.; et al. Selenium and Type 2 Diabetes: Systematic Review. *Nutrients* **2018**, *10*, 1924. [CrossRef] [PubMed]
37. Valea, A.; Georgescu, C.E. Selenoproteins in human body: Focus on thyroid pathophysiology. *Hormones* **2018**, *17*, 183–196. [CrossRef]
38. Vicente-Zurdo, D.; Romero-Sánchez, I.; Rosales-Conrado, N.; León-González, M.E.; Madrid, Y. Ability of selenium species to inhibit metal-induced Aβ aggregation involved in the development of Alzheimer's disease. *Anal. Bioanal. Chem.* **2020**, *412*, 6485–6497. [CrossRef]
39. Zhang, X.; Liu, R.-P.; Cheng, W.-H.; Zhu, J.-H. Prioritized brain selenium retention and selenoprotein expression: Nutritional insights into Parkinson's disease. *Mech. Ageing Dev.* **2019**, *180*, 89–96. [CrossRef] [PubMed]
40. Ellwanger, J.H.; Franke, S.I.; Bordin, D.L.; Pra, D.; Henriques, J.A. Biological functions of selenium and its potential influence on Par-kinson's disease. *An. Da Acad. Bras. De Ciências* **2016**, *88*, 1655–1674. [CrossRef]

41. Wadhwani, S.A.; Shedbalkar, U.U.; Singh, R.; Chopade, B.A. Biogenic selenium nanoparticles: Current status and future prospects. *Appl. Microbiol. Biotechnol.* **2016**, *100*, 2555–2566. [CrossRef] [PubMed]
42. Gunti, L.; Dass, R.S.; Kalagatur, N.K. Phytofabrication of Selenium Nanoparticles From Emblica officinalis Fruit Extract and Exploring Its Biopotential Applications: Antioxidant, Antimicrobial, and Biocompatibility. *Front. Microbiol.* **2019**, *10*, 931. [CrossRef] [PubMed]
43. Quester, K.; Avalos-Borja, M.; Castro-Longoria, E. Biosynthesis and microscopic study of metallic nanoparticles. *Micron* **2013**, *54–55*, 1–27. [CrossRef] [PubMed]
44. Fardsadegh, B.; Jafarizadeh-Malmiri, H. Aloe vera leaf extract mediated green synthesis of selenium nanoparticles and assessment of their in vitro antimicrobial activity against spoilage fungi and pathogenic bacteria strains. *Green Process. Synth.* **2019**, *8*, 399–407. [CrossRef]
45. Ahmadi, O.; Jafarizadeh-Malmiri, H.; Jodeiri, N. Eco-friendly microwave-enhanced green synthesis of silver nanoparticles using Aloe vera leaf extract and their physico-chemical and antibacterial studies. *Green Process. Synth.* **2017**, *7*, 231–240. [CrossRef]
46. Ashraf, H.; Meer, B.; Iqbal, J.; Ali, J.S.; Andleeb, A.; Butt, H.; Zia, M.; Mehmood, A.; Nadeem, M.; Drouet, S.; et al. Comparative evaluation of chemically and green synthesized zinc oxide nanoparticles: Their in vitro antioxidant, antimicrobial, cytotoxic and anticancer potential towards HepG2 cell line. *J. Nanostruct. Chem.* **2022**, *1*, 1–19. [CrossRef]
47. Sawant, V.J.; Sawant, V.J. Biogenic capped selenium nano rods as naked eye and selective hydrogen peroxide spectrometric sensor. *Sens. Bio-Sens. Res.* **2020**, *27*, 100314. [CrossRef]
48. Krishnan, M.; Ranganathan, K.; Maadhu, P.; Thangavelu, P.; Kundan, S.; Arjunan, N. Leaf Extract of *Dillenia indica* as a Source of Selenium Nanoparticles with Larvicidal and Antimicrobial Potential toward Vector Mosquitoes and Pathogenic Microbes. *Coatings* **2020**, *10*, 626. [CrossRef]
49. Sharma, G.; Sharma, A.R.; Bhavesh, R.; Park, J.; Ganbold, B.; Nam, J.-S.; Lee, S.-S. Biomolecule-Mediated Synthesis of Selenium Nanoparticles using Dried *Vitis vinifera* (Raisin) Extract. *Molecules* **2014**, *19*, 2761–2770. [CrossRef] [PubMed]
50. Sadalage, P.S.; Nimbalkar, M.S.; Sharma, K.K.K.; Patil, P.S.; Pawar, K.D. Sustainable approach to almond skin mediated synthesis of tunable selenium microstructures for coating cotton fabric to impart specific antibacterial activity. *J. Colloid Interface Sci.* **2020**, *569*, 346–357. [CrossRef] [PubMed]
51. Anu, K.; Singaravelu, G.; Murugan, K.; Benelli, G. Green-Synthesis of Selenium Nanoparticles Using Garlic Cloves (Allium sativum): Biophysical Characterization and Cytotoxicity on Vero Cells. *J. Clust. Sci.* **2016**, *28*, 551–563. [CrossRef]
52. Huang, B.; Zhang, J.; Hou, J.; Chen, C. Free radical scavenging efficiency of Nano-Se in vitro. *Free. Radic. Biol. Med.* **2003**, *35*, 805–813. [CrossRef] [PubMed]
53. Song, X.; Chen, Y.; Zhao, G.; Sun, H.; Che, H.; Leng, X. Effect of molecular weight of chitosan and its oligosaccharides on antitumor activities of chitosan-selenium nanoparticles. *Carbohydr. Polym.* **2019**, *231*, 115689. [CrossRef]
54. Hosnedlova, B.; Kepinska, M.; Skalickova, S.; Fernandez, C.; Ruttkay-Nedecky, B.; Peng, Q.; Baron, M.; Melcova, M.; Opatrilova, R.; Zidkova, J.; et al. Nano-selenium and its nanomedi-cine applications: A critical review. *Int. J. Nanomed.* **2018**, *13*, 2107. [CrossRef] [PubMed]
55. Chen, J.; Berry, M.J. Selenium and selenoproteins in the brain and brain diseases. *J. Neurochem.* **2004**, *86*, 1–12. [CrossRef]
56. Khandel, P.; Yadaw, R.K.; Soni, D.K.; Kanwar, L.; Shahi, S.K. Biogenesis of metal nanoparticles and their pharmacological applications: Present status and application prospects. *J. Nanostruct. Chem.* **2018**, *8*, 217–254. [CrossRef]
57. Chintamani, R.B.; Salunkhe, K.S.; Chavan, M.J. Emerging use of green synthesis silver nanoparticle: An updated review. *Int. J. Pharm. Sci. Res.* **2018**, *9*, 4029–4055.
58. Guo, L.; Xiao, J.; Liu, H.; Liu, H. Selenium nanoparticles alleviate hyperlipidemia and vascular injury in ApoE-deficient mice by regulating cholesterol metabolism and reducing oxidative stress. *Metallomics* **2020**, *12*, 204–217. [CrossRef] [PubMed]
59. Abdraboh, M.E.; Essa, Z.S.; Abdelrazzak, A.; El-Far, Y.M.; Elsherbini, Y.; El-Zayat, M.M.; Ali, D.A. Radio-sensitizing effect of a cocktail of phytochemicals on HepG2 cell proliferation, motility and survival. *Biomed. Pharmacother.* **2020**, *131*, 110620. [CrossRef] [PubMed]
60. Park, H.; Kim, M.-J.; Ha, E.; Chung, J.-H. Apoptotic effect of hesperidin through caspase3 activation in human colon cancer cells, SNU-C4. *Phytomedicine* **2008**, *15*, 147–151. [CrossRef] [PubMed]
61. Martinelli, C.; Pucci, C.; Battaglini, M.; Marino, A.; Ciofani, G. Antioxidants and Nanotechnology: Promises and Limits of Potentially Disruptive Approaches in the Treatment of Central Nervous System Diseases. *Adv. Health Mater.* **2019**, *9*, e1901589. [CrossRef] [PubMed]
62. Kitts, D.D.; Wijewickreme, A.N.; Hu, C. Antioxidant properties of a North American ginseng extract. *Mol. Cell. Biochem.* **2000**, *203*, 1–10. [CrossRef] [PubMed]
63. Chopade, B.A.; Ghosh, S.; Patil, S.; Ahire, M.; Kitture, R.; Jabgunde, A.; Kale, S.; Pardesi, K.; Cameotra, S.S.; Bellare, J.; et al. Synthesis of silver nanoparticles using Dioscorea bulbifera tuber extract and evaluation of its synergistic potential in combination with antimicrobial agents. *Int. J. Nanomed.* **2012**, *7*, 483–496. [CrossRef] [PubMed]
64. Egorova, E.; Revina, A. Synthesis of metallic nanoparticles in reverse micelles in the presence of quercetin. *Colloids Surf. A: Physicochem. Eng. Asp.* **2000**, *168*, 87–96. [CrossRef]
65. El-Refai, A.A.; Ghoniem, G.A.; El-Khateeb, A.Y.; Hassaan, M.M. Eco-friendly synthesis of metal nanoparticles using ginger and garlic extracts as biocompatible novel antioxidant and antimicrobial agents. *J. Nanostruct. Chem.* **2018**, *8*, 71–81. [CrossRef]

66. Zhang, H.; Wang, T.; Qiu, W.; Han, Y.; Sun, Q.; Zeng, J.; Yan, F.; Zheng, H.; Li, Z.; Gao, M. Monitoring the Opening and Recovery of the Blood–Brain Barrier with Noninvasive Molecular Imaging by Biodegradable Ultrasmall $Cu_{2-x}Se$ Nanoparticles. *Nano Lett.* **2018**, *18*, 4985–4992. [CrossRef] [PubMed]
67. Magaldi, S.; Mata-Essayag, S.; De Capriles, C.H.; Pérez, C.; Colella, M.T.; Olaizola, C.; Ontiveros, Y. Well diffusion for antifungal susceptibility testing. *Int. J. Infect. Dis.* **2004**, *8*, 39–45. [CrossRef]
68. Zhang, J.; Zhou, X.; Yu, Q.; Yang, L.; Sun, D.; Zhou, Y.; Liu, J. Epigallocatechin-3-gallate (EGCG)-Stabilized Selenium Nanoparticles Coated with Tet-1 Peptide To Reduce Amyloid-β Aggregation and Cytotoxicity. *ACS Appl. Mater. Interfaces* **2014**, *6*, 8475–8487. [CrossRef]
69. Zhou, X.; Sun, J.; Yin, T.; Le, F.; Yang, L.; Liu, Y.; Liu, J. Enantiomers of cysteine-modified SeNPs (d/l SeNPs) as inhibitors of met-al-induced Aβ aggregation in Alzheimer's disease. *J. Mater. Chem. B* **2015**, *3*, 7764–7774. [CrossRef]
70. Yang, L.; Sun, J.; Xie, W.; Liu, Y.; Liu, J. Dual-functional selenium nanoparticles bind to and inhibit amyloid β fiber formation in Alzheimer's disease. *J. Mater. Chem. B* **2017**, *5*, 5954–5967. [CrossRef] [PubMed]
71. Sun, D.; Zhang, W.; Yu, Q.; Chen, X.; Xu, M.; Zhou, Y.; Liu, J. Chiral penicillamine-modified selenium nanoparticles enantioselectively inhibit metal-induced amyloid β aggregation for treating Alzheimer's disease. *J. Colloid Interface Sci.* **2017**, *505*, 1001–1010. [CrossRef]
72. Williams, P.; Sorribas, A.; Howes, M.-J.R. Natural products as a source of Alzheimer's drug leads. *Nat. Prod. Rep.* **2010**, *28*, 48–77. [CrossRef]
73. Ramassamy, C. Emerging role of polyphenolic compounds in the treatment of neurodegenerative diseases: A review of their intracellular targets. *Eur. J. Pharmacol.* **2006**, *545*, 51–64. [CrossRef] [PubMed]
74. Vingtdeux, V.; Dreses-Werringloer, U.; Zhao, H.; Davies, P.; Marambaud, P. Therapeutic potential of resveratrol in Alzheimer's disease. *BMC Neurosci.* **2008**, *9*, S6. [CrossRef]
75. Yang, L.; Wang, W.; Chen, J.; Wang, N.; Zheng, G. A comparative study of resveratrol and resveratrol-functional selenium nano-particles: Inhibiting amyloid β aggregation and reactive oxygen species formation properties. *J. Biomed. Mater. Res. Part A* **2018**, *106*, 3034–3041. [CrossRef]
76. Hald, A.; Lotharius, J. Oxidative stress and inflammation in Parkinson's disease: Is there a causal link? *Exp. Neurol.* **2005**, *193*, 279–290. [CrossRef]
77. Jankovic, J. Parkinson's disease: Clinical features and diagnosis. *J. Neurol. Neurosurg. Psychiatry* **2008**, *79*, 368–376. [CrossRef]
78. Tatton, W.G.; Eastman, M.J.; Bedingham, W.; Verrier, M.C.; Bruce, I.C. Defective utilization of sensory input as the basis for bradykinesia, rigidity and decreased movement repertoire in Parkinson's disease: A hypothesis. *Can. J. Neurol. Sci.* **1984**, *11*, 136–143. [CrossRef] [PubMed]
79. Ashraf, H.; Solla, P.; Sechi, L.A. Current Advancement of Immunomodulatory Drugs as Potential Pharmacotherapies for Auto-immunity Based Neurological Diseases. *Pharmaceuticals* **2022**, *15*, 1077. [CrossRef]
80. Fedorova, T.N.; Logvinenko, A.A.; Poleshchuk, V.V.; Illarioshkin, S.N. The state of systemic oxidative stress during Parkinson's disease. *Neurochem. J.* **2017**, *11*, 340–345. [CrossRef]
81. Exner, N.; Lutz, A.K.; Haass, C.; Winklhofer, K.F. Mitochondrial dysfunction in Parkinson's disease: Molecular mechanisms and pathophysiological consequences. *EMBO J.* **2012**, *31*, 3038–3062. [CrossRef] [PubMed]
82. Yue, D.; Zeng, C.; Okyere, S.K.; Chen, Z.; Hu, Y. Glycine nano-selenium prevents brain oxidative stress and neurobehavioral ab-normalities caused by MPTP in rats. *J. Trace Elem. Med. Biol.* **2021**, *64*, 126680. [CrossRef]
83. A framework for human microbiome research. *Nature* **2012**, *486*, 215–221. [CrossRef] [PubMed]
84. Liang, S.; Wu, X.; Jin, F. Gut-brain psychology: Rethinking psychology from the microbiota–gut–brain axis. *Front. Integr. Neurosci.* **2018**, *12*, 33. [CrossRef] [PubMed]
85. Magnusson, K.; Hauck, L.; Jeffrey, B.; Elias, V.; Humphrey, A.; Nath, R.; Perrone, A.; Bermudez, L. Relationships between diet-related changes in the gut microbiome and cognitive flexibility. *Neuroscience* **2015**, *300*, 128–140. [CrossRef] [PubMed]
86. Fung, T.C.; Olson, C.A.; Hsiao, E.Y. Interactions between the microbiota, immune and nervous systems in health and disease. *Nat. Neurosci.* **2017**, *20*, 145–155. [CrossRef] [PubMed]
87. Heijtz, R.D. Fetal, neonatal, and infant microbiome: Perturbations and subsequent effects on brain development and behavior. *Semin. Fetal Neonatal Med.* **2016**, *21*, 410–417. [CrossRef]
88. de Weerth, C. Do bacteria shape our development? Crosstalk between intestinal microbiota and HPA axis. *Neurosci. Biobehav. Rev.* **2017**, *83*, 458–471. [CrossRef]
89. Sudo, N. Microbiome, HPA axis and production of endocrine hormones in the gut. In *Microbial Endocrinology: The Microbio-Ta-Gut-Brain Axis in Health and Disease*; Springer: Berlin/Heidelberg, Germany, 2014; pp. 177–194.
90. Gensollen, T.; Blumberg, R.S. Correlation between early-life regulation of the immune system by microbiota and allergy development. *J. Allergy Clin. Immunol.* **2017**, *139*, 1084–1091. [CrossRef] [PubMed]
91. Cryan, J.F.; O'Riordan, K.J.; Cowan, C.S.; Sandhu, K.V.; Bastiaanssen, T.F.; Boehme, M.; Codagnone, M.G.; Cussotto, S.; Fulling, C.; Golubeva, A.V.; et al. The microbiota-gut-brain axis. *Physiol. Rev.* **2019**, *99*, 1877–2013. [CrossRef]
92. Peng, W.; Yi, P.; Yang, J.; Xu, P.; Wang, Y.; Zhang, Z.; Huang, S.; Wang, Z.; Zhang, C. Association of gut microbiota composition and function with a senes-cence-accelerated mouse model of Alzheimer's Disease using 16S rRNA gene and metagenomic sequencing analysis. *Aging* **2018**, *10*, 4054. [CrossRef]

93. Bell, J.S.; Spencer, J.I.; Yates, R.L.; Yee, S.A.; Jacobs, B.M.; DeLuca, G.C. Invited Review: From nose to gut—The role of the microbiome in neurological disease. *Neuropathol. Appl. Neurobiol.* **2018**, *45*, 195–215. [CrossRef]
94. Kanayama, M.; Danzaki, K.; He, Y.-W.; Shinohara, M.L. Lung inflammation stalls Th17-cell migration en route to the central nervous system during the development of experimental autoimmune encephalomyelitis. *Int. Immunol.* **2016**, *28*, 463–469. [CrossRef]
95. Turnbaugh, P.J.; Hamady, M.; Yatsunenko, T.; Cantarel, B.L.; Duncan, A.; Ley, R.E.; Sogin, M.L.; Jones, W.J.; Roe, B.A.; Affourtit, J.P.; et al. A core gut microbiome in obese and lean twins. *Nature* **2009**, *457*, 480–484. [CrossRef]
96. Ley, R.E.; Peterson, D.A.; Gordon, J.I. Ecological and Evolutionary Forces Shaping Microbial Diversity in the Human Intestine. *Cell* **2006**, *124*, 837–848. [CrossRef]
97. Stocchi, F.; Torti, M. Constipation in Parkinson's disease. *Int. Rev. Neurobiol.* **2017**, *134*, 811–826.
98. Jandhyala, S.M.; Talukdar, R.; Subramanyam, C.; Vuyyuru, H.; Sasikala, M.; Reddy, D.N. Role of the normal gut microbiota. *World J. Gastroenterol. WJG* **2015**, *21*, 8787. [CrossRef]
99. Baquero, F.; Nombela, C. The microbiome as a human organ. *Clin. Microbiol. Infect.* **2012**, *18*, 2–4. [CrossRef]
100. Iwatsubo, T. Aggregation of α-synuclein in the pathogenesis of Parkinson's disease. *J. Neurol.* **2003**, *250*, iii11–iii14. [CrossRef]
101. Ueki, A.; Otsuka, M. Life style risks of Parkinson's disease: Association between decreased water intake and constipation. *J. Neurol.* **2004**, *251*, vii18–vii23. [CrossRef]
102. Savica, R.; Carlin, J.M.; Grossardt, B.R.; Bower, J.H.; Ahlskog, J.E.; Maraganore, D.M.; Bharucha, A.E.; Rocca, W.A. Medical records documentation of constipation pre-ceding Parkinson disease: A case-control study. *Neurology* **2009**, *73*, 1752–1758. [CrossRef] [PubMed]
103. Chandra, R.; Hiniker, A.; Kuo, Y.M.; Nussbaum, R.L.; Liddle, R.A. α-Synuclein in gut endocrine cells and its implications for Par-kinson's disease. *JCI Insight* **2017**, *2*, e92295. [CrossRef] [PubMed]
104. Hilton, D.; Stephens, M.; Kirk, L.; Edwards, P.; Potter, R.; Zajicek, J.; Broughton, E.; Hagan, H.; Carroll, C. Accumulation of α-synuclein in the bowel of patients in the pre-clinical phase of Parkinson's disease. *Acta Neuropathol.* **2013**, *127*, 235–241. [CrossRef]
105. Sampson, T.R.; Debelius, J.W.; Thron, T.; Janssen, S.; Shastri, G.G.; Ilhan, Z.E.; Challis, C.; Schretter, C.E.; Rocha, S.; Gradinaru, V.; et al. Gut microbiota regulate motor deficits and neu-roinflammation in a model of Parkinson's disease. *Cell* **2016**, *167*, 1469–1480.e12. [CrossRef]
106. Minter, M.R.; Zhang, C.; Leone, V.; Ringus, D.L.; Zhang, X.; Oyler-Castrillo, P.; Musch, M.W.; Liao, F.; Ward, J.F.; Holtzman, D.M.; et al. Antibiotic-induced perturbations in gut microbial diversity influences neuro-inflammation and amyloidosis in a murine model of Alzheimer's disease. *Sci. Rep.* **2016**, *6*, 30028. [CrossRef]
107. Harach, T.; Marungruang, N.; Duthilleul, N.; Cheatham, V.; Mc Coy, K.D.; Frisoni, G.; Neher, J.J.; Fåk, F.; Jucker, M.; Lasser, T.; et al. Reduction of Abeta amyloid pathology in APPPS1 transgenic mice in the absence of gut microbiota. *Sci. Rep.* **2017**, *7*, 41802. [CrossRef]
108. Amini, M.E.; Shomali, N.; Bakhshi, A.; Rezaei, S.; Hemmatzadeh, M.; Hosseinzadeh, R.; Eslami, S.; Babaie, F.; Aslani, S.; Torkamandi, S.; et al. Gut microbiome and multiple sclerosis: New insights and perspective. *Int. Immunopharmacol.* **2020**, *88*, 107024. [CrossRef]
109. Cattaneo, A.; Cattane, N.; Galluzzi, S.; Provasi, S.; Lopizzo, N.; Festari, C.; Ferrari, C.; Guerra, U.P.; Paghera, B.; Muscio, C.; et al. Association of brain amyloidosis with pro-inflammatory gut bacterial taxa and peripheral inflammation markers in cognitively impaired elderly. *Neurobiol. Aging* **2017**, *49*, 60–68. [CrossRef]
110. Lee, Y.K.; Menezes, J.S.; Umesaki, Y.; Mazmanian, S.K. Proinflammatory T-cell responses to gut microbiota promote experimental autoimmune encephalomyelitis. *Proc. Natl. Acad. Sci. USA* **2010**, *108*, 4615–4622. [CrossRef]
111. Berer, K.; Gerdes, L.A.; Cekanaviciute, E.; Jia, X.; Xiao, L.; Xia, Z.; Liu, C.; Klotz, L.; Stauffer, U.; Baranzini, S.E.; et al. Gut microbiota from multiple sclerosis patients enables spon-taneous autoimmune encephalomyelitis in mice. *Proc. Natl. Acad. Sci. USA* **2017**, *114*, 10719–10724. [CrossRef]
112. Ochoa-Repáraz, J.; Mielcarz, D.W.; Wang, Y.; Begum-Haque, S.; Dasgupta, S.; Kasper, D.L.; Kasper, L.H. A polysaccharide from the human commensal Bacteroides fragilis protects against CNS demyelinating disease. *Mucosal Immunol.* **2010**, *3*, 487–495. [CrossRef] [PubMed]
113. Navarro-López, V.; Méndez-Miralles, M.Á.; Vela-Yebra, R.; Fríes-Ramos, A.; Sánchez-Pellicer, P.; Ruzafa-Costas, B.; Núñez-Delegido, E.; Gómez-Gómez, H.; Chumillas-Lidón, S.; Picó-Monllor, J.A.; et al. Gut Mi-crobiota as a Potential Predictive Biomarker in Relapsing-Remitting Multiple Sclerosis. *Genes* **2022**, *13*, 930. [CrossRef]
114. Elgendy, S.G.; Abd-Elhameed, R.; Daef, E.; Mohammed, S.M.; Hassan, H.M.; El-Mokhtar, M.A.; Nasreldein, A.; Khedr, E.M. Gut microbiota in forty cases of Egyptian relapsing remitting multiple sclerosis. *Iran. J. Microbiol.* **2021**, *13*, 632. [CrossRef] [PubMed]
115. Kadowaki, A.; Saga, R.; Lin, Y.; Sato, W.; Yamamura, T. Gut microbiota-dependent CCR9+CD4+ T cells are altered in secondary progressive multiple sclerosis. *Brain* **2019**, *142*, 916–931. [CrossRef]
116. Liu, S.; Gao, J.; Zhu, M.; Liu, K.; Zhang, H.-L. Gut Microbiota and Dysbiosis in Alzheimer's Disease: Implications for Pathogenesis and Treatment. *Mol. Neurobiol.* **2020**, *57*, 5026–5043. [CrossRef] [PubMed]
117. Vogt, N.M.; Kerby, R.L.; Dill-McFarland, K.A.; Harding, S.J.; Merluzzi, A.P.; Johnson, S.C.; Carlsson, C.M.; Asthana, S.; Zetterberg, H.; Blennow, K.; et al. Gut microbiome alterations in Alz-heimer's disease. *Sci. Rep.* **2017**, *7*, 13537. [CrossRef]

118. Mancuso, C.; Santangelo, R. Alzheimer's disease and gut microbiota modifications: The long way between preclinical studies and clinical evidence. *Pharmacol. Res.* **2018**, *129*, 329–336. [CrossRef]
119. Kesika, P.; Suganthy, N.; Sivamaruthi, B.S.; Chaiyasut, C. Role of gut-brain axis, gut microbial composition, and probiotic inter-vention in Alzheimer's disease. *Life Sci.* **2021**, *264*, 118627. [CrossRef]
120. Pellegrini, C.; Antonioli, L.; Colucci, R.; Blandizzi, C.; Fornai, M. Interplay among gut microbiota, intestinal mucosal barrier and enteric neuro-immune system: A common path to neurodegenerative diseases? *Acta Neuropathol.* **2018**, *136*, 345–361. [CrossRef]
121. Leitner, G.R.; Wenzel, T.; Marshall, N.; Gates, E.J.; Klegeris, A. Targeting toll-like receptor 4 to modulate neuroinflammation in central nervous system disorders. *Expert Opin. Ther. Targets* **2019**, *23*, 865–882. [CrossRef]
122. Huo, J.-Y.; Jiang, W.-Y.; Yin, T.; Xu, H.; Lyu, Y.-T.; Chen, Y.-Y.; Chen, M.; Geng, J.; Jiang, Z.-X.; Shan, Q.-J. Intestinal Barrier Dysfunction Exacerbates Neuroinflammation via the TLR4 Pathway in Mice with Heart Failure. *Front. Physiol.* **2021**, *12*, 1263. [CrossRef] [PubMed]
123. Akbari, E.; Asemi, Z.; Daneshvar Kakhaki, R.; Bahmani, F.; Kouchaki, E.; Tamtaji, O.R.; Ali Hamidi, G.; Salami, M. Effect of probiotic supplementation on cognitive function and metabolic status in Alzheimer's disease: A randomized, double-blind and controlled trial. *Front. Aging Neurosci.* **2016**, *8*, 256. [CrossRef] [PubMed]
124. Den, H.; Dong, X.; Chen, M.; Zou, Z. Efficacy of probiotics on cognition, and biomarkers of inflammation and oxidative stress in adults with Alzheimer's disease or mild cognitive impairment—A meta-analysis of randomized controlled trials. *Aging* **2020**, *12*, 4010. [CrossRef] [PubMed]
125. Generoso, J.S.; Giridharan, V.V.; Lee, J.; Macedo, D.; Barichello, T. The role of the microbiota-gut-brain axis in neuropsychiatric dis-orders. *Braz. J. Psychiatry* **2020**, *43*, 293–305. [CrossRef] [PubMed]
126. Koc, E.R.; Ilhan, A.; Aytürk, Z.; Acar, B.; Gürler, M.; Altuntaş, A.; Bodur, A.S. A comparison of hair and serum trace elements in patients with Alzheimer disease and healthy participants. *Turk. J. Med. Sci.* **2015**, *45*, 1034–1039. [CrossRef] [PubMed]
127. Tamtaji, O.R.; Heidari-Soureshjani, R.; Mirhosseini, N.; Kouchaki, E.; Bahmani, F.; Aghadavod, E.; Tajabadi-Ebrahimi, M.; Asemi, Z. Probiotic and selenium co-supplementation, and the effects on clinical, metabolic and genetic status in Alzheimer's disease: A randomized, double-blind, controlled trial. *Clin. Nutr.* **2019**, *38*, 2569–2575. [CrossRef]
128. Cardoso, B.R.; Roberts, B.R.; Malpas, C.B.; Vivash, L.; Genc, S.; Saling, M.M.; Desmond, P.; Steward, C.; Hicks, R.J.; Callahan, J.; et al. Supranutritional Sodium Selenate Supplementation Delivers Selenium to the Central Nervous System: Results from a Randomized Controlled Pilot Trial in Alzheimer's Disease. *Neurotherapeutics* **2018**, *16*, 192–202. [CrossRef] [PubMed]
129. Qiao, L.; Dou, X.; Song, X.; Xu, C. Green synthesis of nanoparticles by probiotics and their application. *Adv. Appl. Microbiol.* **2022**, *119*, 83–128. [CrossRef]
130. Qiao, L.; Chen, Y.; Song, X.; Dou, X.; Xu, C. Selenium Nanoparticles-Enriched Lactobacillus casei ATCC 393 Prevents Cognitive Dysfunction in Mice Through Modulating Microbiota-Gut-Brain Axis. *Int. J. Nanomed.* **2022**, *17*, 4807–4827. [CrossRef]
131. Yang, L.; Cui, Y.; Liang, H.; Li, Z.; Wang, N.; Wang, Y.; Zheng, G. Multifunctional selenium nanoparticles with different surface modifi-cations ameliorate neuroinflammation through the gut microbiota-NLRP3 inflammasome-brain Axis in APP/PS1 mice. *ACS Appl. Mater. Interfaces* **2022**, *14*, 30557–30570. [CrossRef]
132. Li, C.; Wang, N.; Zheng, G.; Yang, L. Oral Administration of Resveratrol-Selenium-Peptide Nanocomposites Alleviates Alzheimer's Disease-like Pathogenesis by Inhibiting Aβ Aggregation and Regulating Gut Microbiota. *ACS Appl. Mater. Interfaces* **2021**, *13*, 46406–46420. [CrossRef] [PubMed]
133. Liu, X.; Mao, Y.; Huang, S.; Li, W.; Zhang, W.; An, J.; Jin, Y.; Guan, J.; Wu, L.; Zhou, P. Selenium nanoparticles derived from *Proteus mirabilis* YC801 alleviate oxidative stress and inflammatory response to promote nerve repair in rats with spinal cord injury. *Regen. Biomater.* **2022**, *9*, rbac042. [CrossRef] [PubMed]

Disclaimer/Publisher's Note: The statements, opinions and data contained in all publications are solely those of the individual author(s) and contributor(s) and not of MDPI and/or the editor(s). MDPI and/or the editor(s) disclaim responsibility for any injury to people or property resulting from any ideas, methods, instructions or products referred to in the content.

Article

Miconazole Nitrate Microparticles in Lidocaine Loaded Films as a Treatment for Oropharyngeal Candidiasis

Guillermo Tejada [1], Natalia L. Calvo [2,3], Mauro Morri [4], Maximiliano Sortino [5,6], Celina Lamas [2,7], Vera A. Álvarez [1,*] and Darío Leonardi [2,7,*]

1. Grupo Materiales Compuestos Termoplásticos, Instituto de Investigaciones en Ciencia y Tecnología de Materiales, Av. Colón 10850, Mar Del Plata 7600, Argentina
2. Instituto de Química Rosario, Suipacha 570, Rosario 2000, Argentina
3. Área de Análisis de Medicamentos, Departamento Química Orgánica, Facultad de Ciencias Bioquímicas y Farmacéuticas, Universidad Nacional de Rosario, Suipacha 570, Rosario 2000, Argentina
4. Planta Piloto de Producción de Medicamentos, Facultad de Ciencias Bioquímicas y Farmacéuticas, Universidad Nacional de Rosario, Suipacha 570, Rosario 2000, Argentina
5. Centro de Referencia de Micología, Área Farmacognosia, Departamento Química Orgánica, Facultad de Ciencias Bioquímicas y Farmacéuticas, Universidad Nacional de Rosario, Suipacha 570, Rosario 2000, Argentina
6. Área Farmacognosia, Departamento Química Orgánica, Facultad de Ciencias Bioquímicas y Farmacéuticas, Universidad Nacional de Rosario, Suipacha 570, Rosario 2000, Argentina
7. Área Técnica Farmacéutica, Departamento Farmacia, Facultad de Ciencias Bioquímicas y Farmacéuticas, Universidad Nacional de Rosario, Suipacha 570, Rosario 2000, Argentina
* Correspondence: alvarezvera@gmail.com (V.A.Á.); leonardi@iquir-conicet.gov.ar (D.L.)

Abstract: Oral candidiasis is an opportunistic infection that affects mainly individuals with weakened immune system. Devices used in the oral area to treat this condition include buccal films, which present advantages over both oral tablets and gels. Since candidiasis causes pain, burning, and itching, the purpose of this work was to develop buccal films loaded with both lidocaine (anesthetic) and miconazole nitrate (MN, antifungal) to treat this pathology topically. MN was loaded in microparticles based on different natural polymers, and then, these microparticles were loaded in hydroxypropyl methylcellulose-gelatin-based films containing lidocaine. All developed films showed adequate adhesiveness and thickness. DSC and XRD tests suggested that the drugs were in an amorphous state in the therapeutic systems. Microparticles based on chitosan-alginate showed the highest MN encapsulation. Among the films, those containing the mentioned microparticles presented the highest tensile strength and the lowest elongation at break, possibly due to the strong interactions between both polymers. These films allowed a fast release of lidocaine and a controlled release of MN. Due to the latter, these systems showed antifungal activity for 24 h. Therefore, the treatment of oropharyngeal candidiasis with these films could reduce the number of daily applications with respect to conventional treatments.

Keywords: oropharyngeal candidiasis; miconazole nitrate; lidocaine; polymeric microparticles; polymeric films

Citation: Tejada, G.; Calvo, N.L.; Morri, M.; Sortino, M.; Lamas, C.; Álvarez, V.A.; Leonardi, D. Miconazole Nitrate Microparticles in Lidocaine Loaded Films as a Treatment for Oropharyngeal Candidiasis. *Materials* 2023, *16*, 3586. https://doi.org/10.3390/ma16093586

Academic Editors: Paul Cătălin Balaure and Alexandru Mihai Grumezescu

Received: 3 March 2023
Revised: 27 April 2023
Accepted: 5 May 2023
Published: 7 May 2023

Copyright: © 2023 by the authors. Licensee MDPI, Basel, Switzerland. This article is an open access article distributed under the terms and conditions of the Creative Commons Attribution (CC BY) license (https://creativecommons.org/licenses/by/4.0/).

1. Introduction

Candidiasis is one of the most common human opportunistic fungal infections of the oral cavity. Candida strains reside as a commensal in oral flora; however, it causes fungal infection when the host becomes immunocompromised. Particularly, oral candidiasis is common in the population of patients with hematological malignancies as well as in people with acquired immunodeficiency syndrome (AIDS) [1]. Currently, there are several antifungal drugs (in different presentations) approved for the topical treatment of oral candidiasis [2], among them are ketoconazole gel 2% (dose: 3 times/day), clotrimazole gel 1% (dose: 3 times/day), nystatin suspension (dose: 4–6 mL/6 h) and miconazole nitrate (MN) gel 2% (dose 100 mg/6 h). These formulations allow the eradication of fungal infection but need to be applied several times daily, and the treatment in HIV/AIDS patients may be prolonged (from 4 to 14 days using nystatin rinses) [3]. Thus, the development of an appropriate formulation to treat oropharyngeal candidiasis, which allows for a reduction in the number of applications per day, will improve patient's compliance during treatment. In recent years, several buccal devices containing MN have been developed and analyzed. In 2014, Rai et al. developed a cellulosic polymer-based gel containing MN for buccal delivery. The final formulation, obtained after an optimization procedure, showed an extended residence time in oral mucosa and a broader zone of growth inhibition compared with a marketed antifungal formulation [4]. However, in general, gels have a short residence time on the mucosa, which is a disadvantage that can be solved when buccal films are developed [5]. In 2018 Mady et al. developed polymeric films containing MN for treatment of buccal candidiasis. The authors combined MN with urea in the films and observed an improvement in the inhibition zone diameters for films containing increasing concentrations of both urea and MN. It was concluded that urea acts as a penetration enhancer against *C. albicans* and there is a synergistic effect between MN and urea. Although this finding is highly relevant, the release of the MN from the films was analyzed only for 120 min, and therefore, it is not possible to determine the number of applications per day that will be necessary for treatment [6]. On the other hand, in 2021, De Caro et al. developed and characterized solid and semisolid formulations containing MN-loaded solid lipid microparticles as therapy for oral candidiasis. The developed formulations allowed them to overcome the short retention time and suboptimal drug concentration that are present in the currently available antifungal therapy. The solid lipid microparticles loaded with MN were able to increase up to three-fold MN accumulation in the buccal mucosa compared with Daktarin® 2% oral gel, probably due to the effect of oromucosal penetration enhancers of the microparticles. Finally, these solid lipid microparticles were loaded into a buccal gel (based on trehalose, PVP-K90, and hydroxyethylcellulose) or into a mucoadhesive buccal film based on trehalose, PVP-K90, limonene, and hydroxyethylcellulose. In particular, the buccal film was able to release the MN for an extended period of time and avoid MN permeation, limiting the possibility of adverse side effects [7]. These findings are remarkable and allow for progress in the search for an ideal device for treating buccal candidiasis. Since candidiasis may cause burning pain and pruritus, it is desirable to develop a formulation containing an anesthetic in addition to an antifungal agent. Thus, the purpose of this work was to develop films loaded with both lidocaine (to obtain a fast dissolution rate of the anesthetic) and miconazole loaded microparticles (to achieve controlled release of the antifungal agent) and evaluate the antifungal activity over time in the systems.

2. Materials and Methods

2.1. Materials

Chitosan (CH, MW 230 KDa and 80.6% N-deacetylation) was acquired from Aldrich Chemical Co. (Milwaukee, WI, USA) and hydroxypropyl methylcellulose (HPMC, MW 250 kDa, hydroxypropyl 7–12%, methoxyl 19–24%) was obtained from Eigenmann & Veronelli (Milan, Italy). Gelatine (GEL, type A from pork skin, 125 Bloom value) and sodium alginate (ALG, Sigma-Aldrich Co. Buenos Aires, Argentina) were also used. FMC BioPolymer (Philadelphia, PA, USA) donated samples of food-grade λ and κ-carrageenan (λ-c and κ-C), and sodium lauryl sulphate (SLS) was acquired from Biopack (Buenos Aires, Argentina). Ammonium acetate (HPLC grade) was acquired from Thermo Fisher Scientific (Cleveland, OH, USA), acetonitrile (HPLC grade) and sorbitol (70%) were obtained from Cicarelli (Buenos Aires, Argentina), while lidocaine hydrochloride (LDCH) and miconazole nitrate (MN), both of pharmaceutical grade, were obtained from Parafarm (Buenos Aires, Argentina). All other reagents were of analytical grade.

2.2. Preparation of Microparticles

CH solutions (0.5% w/v) were obtained by dispersing the polymer in acetic acid solution (pH = 2.50, 30% v/v), while SLS, ALG, κ, and λ-carrageenan solutions (0.2% w/v) were prepared by dissolving them in water. The solutions were stirred at 400 rpm and 40 °C for 30 min). Subsequently, MN at 20% w/w (with respect to the mass of polymers) was added to the solution and stirred again at 400 rpm for 10 min at 40 °C. Then, mixtures were sprayed using a Buchi Mini dryer B-290 (Flawil, Switzerland). The composition of the microparticles was 50% CH and 50% SLS, ALG, κ or λ-carrageenan. Some parameters such as airflow rate (38 m^3/h), feed rate (5 mL/min), pump (10%), aspirator (100%), spray-drying (SD) inlet temperature (130 °C), and outlet temperature (70 °C) remained constant during the prosses. After SD, the powders were stored at room temperature.

2.3. Lidocaine Films Loaded Microparticles

Solutions of HPMC and GEL both at 3% w/v, were obtained by dispersing the polymers in water. The dispersion was stirred (for 24 h) and filtered using Miracloth® (Calbiochem-Novabiochem Corp., San Diego, CA, USA). Then, GEL solution was dripped over the HPMC one under stirring (400 rpm at 80 °C). Sorbitol at 20% w/w (as a plasticizer) and LDCH (at 5% w/w with respect to the mass of polymers) were added to the mixture. This mixture was stirred again at 400 rpm for 30 min. Then, different microparticulate systems loaded with MN were added to the mixture (at 2% w/w of the total polymeric mass) and stirred at 400 rpm for 1 h at 50 °C to homogenize). Dispersions were poured into Petri dishes (9 cm diameter) and dried at 40 °C and 58% relative humidity (RH) for 48 h. After that, films were detached from the Petri dishes and conditioned for 72 h (25 °C and 58% RH). Systems presenting lack of physical defects such as cracks, holes, and bubbles were selected and used in the different tests. The different compositions of the systems are shown in Table 1.

Table 1. Composition of different formulation.

Films Matrix Composition (Containing LDCH 2% w/w)	Microparticles Matrix Composition (Containing MN 5% w/w)	Films Loaded Microparticles Abbreviation
HPMC-GEL-LDCH	CH-ALG-MN	HG-CH-ALG
HPMC-GEL-LDCH	CH-κC-MN	HG-CH-κC
HPMC-GEL-LDCH	CH-λC-MN	HG-CH-λC
HPMC-GEL-LDCH	CH-SLS-MN	HG-CH-SLS

2.4. High-Performance Liquid Chromatography (HPLC)

High-Performance Liquid Chromatography was carried out using an Agilent Technologies 1200 Series chromatograph (Santa Clara, CA, USA) comprising a SiliaChrom C18 column (5 μm, 150 × 4.6 mm, S/N S11410) thermostatted at 40 °C [8] and a diode array detector. The amounts of MN and LDCH released were obtained spectrophotometrically by measuring the absorbance at 230 [9] and 265 nm [10] using ammonium acetate buffer (50 mM, pH 4): acetonitrile (35:65) as the mobile phase in isocratic mode with a flow rate of 1.5 mL min^{-1}. The retention time of LDCH was 2.9 min while that of MN was 14.4 min. The construction of the LDCH calibration curve was developed in the range 25–400 mg/mL (equation: y = 1675x + 80,010) showing R^2 = 0.9795, while for MN the range was 20–400 mg/mL (equation: y = 41,953x + 185,070) showing R^2 = 0.998.

2.5. Microparticles Loaded MN—Encapsulation Efficiency

Encapsulation efficiency (EE) of each microparticle system was obtained using the HPLC method described above. For this, 12 mg of microparticles was dispersed in 60 mL of methanol and the suspension was stirred for 2 h at 400 rpm to extract the MN from the microparticles.

2.6. Films Loaded Microparticles: Thickness and Folding Endurance

A digital micrometer (Schwyz, China) was used to determine the thickness (TH) of the films. Six measurements were carried out in the center and around of each system. Folding endurance values were obtained by repeatedly folding each system 300 times at the same place, or until it broke. The assay was carried out in triplicate [11].

2.7. Mechanical Properties

The mechanical strength of the systems was analyzed by using a Universal Testing Machine Instron, Series 3340, single column (Instron, Norwood, MA, USA) with a 50 N load cell. Systems were conditioned (24 h at 25 °C and 58% RH) for each mechanical test, and then cut into strips 7 mm wide and 60 mm long, to evaluate both tensile strength and elongation at break. To carry out the assay it was settled at 30 mm the initial grip distance and at 0.05 mm/s the crosshead speed. From stress/strain curves, the parameters tensile strength and elongation at break were obtained. The assay was carried out in triplicate [12].

2.8. In Vitro Mucoadhesive Strength

The in vitro mucoadhesive strength was obtained following the technique described by Tejada et al. [13]. Briefly, an Instron universal testing machine was employed to analyze the mucoadhesive strength of each system. This parameter was evaluated in vitro by determining the force required to detach each system from a pork gum disc (donated by "Paladini" slaughterhouse, V.G. Galvez, Argentina)

2.9. Swelling Index

Swelling index (SI) values were obtained by immersing a portion of the systems (2.5 cm diameter) in artificial saliva (1 mL, 37 °C, pH = 6.8). At different time intervals, the systems were removed, and the excess adhering moisture was gently blotted off and weighed. Then, artificial saliva was added again (0.5 mL), and the process was repeated until 24 h. Using the Eqaution (1), the SI was calculated. The assay was carried out in triplicate.

$$SI = (W_t - W_0)/W_0 \qquad (1)$$

where W_t is the weight of swollen films and W_0 is the weight of dry film.

2.10. Morphology Analysis and Size Determination by Scanning Electron Microscopy

Scanning electron microscopy (SEM, AMR 1000, Leitz, Wetzlar, Germany) was used to observe the morphology of the systems. Both microparticles and films containing microparticles were placed on an aluminum sample holder using a conductive double-sided adhesive. To make the systems conductive they were coated with a fine gold layer (15 min at 70–80 mTorr). SEM images were obtained using 20 kV accelerating voltage and magnifications of 200× and 500× for films and 10,000× for microparticles.

2.11. Thermal Analysis

Thermogravimetric analysis (TGA) was performed using a TG HI-Res thermal analyzer (TA Instruments) at a 10 °C/min heating rate in the range of room temperature to 800 °C, in air flow. Differential scanning calorimetry (DSC) tests were carried out in a DSC Q2000 (TA Instruments) at a 10 °C/min heating rate in the range of room temperature to 200 °C, under nitrogen.

2.12. X-ray Diffraction

An automated X'Pert Phillips MPD diffractometer (Eindhoven, The Netherlands) was used, and data collection was performed in transmission mode. The patterns of X-ray diffraction (XRD) were recorded using CuKα radiation (λ = 1.540562 Å), 20 mA (current), 40 kV (voltage), and 0.02° (steps) on the interval 2θ from 10° to 50°. The Stoe Visual-Xpow package, Version 2.75 (Germany) was employed for data acquisition and evaluation.

2.13. Dissolution Studies

Dissolution assays were performed in artificial saliva (900 mL, 37 °C), using a USP XXIV apparatus II (Hanson Research, SR8 8-Flask Bath, ON, Canada) with paddles rotating at 50 rpm [10,14]. Portions of films (containing 25 mg LDCH) were placed in the dissolution medium. At different times, 3 samples of 1 mL each were taken, and after each sample collection, an equal volume of the dissolution medium was added. The drug content of the aliquots was determined using HPLC as described above [15].

2.14. Halo Zone Test over Time

The guidelines of the disk diffusion method described in CLSI document M44-A2 were followed to carry out the halo zone test [16]. Agar plates (90 mm diameter) containing Mueller—Hinton agar, supplemented with glucose and methylene blue, were used to perform the assay. The agar surface was inoculated by dipping sterile cotton swabs into a fungal suspension and by streaking the plate surface in three directions. After the plate was dried for 20 min, systems (containing 10 mg of MN) were placed onto the surface of the agar. Each hour, the films were moved to another zone of the culture plate and the process was repeated for 24 h. A paper disk containing MN powder (10 mg) was designed as a control. Finally, the plates were incubated in air (28 °C) and read after 24 h. Using a caliper, the zone of complete growth inhibition (halo diameters) was determined [17]. This assay makes it possible to analyze the release of MN from the systems and their antifungal activity over time.

2.15. Statistical Analysis

An ANOVA test was used, and when it was found that the effect of the factors was significant, the Tukey multiple ranks test was applied to establish significant differences among samples with a 95% confidence ($p < 0.05$).

3. Results and Discussion

3.1. Encapsulation Efficiency of Microparticles

Encapsulation efficiencies values ranged from 48.88% to 97.15%. The highest encapsulation value corresponds to the CH-ALG system, while the lowest value was obtained in the microparticles based on CH-SLS. Systems formed by CH-κC and CH-λC presented intermediate encapsulation values: 67.37 and 57.44%, respectively. This could be because, before the SD process, polymers with a high charge density have a greater number of free sites to interact with the drug, increasing its capacity to encapsulate the MN [18]. Furthermore, the EE values are also probably related to the positive charge densities present in CH and the negative charge densities present in the other polymers. In the preparation of the CH solution (pH = 2.50), the amino groups are positively charged, while the other polymers (pH solution = 5.40) are negatively charged. ALG has in its structure monomeric units of α-L-guluronic acid, β-D-mannuronic units, and the two bonded together, being a polymer with abundant negative charge density which may interact more easily with CH [19]. Following the descending order of EE values, the Carrageenans present sulfate ester groups in the galactose units to interact with the amino groups of the CH [20]; finally, the lowest EE value corresponds to the CH-SLS formulation, which could be due to the fact that in the SLS chain there is only one sulfate group to interact with CH.

3.2. Film Thickness, Folding Endurance and Mechanical Properties

Once the films were loaded with microparticles, film TH, folding endurance, and mechanical properties were determined. TH measurements are essential to evaluate the potential discomfort of the patient in the gingiva. The TH values of the formulations ranged from 0.527 to 0.820 mm with no significant differences ($p > 0.05$) between them (Table 2). These TH values are lesser than 1 mm and therefore are considered adequate to avoid discomfort after application [21].

Table 2. Values obtained in physical measurements and mechanical properties.

Film Formulation	Thickness (mm)	Folding Endurance	Tensile Strength (N)	Elongation (%)
HG-CH-ALG	0.818 ± 0.085	✓	2.30 ± 0.57	12.11 ± 3.37
HG-CH-κC	0.696 ± 0.118	✓	2.20 ± 0.54	15.22 ± 5.07
HG-CH-λC	0.820 ± 0.075	✓	1.53 ± 0.33	14.97 ± 5.60
HG-CH-SLS	0.527 ± 0.076	✓	1.36 ± 0.48	19.04 ± 3.92

The folding endurance is the number of times each film can be folded in the same place without breaking. The analysis was carried out to analyze the flexibility of the films so that they can be easily manipulated, without causing discomfort, and achieve a safe application at the site of action [11]. All formulations were folded at least 300 times, complying with the test (Table 2). The obtained values of elongation at break and tensile strength for the different films are also shown in Table 2. High tensile strength values were observed in the films containing CH-ALG and CH-κC microparticles. Although the matrix of all four polymeric films is constituted by the same polymers, the difference lies in the loaded microparticulate system. Those microparticles that have higher charge density and free charges would interact strongly with the polymers in the films.

The TH of the films varied with respect to the values obtained for the not loaded film (0.517 mm ± 0.04), demonstrating that the incorporation of the microparticles produced a slight increase in the TH of the formulations. This result is probably because the presence of microparticles disrupts the polymeric matrix of the films, preventing an adequate compaction, which is also reflected in the tensile strength and elongation values of the formulations [22–24]. The tensile strength of the not loaded film (3.28 N) was diminished significantly with the incorporation of microparticles. When the microparticles based on CH-ALG and CH-κC were added, the tensile strength of the films was reduced from 3.28 N to 2.30 N and 2.20 N, respectively. Additionally, the incorporation of the microparticles based on CH-λC and CH-SLS reduced even more this parameter, reaching values of 1.53 N and 1.36 N, respectively. In addition, the elongation of films containing microparticles based on CH-ALG, CH-κC, and CH-λC decreased significantly with respect to the not load film (18.54 ± 4.73%). Conversely, this parameter was not altered when the film was loaded with the CH-SLS microparticles. As mentioned, CH is a cationic polymer presenting amino groups, which can interact with anionic and neutral compounds, while ALG, C, and SLS are anionic polymers which present different groups (carboxylic, sulphate ester, and sulphate groups, respectively) to interact with the amino groups of CH. Yuhua C. et al. reported that the reactivity of the substituent anionic groups of the mentioned anionic polymers increases in the following order: (1) sulfate group, (2) sulfate ester group, (3) carboxylic group [25]. Thus, the microparticles based on CH and ALG present the strongest interactions [26]. Between the loaded films, those containing CH-ALG microparticles were more rigid while those loaded with CH-SLS showed the smallest tensile strength and the highest elongation at break. This fact shows a direct relationship between the interactions that exist in the microparticles with the mechanical properties of the loaded films.

3.3. Swelling Index

The swelling index values are shown in Figure 1. The formulations loaded with CH-ALG, CH-κC, and CH-λC microparticles were able to complete the test after 24 h. Systems loaded with CH-SLS and CH-λC microparticles achieved maximum in swelling at approximately at 6 h assay. After this time, the weight of both films decreased due to the partial (CH-λC) and complete (CH-SLS) disintegration of the matrices. Polymers presenting free groups with hydrophilic nature play an important role in water uptake [27]. In film containing microparticles that present strong interactions between their oppositely charged polymers (CH-ALG), the amino groups of CH interact with the carboxylic groups of ALG. These interactions reduced the number of amino and carboxylic free groups, diminishing the water retention capacity of this matrix. Oppositely, microparticles based on CH-SLS are able of capturing a greater amount of fluid, generating greater swelling, which finally causes the disintegration of the matrix.

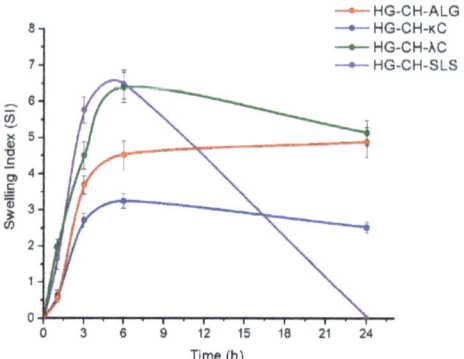

Figure 1. Swelling index of different systems in artificial saliva.

3.4. In Vitro Mucoadhesive Strength

Mucoadhesivity values of the systems are shown in Figure 2. Films loaded with microparticles significantly decreased the adhesiveness of the unloaded film (1.52 N ± 0.18). On the other hand, no significant differences were found between the adhesiveness of the different loaded films. This fact could be due to the presence of the microparticles; when they interact with the matrix, the number of free surface charges available to interact with the mucin of the gingiva is reduced [28]. The films loaded with microparticles based on SLS were the most flexible, presenting the highest swelling index, so they could adhere more easily to the gingiva than the other formulations [29].

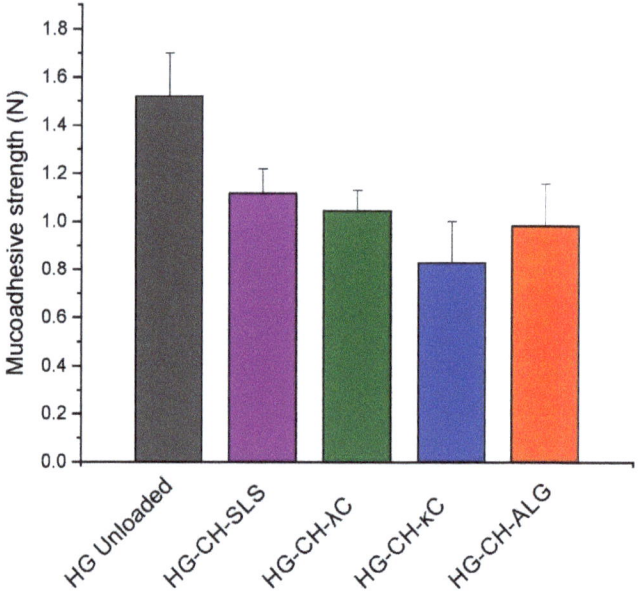

Figure 2. Film mucoadhesive strength values.

3.5. Morphology Analysis and Size Determination by Scanning Electron Microscopy

Images taken by SEM of the microparticulate systems and the films loaded with microparticles are shown in Figure 3. CH-ALG microparticles (Figure 3a) showed a smooth and uniform surface (size 2.973 ± 0.886 µm), while microparticles based on CH-κC and CH-λC (Figure 3b,c) presented nonuniform and irregular surface (size 2.239 ± 0.574 and 1.734 ± 0.382 µm, respectively), whereas microparticles based only on CH-SLS (size 2.774 ± 1.035 µm) were irregular with a smooth surface (Figure 3d). All films loaded with microparticles showed a nonhomogeneous surface with cavities, while the presence of discontinuous areas and rugosities were observed in the cross section of films (Figure 3i–l). A probable incompatibility between HPMC and GEL has previously been reported, which may be the cause of this morphology [30].

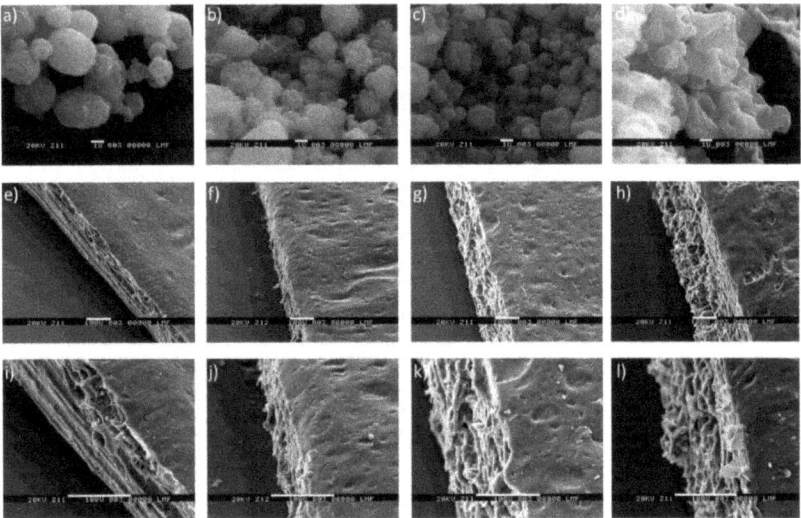

Figure 3. Scanning electron microscopy of microparticles (**a**–**d**) at 10,000× and films loaded with microparticles at 200× (**e**–**h**) and at 500× (**i**–**l**). (**a**) CH-ALG, (**b**) CH-κC, (**c**) CH-λC, (**d**) CH-SLS, (**e**) HG-CH-ALG, (**f**) HG-CH-κC, (**g**) HG-CH-λC, (**h**) HG-CH-SLS, (**i**) HG-CH-ALG, (**j**) HG-CH-κC, (**k**) HG-CH-λC, (**l**) HG-CH-SLS.

3.6. Thermal Characterization

Figure 4 shows the derivate of TGA curves of the systems. Regarding the MN (Figure 4a), it has a broad peak centered at 305 °C corresponding to its degradation [13]. In the microparticles loaded with MN, the characteristic peaks of the degradations step of the forming polymers were observed, but not the peaks of MN [31–35]. As observed in Figure 4b, LDCH presented a degradation peak centered at around 250 °C [36]. The different film formulations exhibit a central peak around 300 °C, corresponding to the degradation of the film forming polymers, HPMC, and GEL [37]. The absence of the characteristic peaks of MN and LDCH loaded in the systems is due to the ability of the polymeric matrices to protect both drugs against thermal degradation [38].

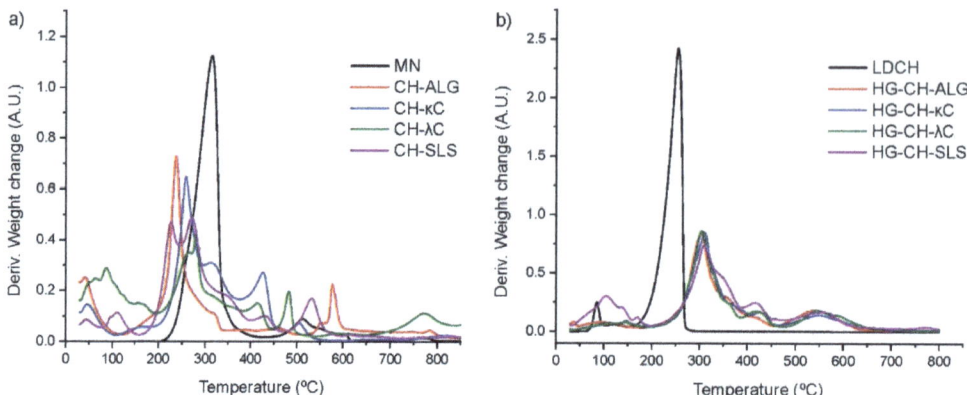

Figure 4. (a) DTGA curves of MN and microparticles loaded with MN. (b) DTGA curves of LDCH and different film formulations.

The melting temperatures of the systems were observed by DSC (Figure 5). As shown in Figure 5a, the endothermic peak for MN was at 186 °C [39]. This endothermic peak was absent in MN loaded microparticles. Figure 5b shows the endothermic peak of LDCH at 80 °C, corresponding to the fusion of this drug [40]. As in the case of MN, the same performance of the drug in the different formulations was observed in this curve. The absence of the characteristic melting peaks of MN and LDCH is due to: (1) the drug reduced its crystallinity below the detection limit, or (2) the drug changed to an amorphous state [41,42].

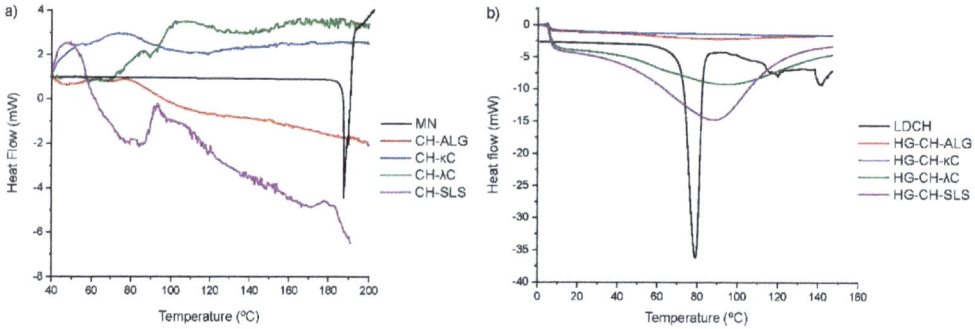

Figure 5. (a) DSC curves of MN and microparticles loaded with MN. (b) DSC curves of LDCH and films loaded with microparticles and LDCH.

3.7. X-ray Diffraction

Figure 6a shows the diffractogram of MN and microparticles loaded with MN and Figure 6b shows the patterns of LDCH and films loaded with the microparticles and LDCH. The diffractogram of the MN presented the typical crystalline pattern [43]. On the other hand, the XRD patterns of the loaded microparticles did not show any defined peak. This could be due to multiple factors such as the fact that, in the SD process, the MN partially lost its crystallinity or, that the MN loaded in the microparticulate systems is in an amorphous state [42].

As observed in Figure 6b, the diffractogram of the LDCH showed peaks at diffraction angles (2θ): 6.9°, 14.2°, 20.8°, 25.2°, and 27.8°, with a typical crystalline pattern [44]. The characteristic main peaks of LDCH were absent in the XRD pattern of the films, suggesting that LDCH was in an amorphous state.

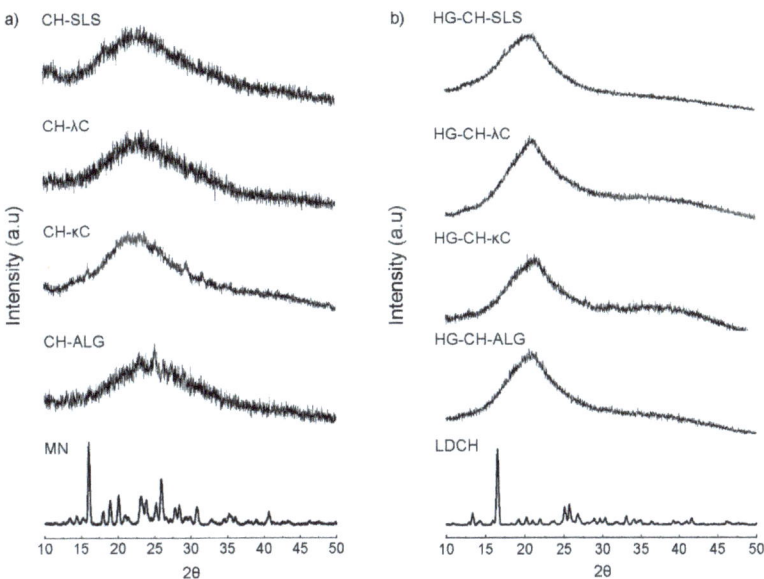

Figure 6. (a) Diffractogram of MN and microparticles loaded with MN. (b) Diffractogram of LDCH and films loaded with microparticles and LDCH.

3.8. Dissolution Studies

The dissolution of LDCH (as raw material), and the different systems in artificial saliva are shown in Figure 7. LDCH raw material showed a fast dissolution rate (100% after 1 h assay), while when the drug was loaded in the films, its dissolution rate was reduced. The release of LDCH from the films was in agreement with the swelling results. Among the formulations, films containing microparticles based on CH-SLS allowed for the fastest dissolution rate of the drug (60% after 1 h assay), while films containing microparticles based on CH-ALG released only 20% of LDCH at this time. This result was predictable, since LDCH is a drug that is highly soluble in water; the more swelling the matrix presents, the greater the dissolution of the drug there will be.

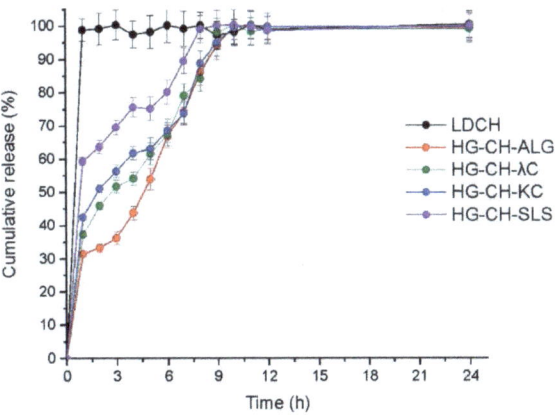

Figure 7. Dissolution profiles of LDCH raw material and films loaded with microparticles and LDCH.

3.9. Halo Zone Test over Time

All systems produced different inhibition halos at different times (Figure 8).

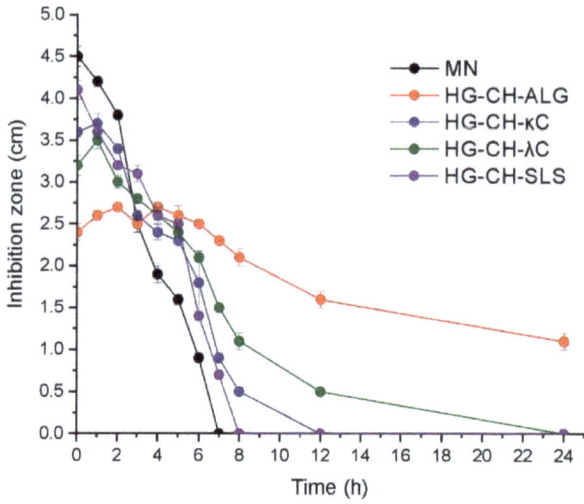

Figure 8. Inhibition halos produced by all formulations in cultures of *C. albicans*.

After 1 h assay, the highest inhibition halo corresponds to MN powder (4.2 cm) followed by the formulations HG-CH-SLS, HG-CH-κC, and HG-CH-λC (around 3.6 cm). This could be due to the fact that the interactions in the microparticles loaded in these films are weaker than those generated in the CH-ALG microparticles. Thus, the MN is more exposed to the culture medium, resulting in higher inhibition halos at the beginning of the study. Contrarily, after 1 h assay, the formulation HG-CH-ALG produced the smallest halo (2.6 cm). In the same way, both the strongest interaction between CH and ALG and the reduced swelling of this film probably cause less exposition of MN to the culture medium. The activity of the systems was decreasing over time, the paper disk containing MN powder lost its activity after 7 h assay, followed by the film HG-CH-SLS which did not show any halo after 8 h assay, and then HG-CH-κC (which lost its activity after 12 h). On the other hand, the formulation HG-CH-λC presented activity for 12 h, and finally, the film containing microparticles based on CH-ALG showed continuous antifungal activity until 24 h assay, demonstrating that this system (containing MN microencapsulated and added to the HPMC-GEL matrix) produced a sustained release of this drug.

4. Conclusions

In this work, microparticulate systems containing miconazole nitrate were developed by the spray-drying method. These systems were loaded in films based on HMPC and GEL which also contain lidocaine as an anesthetic drug to complement the treatment of oropharyngeal candidiasis. The MN encapsulation efficiency was higher in the CH-ALG system, possibly due to the high reactivity of the amino and carboxylic groups of CH and ALG, respectively, which may interact with the drug. These strong interactions between CH-ALG generated films with the highest tensile strength and the lowest elongation at break. All systems presented adequate values in terms of thickness to not produce discomfort to the gingiva of the potential patient. In the DSC and XRD tests, the characteristic peaks of the drugs were not evident, possibly because the drugs changed to an amorphous state in the systems. In vitro activity of the formulations evaluated by the halo zone test showed that the formulation containing microparticles based on CH-SLS (which allowed the fastest dissolution of MN) lost its antifungal activity after 8 h, while the formulation containing CH-ALG microparticles presented activity for 24 h. Therefore, the film loaded with both lidocaine and microparticles based on CH-ALG containing MN would cause an anesthetic effect and produce a controlled release of MN in the patient's gums for 24 h. Thus, this formulation could be applied once a day, increasing the patients' compliance.

Author Contributions: Conceptualization: G.T.; Methodology: G.T. and N.L.C. and M.M.; Software: G.T. and N.L.C.; Data curation: G.T.; Writing—original draft presentation: G.T., N.L.C., C.L. and M.S.; Visualization: C.L. and D.L.; Investigation: G.T., N.L.C., M.M., M.S., C.L., D.L. and V.A.Á.; Supervision: C.L., D.L. and V.A.Á.; Validation: G.T; Writing—Reviewing and Editing: D.L. and V.A.Á. All authors have read and agreed to the published version of the manuscript.

Funding: This research was funded by the Agencia Nacional de Promoción Científica y Tecnológica of Argentina (PICT 2017-1276).

Institutional Review Board Statement: Not applicable.

Informed Consent Statement: Not applicable.

Data Availability Statement: The data presented in this study are available on request from the corresponding authors.

Conflicts of Interest: The authors declare no conflict of interest.

References

1. Spampinato, C.; Leonardi, D. Candida Infections, Causes, Targets, and Resistance Mechanisms: Traditional and Alternative Antifungal Agents. *BioMed Res. Int.* **2013**, *2013*, 204237. [CrossRef]
2. Garcia-Cuesta, C.; Sarrion-Pérez, M.G.; Bagán, J.V. Current Treatment of Oral Candidiasis: A Literature Review. *J. Clin. Exp. Dent.* **2014**, *6*, e576. [CrossRef]
3. Oji, C.; Chukwuneke, F. Evaluation and Treatment of Oral Candidiasis in HIV/AIDS Patients in Enugu, Nigeria. *Oral Maxillofac. Surg.* **2008**, *12*, 67–71. [CrossRef] [PubMed]
4. Rai, V.K.; Yadav, N.P.; Sinha, P.; Mishra, N.; Luqman, S.; Dwivedi, H.; Kymonil, K.M.; Saraf, S.A. Development of Cellulosic Polymer Based Gel of Novel Ternary Mixture of Miconazole Nitrate for Buccal Delivery. *Carbohydr. Polym.* **2014**, *103*, 126–133. [CrossRef]
5. De Caro, V.; Murgia, D.; Seidita, F.; Bologna, E.; Alotta, G.; Zingales, M.; Campisi, G. Enhanced In Situ Availability of Aphanizomenon Flos-Aquae Constituents Entrapped in Buccal Films for the Treatment of Oxidative Stress-Related Oral Diseases: Biomechanical Characterization and In Vitro/Ex Vivo Evaluation. *Pharmaceutics* **2019**, *11*, 35. [CrossRef] [PubMed]
6. Mady, O.Y.; Donia, A.M.; Al-Madboly, L.A. Miconazole-Urea in a Buccal Film as a New Trend for Treatment of Resistant Mouth Fungal White Patches. *Front. Microbiol.* **2018**, *9*, 837. [CrossRef]
7. De Caro, V.; Giannola, L.I.; Di Prima, G. Solid and Semisolid Innovative Formulations Containing Miconazole-Loaded Solid Lipid Microparticles to Promote Drug Entrapment into the Buccal Mucosa. *Pharmaceutics* **2021**, *13*, 1361. [CrossRef]
8. Morri, M.; Castellano, P.; Leonardi, D.; Vignaduzzo, S. First Development, Optimization, and Stability Control of a Pediatric Oral Atenolol Formulation. *AAPS PharmSciTech* **2018**, *19*, 1781–1788. [CrossRef] [PubMed]
9. Hermawan, D.; Yatim, I.M.; Rahim, K.A.; Sanagi, M.M.; Ibrahim, W.A.W.; Aboul-Enein, H.Y. Comparison of HPLC and MEEKC for Miconazole Nitrate Determination in Pharmaceutical Formulation. *Chromatographia* **2013**, *76*, 1527–1536. [CrossRef]
10. Tejada, G.; Barrera, M.G.; García, P.; Sortino, M.; Lamas, M.C.; Lassalle, V.; Alvarez, V.; Leonardi, D. Nanoparticulated Systems Based on Natural Polymers Loaded with Miconazole Nitrate and Lidocaine for the Treatment of Topical Candidiasis. *AAPS PharmSciTech* **2020**, *21*, 278. [CrossRef]
11. Takeuchi, Y.; Ikeda, N.; Tahara, K.; Takeuchi, H. Mechanical Characteristics of Orally Disintegrating Films: Comparison of Folding Endurance and Tensile Properties. *Int. J. Pharm.* **2020**, *589*, 119876. [CrossRef] [PubMed]
12. Tejada, G.; Piccirilli, G.N.; Sortino, M.; Salomón, C.J.; Lamas, M.C.; Leonardi, D. Formulation and In-Vitro Efficacy of Antifungal Mucoadhesive Polymeric Matrices for the Delivery of Miconazole Nitrate. *Mater. Sci. Eng. C* **2017**, *79*, 140–150. [CrossRef] [PubMed]
13. Tejada, G.; Lamas, M.C.; Svetaz, L.; Salomón, C.J.; Alvarez, V.A.; Leonardi, D. Effect of Drug Incorporation Technique and Polymer Combination on the Performance of Biopolymeric Antifungal Buccal Films. *Int. J. Pharm.* **2018**, *548*, 431–442. [CrossRef] [PubMed]
14. Oza, N.A.; Makwana, A.; Gohil, T.A.; Shukla, P. Statistical Optimization of Miconazole Nitrate Microemulgel by Using 23 Full Factorial Desing. *Int. J. Pharm. Sci. Drug Res.* **2021**, *13*, 15–23. [CrossRef]
15. Raj, V.; Prava, K.; Seru, G. RP-HPLC Method Development and Validation for the Simultaneous Determination of Clindamycin and Miconazole in Pharmaceutical Dosage Forms. *Pharm. Methods* **2014**, *5*, 56–60. [CrossRef]
16. CLSI M44-A2; Method for Antifungal Disk Diffusion Susceptibility Testing of Yeasts. Approved Guideline. 2nd ed. Clinical and Laboratory Standards Institute: Wayne, PA, USA, 2009; ISBN 1-56238-703-0.

17. Calvo, N.L.; Svetaz, L.A.; Alvarez, V.A.; Quiroga, A.D.; Lamas, M.C.; Leonardi, D. Chitosan-Hydroxypropyl Methylcellulose Tioconazole Films: A Promising Alternative Dosage Form for the Treatment of Vaginal Candidiasis. *Int. J. Pharm.* **2019**, *556*, 181–191. [CrossRef]
18. Yeo, Y.; Park, K. Control of Encapsulation Efficiency and Initial Burst in Polymeric Microparticle Systems. *Arch. Pharmacal Res.* **2004**, *27*, 1–12. [CrossRef]
19. Król, Ż.; Malik, M.; Marycz, K.; Jarmoluk, A. Characteristic of Gelatine, Carrageenan and Sodium Alginate Hydrosols Treated by Direct Electric Current. *Polymers* **2016**, *8*, 275. [CrossRef]
20. Kalsoom Khan, A.; Saba, A.U.; Nawazish, S.; Akhtar, F.; Rashid, R.; Mir, S.; Nasir, B.; Iqbal, F.; Afzal, S.; Pervaiz, F.; et al. Carrageenan Based Bionanocomposites as Drug Delivery Tool with Special Emphasis on the Influence of Ferromagnetic Nanoparticles. *Oxid. Med. Cell. Longev.* **2017**, *2017*, 8158315. [CrossRef]
21. Nair, A.B.; Kumria, R.; Harsha, S.; Attimarad, M.; Al-Dhubiab, B.E.; Alhaider, I.A. In Vitro Techniques to Evaluate Buccal Films. *J. Control. Release* **2013**, *166*, 10–21. [CrossRef]
22. Nogueira, G.F.; Fakhouri, F.M.; Velasco, J.I.; de Oliveira, R.A. Active Edible Films Based on Arrowroot Starch with Microparticles of Blackberry Pulp Obtained by Freeze-Drying for Food Packaging. *Polymers* **2019**, *11*, 1382. [CrossRef] [PubMed]
23. Wu, H.; Tang, B.; Wu, P. Optimizing Polyamide Thin Film Composite Membrane Covalently Bonded with Modified Mesoporous Silica Nanoparticles. *J. Membr. Sci.* **2013**, *428*, 341–348. [CrossRef]
24. Lu, R.; Sameen, D.E.; Qin, W.; Wu, D.; Dai, J.; Li, S.; Liu, Y. Development of Polylactic Acid Films with Selenium Microparticles and Its Application for Food Packaging. *Coatings* **2020**, *10*, 280. [CrossRef]
25. Chang, Y.; McLandsborough, L.; McClements, D.J. Interaction of Cationic Antimicrobial (ε-Polylysine) with Food-Grade Biopolymers: Dextran, Chitosan, Carrageenan, Alginate, and Pectin. *Food Res. Int.* **2014**, *64*, 396–401. [CrossRef] [PubMed]
26. Ouellette, R.J.; Rawn, J.D. Carboxylic Acid Derivatives. In *Organic Chemistry*; Academic Press: Cambridge, MA, USA, 2018; pp. 665–710. [CrossRef]
27. Pal, K.; Banthia, A.K.; Majumdar, D.K. Preparation and Characterization of Polyvinyl Alcohol-Gelatin Hydrogel Membranes for Biomedical Applications. *AAPS PharmSciTech* **2007**, *8*, 21. [CrossRef] [PubMed]
28. Morales, J.O.; McConville, J.T. Manufacture and Characterization of Mucoadhesive Buccal Films. *Eur. J. Pharm. Biopharm.* **2011**, *77*, 187–199. [CrossRef]
29. Karki, S.; Kim, H.; Na, S.-J.J.; Shin, D.; Jo, K.; Lee, J. Thin Films as an Emerging Platform for Drug Delivery. *Asian J. Pharm. Sci.* **2016**, *11*, 559–574. [CrossRef]
30. Tedesco, M.P.; Monaco-Lourenço, C.A.; Carvalho, R.A. Gelatin/Hydroxypropyl Methylcellulose Matrices—Polymer Interactions Approach for Oral Disintegrating Films. *Mater. Sci. Eng. C* **2016**, *69*, 668–674. [CrossRef]
31. Sakurai, K.; Maegawa, T.; Takahashi, T. Glass Transition Temperature of Chitosan and Miscibility of Chitosan/Poly(N-Vinyl Pyrrolidone) Blends. *Polymer* **2000**, *41*, 7051–7056. [CrossRef]
32. Patel, N.; Lalwani, D.; Gollmer, S.; Injeti, E.; Sari, Y.; Nesamony, J. Development and Evaluation of a Calcium Alginate Based Oral Ceftriaxone Sodium Formulation. *Prog. Biomater.* **2016**, *5*, 117–133. [CrossRef]
33. Rane, L.R.; Savadekar, N.R.; Kadam, P.G.; Mhaske, S.T. Preparation and Characterization of K-Carrageenan/Nanosilica Biocomposite Film. *J. Mater.* **2014**, *2014*, 736271. [CrossRef]
34. Sadeghi, M. Synthesis of a Biocopolymer Carrageenan-g-Poly(AAm-Co-IA)/ Montmorillonite Superabsorbent Hydrogel Composite. *Braz. J. Chem. Eng.* **2012**, *29*, 295–305. [CrossRef]
35. Bang, J.H.; Jang, Y.N.; Song, K.S.; Jeon, C.W.; Kim, W.; Lee, M.G.; Park, S.J. Effects of Sodium Laurylsulfate on Crystal Structure of Calcite Formed from Mixed Solutions. *J. Colloid Interface Sci.* **2011**, *356*, 311–315. [CrossRef] [PubMed]
36. Wei, Y.; Nedley, M.P.; Bhaduri, S.B.; Bredzinski, X.; Boddu, S.H.S. Masking the Bitter Taste of Injectable Lidocaine HCl Formulation for Dental Procedures. *AAPS PharmSciTech* **2015**, *16*, 455–465. [CrossRef]
37. Tejada, G.; Lamas, M.C.; Sortino, M.; Alvarez, V.A.; Leonardi, D. Composite Microparticles Based on Natural Mucoadhesive Polymers with Promising Structural Properties to Protect and Improve the Antifungal Activity of Miconazole Nitrate. *AAPS PharmSciTech* **2018**, *19*, 3712–3722. [CrossRef] [PubMed]
38. Moseson, D.E.; Jordan, M.A.; Shah, D.D.; Corum, I.D.; Alvarenga, B.R.; Taylor, L.S. Application and Limitations of Thermogravimetric Analysis to Delineate the Hot Melt Extrusion Chemical Stability Processing Window. *Int. J. Pharm.* **2020**, *590*, 119916. [CrossRef]
39. Umeyor, C.E.; Okoye, I.; Uronnachi, E.; Okeke, T.; Kenechukwu, F.; Attama, A. Repositioning Miconazole Nitrate for Malaria: Formulation of Sustained Release Nanostructured Lipid Carriers, Structure Characterization and in Vivo Antimalarial Evaluation. *J. Drug Deliv. Sci. Technol.* **2021**, *61*, 102125. [CrossRef]
40. Zhou, G.; Dong, J.; Wang, Z.; Li, Z.; Li, Q.; Wang, B. Determination and Correlation of Solubility with Thermodynamic Analysis of Lidocaine Hydrochloride in Pure and Binary Solvents. *J. Mol. Liq.* **2018**, *265*, 442–449. [CrossRef]
41. Trivino, A.; Gumireddy, A.; Meng, F.; Prasad, D.; Chauhan, H. Drug-Polymer Miscibility, Interactions, and Precipitation Inhibition Studies for the Development of Amorphous Solid Dispersions for the Poorly Soluble Anticancer Drug Flutamide. *Drug Dev. Ind. Pharm.* **2019**, *45*, 1277–1291. [CrossRef]
42. Leuner, C.; Dressman, J. Improving Drug Solubility for Oral Delivery Using Solid Dispersions. *Eur. J. Pharm. Biopharm.* **2000**, *50*, 47–60. [CrossRef]

43. Ribeiro, A.; Figueiras, A.; Santos, D.; Veiga, F. Preparation and Solid-State Characterization of Inclusion Complexes Formed between Miconazole and Methyl-β-Cyclodextrin. *AAPS PharmSciTech* **2008**, *9*, 1102–1109. [CrossRef] [PubMed]
44. Khan, H.U.; Nasir, F.; Maheen, S.; Shafqat, S.S.; Shah, S.; Khames, A.; Ghoneim, M.M.; Abbas, G.; Shabbir, S.; Abdelgawad, M.A.; et al. Antibacterial and Wound-Healing Activities of Statistically Optimized Nitrofurazone-and Lidocaine-Loaded Silica Microspheres by the Box–Behnken Design. *Molecules* **2022**, *27*, 2532. [CrossRef] [PubMed]

Disclaimer/Publisher's Note: The statements, opinions and data contained in all publications are solely those of the individual author(s) and contributor(s) and not of MDPI and/or the editor(s). MDPI and/or the editor(s) disclaim responsibility for any injury to people or property resulting from any ideas, methods, instructions or products referred to in the content.

Article

Green Synthesized sAuNPs as a Potential Delivery Platform for Cytotoxic Alkaloids

Byron Mubaiwa [1], Mookho S. Lerata [1], Nicole R. S. Sibuyi [2], Mervin Meyer [2], Toufiek Samaai [3,4], John J. Bolton [5], Edith M. Antunes [6,*] and Denzil R. Beukes [1,*]

1. School of Pharmacy, University of the Western Cape, Robert Sobukwe Road, Bellville 7535, South Africa
2. Department of Science and Innovation/Mintek Nanotechnology Innovation Centre (DST/Mintek NIC), Bio-Labels Node, Department of Biotechnology, University of the Western Cape, Robert Sobukwe Road, Bellville 7535, South Africa
3. Department of Environmental Affairs (Oceans and Coasts), Cape Town 8000, South Africa
4. Department of Biodiversity and Conservation Biology, University of the Western Cape, Robert Sobukwe Road, Bellville 7535, South Africa
5. Department of Biological Sciences, University of Cape Town, Rondebosch 7701, South Africa
6. Department of Chemistry, University of the Western Cape, Robert Sobukwe Road, Bellville 7535, South Africa
* Correspondence: ebeukes@uwc.ac.za (E.M.A.); dbeukes@uwc.ac.za (D.R.B.); Tel.: +27-21-959-4020 (E.M.A.); +27-21-959-2352 (D.R.B.)

Abstract: The use of natural products as chemotherapeutic agents is well established. However, many are associated with undesirable side effects, including high toxicity and instability. Previous reports on the cytotoxic activity of pyrroloiminoquinones isolated from Latrunculid sponges against cancer cell lines revealed extraordinary activity at IC_{50} of 77nM for discorhabdins. Their general lack of selectivity against the cancer and normal cell lines, however, precludes further development. In this study, extraction of a South African Latrunculid sponge produced three known pyrroloiminoquinone metabolites (14-bromodiscorhabdin C (**5**), Tsitsikammamine A (**6**) and B (**7**)). The assignment of the structures was established using standard 1D and 2D NMR experiments. To mitigate the lack of selectivity, the compounds were loaded onto gold nanoparticles synthesized using the aqueous extract of a brown seaweed, *Sargassum incisifolium* (sAuNPs). The cytotoxicity of the metabolites alone, and their sAuNP conjugates, were evaluated together with the known anticancer agent doxorubicin and its AuNP conjugate. The compound-AuNP conjugates retained their strong cytotoxic activity against the MCF-7 cell line, with >90% of the pyrroloiminoquinone-loaded AuNPs penetrating the cell membrane. Loading cytotoxic natural products onto AuNPs provides an avenue in overcoming some issues hampering the development of new anticancer drugs.

Keywords: pyrroloiminoquinones; gold nanoparticles; *Sargassum incisifolium*; Latrunculid sponges; *Tsitsikamma favus*; cytotoxicity

1. Introduction

Cytotoxic natural products such as vincristine, vinblastine, paclitaxel, doxorubicin, podophyllotoxin, camptothecin, and ecteinascidin-743 have been the mainstay of cancer chemotherapy for decades [1]. However, one of the major drawbacks of these drugs is their non-selective action in killing both cancer and normal cells [2,3]. In recent years, several nano-delivery systems have been developed for anticancer drugs in the hope of improving selectivity and thus reducing toxicity [4].

Of particular interest to this study was nano-delivery systems based on gold nanoparticles (AuNPs). The advantages of the AuNP-based delivery platforms include: (1) their biocompatibility, (2) ability to passively target tumors through the leaky vasculature (enhanced permeability and retention effect), (3) controlled release in response to internal or external stimuli, and (4) the ability to modify the AuNPs surface with targeting ligands to enhance tumor selective accumulation compared to free drugs [5]. Several studies have demonstrated the effective delivery and/or improved activity of several anticancer drugs using this approach. Two main strategies have been developed to load the AuNP with the anticancer drug. Firstly, through covalent linking via thiolated derivatives, and secondly, by loading the drug onto polysaccharide-capped AuNPs. These studies have demonstrated more selective targeting [6], intracellular delivery of the drug [7], in vivo stability and efficacy [8,9], lower plasma levels of the free drug (cf. AuNP bound) and reduced general toxicities in normal tissues [9].

Biological materials employed in synthesizing nanoparticles possess excellent reducing (e.g., polyphenols) and stabilizing agents (e.g., polysaccharides) [10]. The main advantage of loading the anticancer drug onto polysaccharide-capped AuNPs is the simple, green synthetic methods employed using biodegradable polymers compared to the more complex synthetic steps, and the toxic coupling reagents currently used to covalently link the anticancer drug to the AuNP [10]. Several different types of polysaccharide-based capping agents have been used to deliver anticancer drugs, including fucoidan [11], carboxymethyl xanthan gum [12], carrageenan [13] and pectin [14]. These studies also demonstrated improved cytotoxic activities in vitro, high drug-loading and good stability under varying pH and electrolytic conditions as well as pH-triggered drug release.

We have previously reported on the excellent redox properties of *Sargassum incisifolium* extracts [15] and demonstrated the simple, room temperature synthesis and characterization (FTIR, TEM, XRD etc.) of AuNPs (sAuNPs) using aqueous extracts of *S. incisifolium* (SiAE) [16]. The brown seaweed, *S. incisifolium* is widely distributed along the South African coastline from Cape Point to Mozambique [17]. Metabolites isolated from *S. incisifolium* organic extracts, such as fucoxanthin and sargaquinal, have demonstrated promising antimalarial bioactivities against chloroquine-sensitive strains of *P. falciparum* [18]. The aqueous extracts of *S. incisifolium*, however, is rich in antioxidant molecules which are responsible for the reduction of Au^{3+} to Au^0 (AuNPs), while the water-soluble polysaccharides serve as capping agents [10,16]. Phlorotannins, e.g., fucodiphlorethol (**1**) and tetraphlorethol (**2**), are the main antioxidants in marine brown algae and are complex polymers of phloroglucinol (1,3,5-trihydroxybenzene) (Figure 1). The main water-soluble polysaccharides in the *Sargassum* spp. are alginic acid, fucoidan and laminaran. Alginic acid is a carboxyl-containing polysaccharide formed by joining β-D-mannuronic acid and α-L-guluronic acid through β-(1→4)/α-(1→4) glycosidic bonds. Fucoidan (Figure 1) is a water-soluble sulphated heteropolysaccharide with L-fucose groups linked through α-(1→3) or (1→4) glycosidic bonds [19], while laminaran is mainly composed of β-D-glucose bonded through β-(1→3) glycosidic bond links [20,21] (Figure 1).

The selective delivery and anticancer effects of AuNPs loaded with anticancer drugs can only be confirmed by in vivo studies. However, in this preliminary study we investigated the synthesis, characterization, and preliminary cytotoxic activity of *S. incisifolium* synthesized AuNPs (sAuNPs) loaded with doxorubicin (**3**) and the pyrroloiminoquinone alkaloids 14-bromodiscorhabdin C (**5**), and tsitsikammamine A (**6**) and B (**7**). If the AuNP-drug conjugate has the same, or improved activity, against a cancer cell line in vitro, then it may also demonstrate reduced toxic side effects due to the selective delivery of the conjugates to the cancer tissues. The pyrroloiminoquinones isolated from Latrunculid sponges have shown potent but non-selective cytotoxicity against both normal and cancer cell lines, which precluded their further development as anti-tumor drugs [22]. The targeted delivery, and increased selectivity, of these pyrroloiminoquinone alkaloids to tumor cells using nanoparticles as a delivery vehicle presents a highly attractive endeavor. In addition, the development of advanced targeted drug delivery system platforms for doxorubicin also remains a worthwhile research effort since an ideal, targeted, doxorubicin delivery platform decreases the required concentrations needed as well as the prevalence and intensity of side effects associated with the drug, while still utilizing its potent anti-cancer properties. In this work, we intend to show that (a) the sAuNPs are biocompatible and highly stable under physiological conditions, (b) that conjugation of the sAuNPs with doxorubicin

(3) and the pyrroliminoquinones (5–7) can be achieved, (c) that the 'drug' payload is released, and (d) that the AuNPs bound to cytotoxic alkaloids are able to traverse the MCF-7 cell membrane to exert their cytotoxic activity.

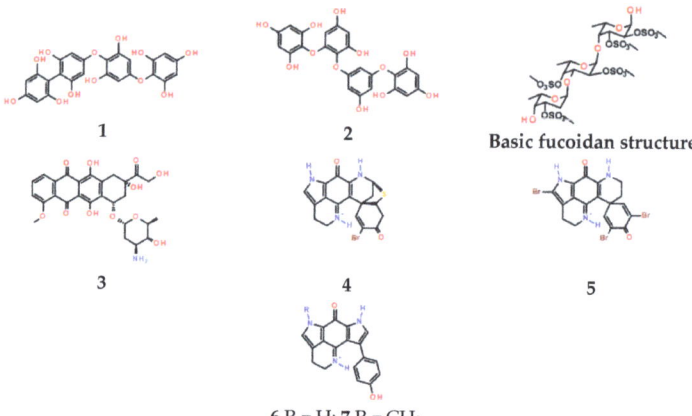

Figure 1. Structures of the compounds referred to or used in this study: fucodiphlorethol (**1**), tetraphlorethol (**2**), Doxorubicin (**3**), Discorhabdin A (**4**), 14-bromodiscorhabdin C (**5**), Tsitsikammamine A (**6**) and Tsitsikammamine B (**7**).

2. Materials and Methods

This section, as shown in Scheme 1, describes the preparation of the aqueous extracts of *S. incisifolium* (steps 1.1 and 1.2), the synthesis of sAuNPs (steps 1.3 and 1.4), the extraction and purification (steps 2.1 and 2.2) of the pyrroliminoquinone (Pq) alkaloids from *T. favus* latrunculid sponge, and the preparation of the sAuNP conjugates (Pq-sAuNPs, step 3.1) with the pyrroliminoquinone metabolites (and doxorubicin) using centrifugation (step 3.2, (Pq)-NP) or no centrifugation (step 3.3, NP(m)).

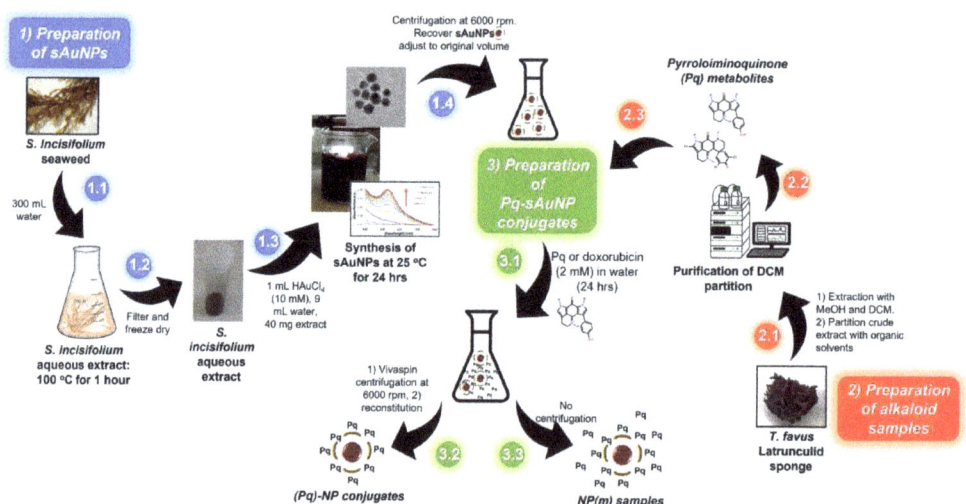

Scheme 1. Preparation of the (1) *S. incisifolium* aqueous extract capped AuNPs (sAuNPs), (2) purification of the pyrroliminoquinone (Pq) alkaloid metabolites and (3) sAuNP conjugates.

2.1. General Experimental/Materials

Gold(III) chloride (HAuCl$_4$.3H$_2$O), sodium citrate, Sartorius® Vivaspin™ Centrifugal Concentrators (Polyethersulfone (PES) membrane, MWCO = 30 kDa), and Dialysis sacks were obtained from Merck-Sigma-Aldrich (St. Louis, MO, USA). The solid phase extractions (SPE) of the Latrunculid sponge were carried out on pre-packed Phenomenex® Sep-Pak® C18 cartridges.

Semi-preparative high performance liquid chromatography (HPLC) was carried out using an Agilent Technologies HPLC equipped with a Phenomenex® Luna C18 (2), 10 μm (250 × 10 mm i.d.), a refractive index detector at 35 °C as well as a variable wavelength (at 245 nm) detector. NMR samples were prepared in deuterated solvents, and all experiments were acquired on a Bruker Avance III HD 400 MHz spectrometer equipped with a 5 mm BBO probe at 298 K using standard 1D and 2D pulse programs. Chemical shifts were referenced to residual undeuterated solvent signals in deuterated solvents peaks (DMSO-d_6: δ_H 2.5, δ_C 39.51; CD$_3$OD: δ_H 3.31, δ_C 49.15; D$_2$O: δ_H 4.70) and reported in ppm. DOSY experiments were carried out using the 2D Bruker DOSY bipolar pulse program with longitudinal eddy current delay, ledbpgp2s at 298 K using a gas flow rate of 400 lph without sample spinning. The diffusion time δ was kept at 100 ms, while the pulse field gradient (δ/2) was adjusted to 2000 μs to acquire 2–5% residual signals with maximum gradient strength. The eddy current delay was 5 ms and the delay for gradient recovery was set to 0.2 ms. The gradient strength was incremented in 25 steps from 2% to 98% of its maximum value in a linear ramp. The data were processed using the Bruker TopSpin 3.6.2 software. An Agilent Technologies Cary-60 UV-Visible spectrophotometer was used for UV-Visible measurements using quartz cuvettes (10 mm). The size, shape, crystallinity, and elemental composition of the AuNPs produced were determined using TEM measurements, coupled with EDX and SAED detectors. The TEM images, EDX data and SAED patterns were obtained using a Tecnai F20 TEM with a field-emission gun (FEG) operating in bright field mode at 200 kV, using a lacy carbon mesh on top of a copper grid and dried under an electric bulb. ImageJ was used to measure the AuNP sizes in the TEM images. A Malvern® Zetasizer Nano ZS™ was for all zeta potential and hydrodynamic particle size, while the powdered X-ray Diffraction patterns were acquired on a Bruker AXS (Germany) D8 Advance diffractometer (voltage 40 kV; current 40 mA). The XRD spectra were recorded in the range 30–90° using a CuKα (λ = 0.154 nm) monochromatic radiation X-ray source. Determination of the gold concentrations in the solutions was then accomplished using a Thermo ICap 6200 inductively coupled plasma optical emission spectroscopy (ICP-OES) utilizing a wavelength of 242.4 nm for Au at the Central Analytical Facility at Stellenbosch University. The instrument was calibrated and validated using NIST (National Institute of Standards and Technology, Gaithersburg, MD, USA) traceable standards to quantify the Au. Cell studies were performed under a class II biological safety cabinet. To visualize the cells, a Nikon light microscope with 20× magnification was used together with a Leica EC digital camera.

2.2. Extraction and Isolation of Pyrroloiminoquinones from Tsitsikamma Favus

The *Tsitsikamma favus* Latrunculid sponge, collected at the Tsitsikamma Marine Reserve, Eastern Cape, South Africa (GPS coordinates: −34.01515185009414, 23.81832453533402) and stored in a freezer at −20 °C, was collected (by SCUBA diving) and identified by Dr. Toufiek Samaai. A voucher specimen for the sample is housed in the School of Pharmacy, University of the Western Cape, South Africa. The mass of the sponge, after extraction and drying, was 240 g. The diced, thawed sponge was initially soaked in water (~2 L), the water decanted, and the sponge sequentially extracted with methanol (~2 L) and dichloromethane-methanol (1:1, ~2 L), for 48 h. The DCM-MeOH extract was dried and re-dissolved in MeOH and sequentially extracted with hexane, DCM and EtOAc giving a hexane fraction (Fr. A, 2.73 g), a DCM fraction (Fr. B, 269 mg), an EtOAc fraction (Fr. C, 101 mg) and an aqueous fraction (Fr. D, 620 mg). Fractions B, C and D revealed the characteristic ^{1}H NMR signals for discorhabdins and tsitsikammamines and were fractionated therefore subjected to further isolation. Fractions

B, C and D were further fractionated on C18 SPE cartridges using a step-gradient from 100% water to 100% MeOH. Final purification was achieved by repeated semi-preparative reversed-phase HPLC using a flow rate of 3 mL/min and 20% MeOH (0.05% TFA) as the eluent to give compounds **5–7**.

Compounds isolated:

14-Bromodiscorhabdin C (**5**): bright-red solid, isolated as a TFA salt; NMR data consistent with published data [23] and available in the supplementary document. HRESIMS m/z 539.8521 [M]$^+$ (calcd. for $C_{18}H_{13}N_3{}^{79}Br_3O_2$, 540.8636). Yield: 15.5 mg.

Tsitsikammamine A (**6**): reddish-brown solid isolated as a TFA salt; NMR data consistent with published data [23] and is available in the supplementary document. LRESIMS m/z 304. 7 [M]$^+$ (calcd. for $C_{18}H_{14}N_3O_2$, 304.1086). Yield: 9.0 mg.

Tsitsikammamine B (**7**): dark red solid, isolated as a TFA salt; NMR data consistent with published data [23] and available in the supplementary document. HRESIMS m/z 318.1237 [M]$^+$ (calcd. for $C_{19}H_{16}N_3O_2$, 318.1242). Yield: 23.2 mg.

2.3. Preparation of Sargassum Incisifolium Aqueous Extracts (SiAE)

Sargassum incisifolium C Agardh, identified and collected by Prof. John Bolton, was attached on the intertidal rocky seashore at Noordhoek, Gqeberha (formerly Port Elizabeth), South Africa (GPS coordinates: −34.03967398674499, 25.63982817957248) and stored in a freezer at −20 °C. A voucher specimen is housed in the School of Pharmacy, University of the Western Cape, South Africa. The thawed seaweed was washed with Milli-Q water, frozen with liquid nitrogen and ground to a fine powder and freeze-dried. The powder (10.4 g) was extracted with methanol (200 mL) for 1 h at room temperature followed by extractions with dichloromethane-methanol (200 mL × 3) at 35 °C for 30 min. The extracted biomass was air-dried and finally extracted with water (300 mL, at 100 °C for 1 h). The extract was filtered and the filtrate freeze-dried to obtain a fine brown powder (SiAE) and stored at room temperature. The total phenolic content, reducing power and the radical scavenging power of the SiAE was determined by the methods described by Tobwala et al. [24] and Mmola et al. [16].

2.4. Synthesis of AuNPs

Systematic studies of the ratio of SiAE to HAuCl$_4$ and reaction times led to the following optimized procedure: SiAE powder (20 mg) was added to HAuCl$_4$ (1 mM, 10 mL). Aliquots (500 µL) were drawn from the reaction mixture and analyzed using UV-Visible spectroscopy every 30 s for 5 min and thereafter every 30 min for 5 h. Details on the optimization studies are included in the supplementary material.

Citrate capped AuNPs (cAuNPs) were also prepared according to well established literature methods [25] to enable comparison to sAuNPs.

2.5. Stability Studies

Measurements were acquired using a UV/Visible spectrometer scanning from 800 nm to 200 nm every 2.5 min. The effect of the NaCl concentration, pH, temperature, and freeze drying, interaction with HSA were all determined. The experimental details are given in the Supplementary Section.

2.6. Loading of Doxorubicin (3) and the Isolated Pyrroloiminoquinone Alkaloids (5–7) onto sAuNPs

Doxorubicin and the pyrroloiminoquinones (2.0 mM) were added to 2 mL of the prepared sAuNPs solution. This solution was divided into two portions. With the first portion, a 2 mL Vivaspin™ concentrator (MWCO =30 kDa) was used to filter off unbound pyrroloiminoquinone metabolites, trapping 200 µL of the pyrroloiminoquinone bound sAuNP conjugates (denoted **5**-NP, for example). The second portion was not applied to a Vivaspin concentrator, thus consisting of a mixture of bound and unbound pyrroloiminoquinones (denoted **5**-NP(m), for example). The preparation of the doxorubicin loaded NPs was similarly done, however, a drop of 0.1 M NaOH was added to assist in solubility of the doxorubicin conjugate and the mixture incubated for 24–48 h. A cellulose-based dialysis sack (MWCO = 12 kDa) was used to separate the unloaded or free doxorubicin from the sAuNP conjugates. The compound loaded sAuNPs were stored in the dark until required.

2.7. Drug Entrapment and Loading Efficiency

To determine the loading efficiency of doxorubicin, the unloaded drug concentration for the doxorubicin-sAuNPs was determined by UV/Visible spectroscopy at a wavelength of 290 nm and compared to a similar study carried out by Manivasagan et al. [11,26].

2.8. Biological Studies

A stock solution of 5 mg/mL of each alkaloid was prepared by incubating an appropriate quantity of alkaloid with the previously prepared sAuNP solution (Section 2.4) for 24 h (sAuNP-1). A portion of this incubated solution was centrifuged at 6000 rpm to remove unbound alkaloid and reconstituted to the original volume (sAuNP-2). These stock solutions were further diluted to give final concentrations in the range 1–50 µg/mL. The antiproliferative effects of the pure natural product, a mixture of natural product and sAuNP (sAuNP-1) and the natural product sAuNP mixture after removal of the unbound natural product by centrifugation (sAuNP-2) were assessed using the water-soluble tetrazolium salt (WST-1). MCF-7 cells were seeded in 96 well plates (100 µL of 1×10^4 cells per well) and incubated for 24 h at 37 °C in a humidified incubator (maintained at 5% CO_2). Fresh media containing 1–50 µg/mL of compounds **3, 5, 6** and **7** and their nanoconjugates, were added to their respective well and incubated for another 48 h. Post incubation, 50 µL of WST-1 dye was added to the cells and incubated for a further three hours. The absorbance in the wells were measured at a wavelength of 440 nm using a microplate reader. Cell viability (as a percentage) was then calculated according to Equation (1):

$$\text{Cell viability (\%)} = \frac{(\text{Absorbance of treated cells})}{(\text{Absorbance of untreated cells})} \times 100 \quad (1)$$

Average values for triplicates were used to calculate IC_{50} concentrations using QuestGraph™ IC_{50} Calculator: https://www.aatbio.com/tools/ic50-calculator (accessed on 8 January 2023).

2.9. Determination of sAuNP Uptake in MCF-7 Cells Using ICP-OES

sAuNP uptake was evaluated using ICP-OES. The MCF-7 cells were seeded in 12-well culture plates at a density of 1×10^5 cells/mL (1 mL per well) and treated with sAuNPs-pyrroloiminoquinones/doxorubicin in triplicate for 48 h. The untreated samples were used as a negative control. Following treatment, the cells were washed twice with PBS. The cells were trypsinized and spun at 3000 rpm. The cell lysates were digested with 2 mL of *aqua regia* (HCl: HNO_3, 3:1) at 90 °C for 2 h. The samples were then allowed to cool to room temperature. The volume of the digested material was adjusted to 10 mL with 2% HCl. The amount of Au taken up by the samples was analyzed using ICP-OES. The amount of AuNPs was calculated based on the concentration of Au found in each cell sample, and compared to the concentration used for treatment to obtain the percentage of AuNPs taken up by the cells (Equation (2)).

$$\text{Au uptake (\%)} = \frac{[Au]_{cell}}{[Au]_{total}} \times 100 \quad (2)$$

$[Au]_{cell}$ is the Au content in the cell pellet and $[Au]_{total}$ is the total Au used in the assay.

3. Results and Discussion

3.1. Characterization of the Pyrroloiminoquinone Alkaloids Isolated from Tsitsikamma Favus

The pyrroloiminoquinone natural products were extracted and purified from *T. favus* (Section 2.2) using C18 solid phase extraction and reversed phased HPLC. The structures of the natural products were confirmed by comparison of their one and two-dimensional NMR data with those reported in the literature [23]. Spectra and tabulated spectroscopic data can be found in the supplementary data.

3.2. Characterization of the Seaweed Aqueous Extracts

The aqueous extract of *S. incisifolium* (SiAE) containing polysaccharides and polyphenols, as the major and minor constituents, was prepared and analyzed using UV/Visible, IR as well as NMR spectroscopies, as previously reported [16]. The presence of the main constituents was confirmed using a multiplicity-edited HSQC NMR experiment (Figure S7). The HSQC data revealed a methyl signal at $\sim\delta_C$ 18, the characteristic sugar oxymethine carbons between δ_C 60 and 80 and the anomeric carbons of the sugar moieties δ_C 100. The intensity of these signals together with the characteristic chemical shifts confirm the presence of a polysaccharide, namely the fucoidan, constituent. Additional, low intensity signals characteristic of phlorotannins/polyphenols were found at δ_C/δ_H 102/6.2, suggesting that phlorotannins are also present, albeit as minor constituents in the *S. incisifolium* crude aqueous extract. The results obtained for the antioxidant and free radical scavenging assays were consistent with the findings obtained with Mmola et al. [16]. The authors reported the total phenolic content and total reducing power for the aqueous extracts to be 235 µg/mg GAE (Gallic Acid Equivalents) and 95 µg/mg AAE (Ascorbic Acid Equivalents) per mg of dried seaweed, respectively [16]. Diffusion ordered spectroscopy (DOSY) was also used to analyze the sample mixture by measuring the differences in the diffusion coefficient (Δ) of the molecules. The Δ value is related to the molecular weight, size, and shape of the molecule and is relevant to the surrounding environment, including temperature and solvent viscosity. The DOSY spectrum (Figure S8) obtained for the SiAE sample showed that the Δ values obtained were remarkably similar (Δ of 6.25×10^{-10} m^2/s and 5.79×10^{-10} m^2/s for the polyphenol and polysaccharide, respectively) indicating that both components in the mixture are large macromolecules. Water was observed with a Δ value of 2.32×10^{-9} m^2/s.

3.3. Synthesis and Characterization of cAuNPs, sAuNPs and Drug Loaded AuNPs

The rate of AuNP formation, stability, and morphology of the AuNPs synthesized were compared using citric acid as the control sample [25], and SiAE as the capping (and reducing) agents. The metal salt to citrate (cAuNPs) or metal salt to extract (sAuNPs) ratios were optimized and the stability of the optimal AuNPs produced for each method compared (Figures 2 and 3 for cAuNPs and sAuNPs, respectively).

The citrate capped AuNPs formed only at higher temperatures (100 °C), and after 5 min an SPR band was observed at 524 nm. Due to the discrepancies in the literature with regards to the metal salt to capping ratio, the reaction conditions were optimized. The TEM images of the cAuNPs using a metal salt to citrate ratio at 2.5:1, revealed a wide range of nanoparticle shapes and sizes, including spherical and cuboid shapes ranging from 8 nm to 168 nm (Figure 2A,B), indicating that the metal salt concentration should be kept to a minimum. The cAuNPs formed at an optimized metal salt to citrate ratio of 0.022:1, revealed spherical, monodispersed NPs which were on average 14.5 nm in size (Figure 2C,D). The Selected Area Electron Diffraction (SAED) pattern revealed the cAuNPs to be crystalline (Figure 2E) in nature.

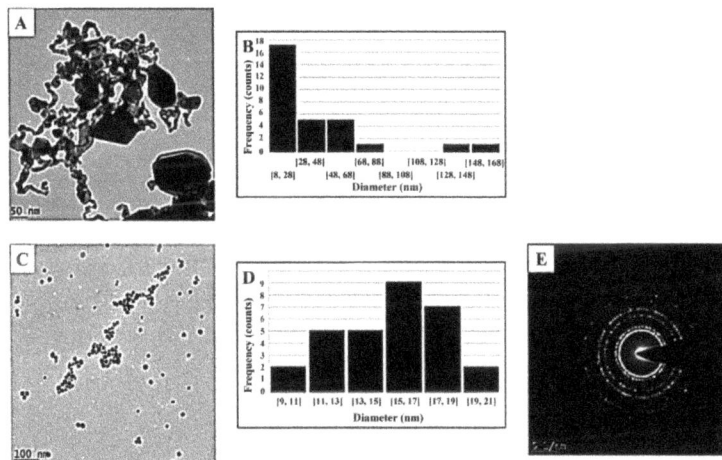

Figure 2. TEM images (**A,C**) and NP size distribution (**B,D**) for the cAuNPs prepared using a gold salt:citrate ratio of 2.5:1 (**A,B**) and the optimized ratio at 0.022:1 (**C,D**). The SAED (**E**) pattern obtained for the cAuNPs synthesized with a metal salt: citrate 0.022:1 ratio.

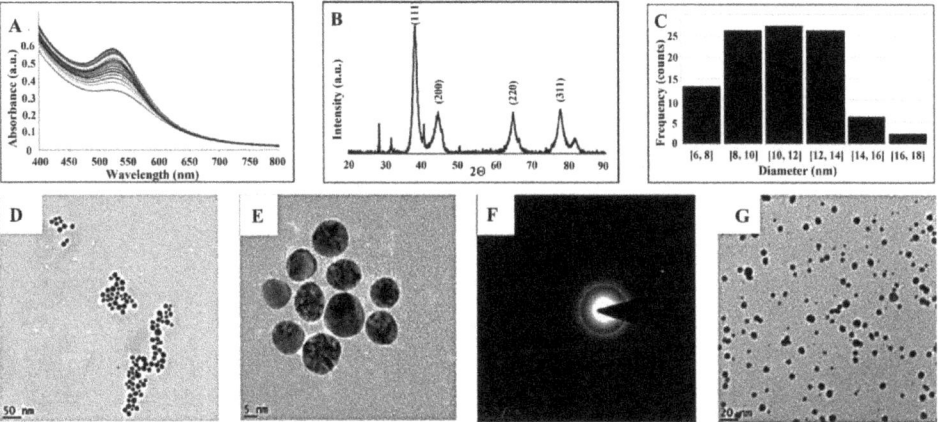

Figure 3. UV/Visible spectra (**A**) collected (in water) every 5 s for 24 h showing the sAuNP formation using SiAE, and the XRD powder pattern (**B**). The sAuNP size distribution (**C**) and TEM images (**D,E**), together with the SAED (**F**) pattern obtained for sAuNPs are also shown. The TEM image obtained for the sAuNPs following the processes of ultrafiltration/centrifugation and reconstitution is given in (**G**).

The sAuNPs were easily prepared at room temperature using the *S. incisifolium* aqueous extract (SiAE). UV/Visible absorption spectra revealed the characteristic surface plasmon resonance band at 525 nm (Figure 3A). The sAuNP formation proceeded at a faster rate with an increasing amount of extract (Figure S9), and optimal conditions were determined to be at a 0.167:1 ratio (metal salt: capping agent) for the sAuNPs. The TEM images for the sAuNPs, the size dispersion histogram and the SAED patterns are shown in Figure 3, together with the X-ray powder patterns. The TEM images (Figure 3D,E) showed spherically shaped, monodispersed sAuNPs which were 10.5 nm in size on average, with sizes ranging from 4 to 18 nm (Figure 3C). Inspection of the SAED pattern (Figure 3F) showed the sAuNPs to be polycrystalline, while the X-ray powder pattern (Figure 3B) confirmed the crystalline nature and face-centered cubic structure of the sAuNPs ($2\theta = 38.23°$ (111), 44.41° (200), 64.76° (220) and 77.77° (311)). Reflections due to NaCl and KCl were attributed to the presence of these salts in the seaweed. These data are consistent with that reported by Mmola et al. [16]. Dynamic light scattering (DLS) measurements, used to compliment the data obtained from the TEM analyses, helped to determine the hydrodynamic (Hd) diameter of the AuNPs suspended in solution and revealed information about the capping of the nanoparticles (Table 1). The DLS measurements for the sAuNPs (Table 1) revealed a hydrodynamic radius of 28.4 nm, a polydispersity index (PdI) of 0.242 and a zeta potential of −47.2 mV. The PdI indicates a broad size distribution, while the negative zeta potential indicated a negative surface charge, as expected for the polysaccharide capping on the NP surface [27]. The large negative charge is also associated with the inherent stability of the NP [28]. The cAuNPs, on the other hand, were found to be unstable and only the hydrodynamic radius (10.4 nm) was obtained.

Table 1. Stability of sAuNPs and cAuNPs (in parentheses) under varying NaCl concentrations, human serum albumin (HSA) concentrations, temperature, and pH.

(mM)			SPR (nm)		Hd (nm)		HSA (mg/mL)			T (°C)			pH		
0	√	(√)	524	(524)	28.4	(10.4)	0	√	(√)	37	√	(X)	2.0	√	(X)
50	-	(√)	-	(524)	-	(28.3)	1.25	√	(X)	25	√	(√)	4.0	√	(X)
75	-	(√)	-	(524)	-	(-)	2.50	√	(X)	4	√	(√)	7.0	√	(√)
100	√	(X)	524	(703)	46.1	(811)	6.25	√	(X)	−20	√	(X)	9.0	√	(X)
1000	√	(√)	524		67.4		12.50	√	(X)	−50	√	(X)	12.0	√	(X)
5000	√	(√)	703		152.7		25.00	√	-						
6000	√	(X)	-				50.00	√	-						

cAuNPs data given in parentheses; nd: not determined; Hd: hydrodynamic diameter; X = not stable; √ = stable.

3.4. Stability of the cAuNPs and sAuNPs

The stability of the synthesized sAuNPs were assessed against the standard citrate capped AuNP (control) for four weeks. The retention of the SPR band, the hydrodynamic radii as well as visual observations were recorded, and the stability of the NPs at various temperature and pH conditions were assessed (Table 1). The experimental methods are given in the Supplementary Materials.

As potential drug carriers, the AuNPs are expected to be stable and robust in the concentrations considered to be isotonic to blood plasma and red blood cells (to avoid lysis). Shukla et al. described a variety of biocompatible conditions requiring stability in the human body [28]. These include challenging the AuNPs at various NaCl concentrations (normal saline concentrations are isotonic at 154 mM), observing the interaction with human serum albumin (HSA) (one of the major plasma proteins involved in drug binding), temperature and pH [29]. Increasing the ionic strength (to 100 mM NaCl) resulted in a shift to longer wavelengths for the SPR band in the cAuNPs (Table 1 in parentheses) which is attributed to the aggregation of the cAuNPs. Changes in the zeta potential from −1.8 mV to more positive values, and the hydrodynamic diameters increasing to 811 nm (from 10.4 nm), all point to a decrease in the stability of the AuNP [30]. This contrasts sharply with the sAuNPs, which were found to be stable up to 5000 mM NaCl, i.e., 30 times the concentration of the medium isotonic. The sAuNPs retained their color and the SPR wavelength, with a slight increase in their hydrodynamic diameters observed. The ionic adsorption of Na^+ and Cl^- ions are therefore not likely to destabilize the NP (see Figure S11). Observations (including UV/Visible spectra) revealed that the sAuNPs, were stable at low HSA concentrations. It is expected that the proteins will interact via electrostatic interactions, with the polysaccharide's sulfate ions coating the sAuNP. Conversely, the cAuNPs were observed to immediately destabilize, coalesce, and flocculate. TEM images (Figure S12) revealed that the sAuNPs retained their structure, while the cAuNPs did not (Table 1). The sAuNPs also proved to be more stable at all temperatures and pH conditions, while the cAuNPs were not stable at low temperature and only stable at neutral pH (Table 1). The response of the sAuNPs to the various conditions was encouraging, as these conditions are related to the human body i.e., pH 7.4 (blood), pH 2 (stomach), pH 5.6–6.8 (intestines) and pH > 6.8 (tissues) [31].

3.5. Preparation of the Pyrroloiminoquinone (5–7) Loaded sAuNPs

The sAuNP-Pyrroloiminoquinone conjugates were prepared by loading the pyrroloiminoquinone metabolites (**5, 6** and **7**) onto the sAuNPs as a DMSO solution (~2 mM). A 2 mL Vivaspin™ concentrator (MWCO: 30 kDa) filtered off unbound metabolites, while trapping 200 µL of the sAuNPs. The Vivaspin™ concentrator was used to separate the un-loaded or free pyrroloiminoquinones from the sAuNP-pyrroloiminoquinone conjugates, blocking molecules more than 5 nm in diameter. The cAuNPs turned blue and became insoluble when subjected to the same process. Representative UV/Visible spectra for the highly colored, unbound 14-bromodiscorhabdin C (**5**), the sAuNPs alone and the pyrroloiminoquinone loaded sAuNP are shown in Figure 4. The spectra for the conjugate reveal a typical profile which is the sum of the individual absorption spectra of the discorhabdin and the sAuNPs, as expected. No real shifts in the absorption maxima for the AuNP or the pyrroloiminoquinone were observed, and the materials are thus expected to be intact.

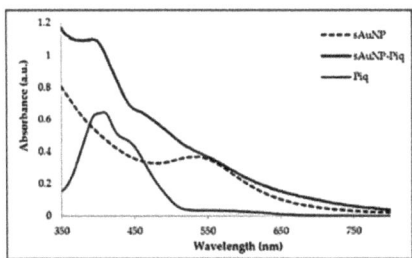

Figure 4. UV/Visible absorption spectra of the sAuNP, a representative pyrroloiminoquinone metabolite (BDC (**5**)), and their conjugate with the sAuNP in water.

3.6. Preparation of the Doxorubicin (3) Loaded sAuNPs

Loading of doxorubicin onto the sAuNPs required a few modifications, as adding doxorubicin as an aqueous solution (~2 mM) resulted in precipitation due to the acidic environment of the aqueous extract (sAuNPs have pH of ~4). A cellulose-based dialysis sack (MWCO: 12 kDa) was used to separate the unloaded or free compound from the sAuNP conjugates. A total of 20 mg of aqueous extract was used to produce the sAuNPs to enable comparison with a previously reported fucoidan-capped doxorubicin loaded AuNP [27], where the authors reported a loading as high as 90%. The UV/Visible absorption and ^{1}H NMR spectra obtained for the sAuNPs-doxorubicin conjugate is shown in Figures 5 and 6, respectively, together with the individual components. No obvious changes were observed for the conjugate upon comparison with the individual constituents.

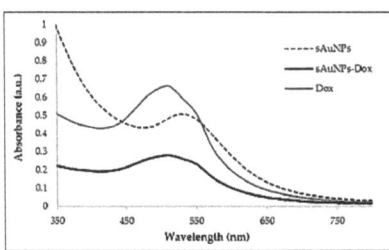

Figure 5. UV/Visible absorption spectra of doxorubicin alone, the sAuNPs and the sAuNP-doxorubicin (sAuNPs-Dox) conjugate in water.

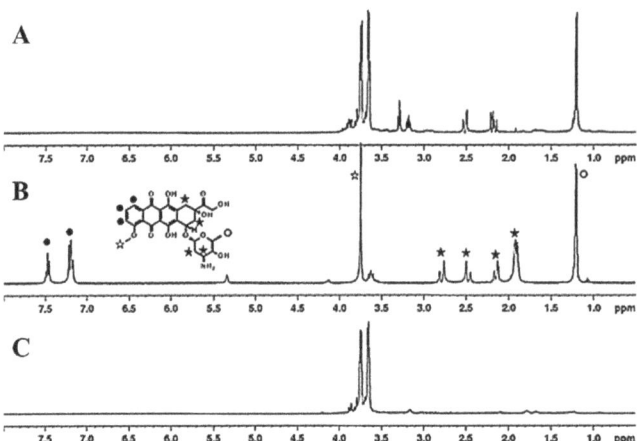

Figure 6. ^{1}H NMR spectra (400 MHz, D$_2$O) of the sAuNP-doxorubicin conjugate (**A**), doxorubicin (**B**) and the sAuNPs alone (**C**) at 298 K using water suppression.

The ^{1}H NMR spectra obtained for doxorubicin, the sAuNPs and the conjugate are shown in Figure 6. The NPs alone (Figure 6C) show signals between δ_H 3.5 and 4 which are characteristic of polysaccharides, as expected.

Figure 6B shows the spectrum obtained for doxorubicin alone, with aromatic signals present between δ_H 7 and 7.5, an oxy-methyl moiety at 3.8, the diastereotopic proton signals of the two aliphatic rings between δ_H 1.8 and 2.8, and a methyl signal at δ_H 1.28. Interestingly, while the signals pertaining to the methyl at δ_H 1.28 and some of the diastereotopic protons between δ_H 1.8 and 2.8 remain upon conjugation, albeit with some significant changes in chemical shifts for the latter, the aromatic signals are noticeably absent. The shifts (Figure 6) in the methylene signals suggest a meaningful change in the ^{1}H chemical environment for the aliphatic rings, perhaps through hydrogen bonding and various intermolecular forces, which is expected due to the polysaccharide capping on the AuNP surface. The absence of the aromatic protons suggests some binding to the gold nanoparticle surface through the aromatic rings, perhaps through cation -π stacking [32]. Curry et al. provided substantial practical and theoretical evidence, where the main absorption process for doxorubicin loading on citrate capped AuNPs were driven firstly through hydrophobic forces towards the AuNPs, followed by cation-π interaction primarily through coordination of the doxorubicin anthracene moiety to the cations on the AuNP (through diffusion-limited kinetics) [32]. The authors found that doxorubicin displaced the citrate capping agent easily, however, in this work, the ^{1}H NMR signals due to fucoidan are still observed. This would thus confirm our observations for the absence of the doxorubicin aromatic proton signals. Curry et al. found that the contribution of the amine functional group to be negligible; this was confirmed through high resolution XPS measurements and fluorescence-based measurements [32]. The authors were also able to show that there was no evidence to support electrostatic interactions or salt-bridging between the positively charged amine moiety on doxorubicin, and the negatively charged citrate capping as the main mechanisms of doxorubicin loading. ^{1}H NMR of the pyrroloiminoquinone-loaded-sAuNP conjugates were not obtained due to the paucity of material available.

3.7. Doxorubicin Release/Desorption

The release or desorption of doxorubicin from the doxorubicin loaded sAuNPs was also studied. Dialysis prevented the AuNPs from traversing the cellulose-based dialysis sack (MWCO: 12 kDa) but allowed the free movement of the unbound drug (i.e., the released doxorubicin). Periodic sampling up until 72 h was carried out and followed by UV/Visible absorption at 290 nm, which allowed the determination of the doxorubicin payload release as a function of time. The amount of doxorubicin released from sAuNPs-Dox at 20 °C and pH 7.4 after 72 h was determined to be 17%, as summarized in Table 2. This was comparable to the amount of doxorubicin released by fucoidan-capped AuNPs-doxorubicin (Dox-Fuc-AuNPs) reported by Manivasagan et al. [26].

Table 2. Comparison of the doxorubicin-loaded sAuNPs (sAuNPs-Dox) vs. the fucoidan-capped AuNPs (Dox-Fuc-AuNPs) [26].

Parameter	sAuNPs-Dox *	Dox-Fuc-AuNPs [26]
NP reaction temperature	20 °C	80 °C
NP pH	4.1	Not reported
Payload concentration	1 mM	0.1 mM
NP conjugation method	Adjusted with NaOH [1]	Direct
Release of payload	17% at pH 7.4 (after 72 h)	10% at pH 7.4 (after 72 h)

[1] to facilitate the solubility of doxorubicin; * this work.

3.8. sAuNP Cellular Uptake Analysis Using ICP-OES

It was not clear whether the AuNP-pyrroloiminoquinone conjugates (with its drug payload) could penetrate the MCF-7 cell membrane to exert its activity. ICP-OES was therefore chosen to determine the amount of gold present inside the MCF-7 cells. Following incubation and cell digestion of the nanoconjugates, centrifugation at low speeds (allowing the cells to sediment into pellets, but not the AuNPs themselves) enabled a comparison of the total amount of gold dosed to the cells, against the amount of gold found inside the cells. This permitted the estimation of the percentage penetration of the alkaloid loaded sAuNPs able to enter the cancer cell. Only the Tsitsikammamine B (**7**) and doxorubicin (**3**) sAuNP nanoconjugates were evaluated, and the results are presented in Figure 7.

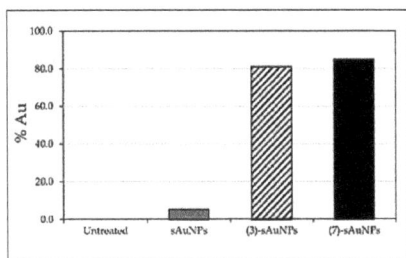

Figure 7. ICP-OES of cellular uptake of sAuNPs and its nanoconjugates by the MCF-7 cells.

The sAuNPs alone revealed poor uptake (5%) by the cancer cells (Figure 7), while the doxorubicin (**3**)- or pyrroloiminoquinone (**7**)-AuNP conjugates showed 80–85% penetration.

3.9. Biological Studies

Pyrroloiminoquinone metabolites have been previously shown to be highly cytotoxic, and were prospective candidates for development as anticancer drugs [21]. Discorhabdin A, the most cytotoxic pyrroloiminoquinone, revealed anti-tumor activity (IC$_{50}$) in the range of 77 nM against the Human Colon Tumor cell line (HCT-116), while the cytotoxic data (IC$_{50}$) obtained for compounds **5**, **6** and **7** were revealed to be 0.08, 1.4 and 2.4 µM, respectively. The role of the pyrroloiminoquinone structural motif plays in the high cytotoxicity was demonstrated by Antunes and co-workers (2004) though isolation and comparison of the relative activities of additional compounds such as Makaluvic acid A and Damirone B [22]. However, the cytotoxic activity of these compounds against cancer cell lines was also demonstrated in subsequent studies in normal cell lines, precluding their use in cancer treatment [22]. The possibility of selective targeting of cancer cells using pyrroloiminoquinone-loaded AuNPs as a carrier was therefore enticing. This could be achieved through selective uptake by cancer cells, or by selective delivery to tumor tissues via leaky vasculature. The antiproliferative activities of the pyrroloiminoquinone metabolites (5–7), doxorubicin (3) and their AuNPs conjugates (e.g., (5) NP) were assessed against the breast cancer (MCF-7) cancer cell line. In addition, the effect of any unbound drug was assessed by its removal through centrifugation. The pyrroloiminoquinone metabolites demonstrated dose dependent antiplroliferative activities in the range 1.5–50 µg/mL as shown in Figure 8 (Table 3, Figure S14).

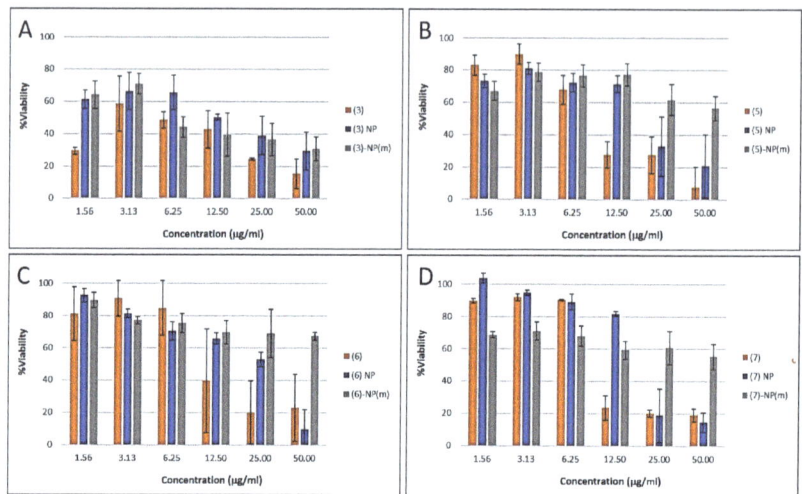

Figure 8. Cytotoxicity of doxorubicin (3), 14-bromo discorhabdin C (5) and the tsitsikammamines (6 and 7) alone, their as-synthesized sAuNP conjugates (NP) and the mixture (NP(m)) against the MCF-7 cancer cells. Dose: 1.5 µg/mL to 50 µg/mL. Graphs are labelled (**A**) doxorubicin (3) and conjugates, (**B**) 14-bromo discorhabdin C (5) and its conjugates, (**C**) Tsitsikammamine A (6) and its conjugates, (**D**) Tsitsikammamine B (7) and its conjugates.

Table 3. IC_{50} values of compounds 3, 5–7 and their sAuNP conjugates, as well as their mixtures (m) of bound and unbound AuNP conjugates, against MCF-7 cells.

Compound	IC_{50} (µg/mL)		
	Compound Alone	AuNP-1 [a]	AuNP-2 [b]
Doxorubicin (3)	17.9	15.9	>50
Tsitsikammamine A (6)	11.1	>50 [c]	>50 [c]
Tsitsikammamine B (7)	9.9	16.0	>50 [c]
14-bromodiscorhabdin C (5)	8.3	19.9	>50 [c]

[a] mixture of natural product and sAuNP. [b] mixture of natural product and sAuNP after removal of unbound natural products. [c] IC_{50} > maximum treatment dose used [50 µg/mL].

The "unloaded" sAuNPs showed very low cytotoxicity, however, the pure natural products displayed IC_{50} values between 8 and 18 µg/mL. In general, the combination of sAuNP and natural products showed either no change (sAuNP-doxorubicin, 3), half the activity of the natural product alone (tsitsikammamine B and 14-bromodiscorhabdin C) or significantly reduced activity (tsitsikammamine A) as shown in Table 3. Interestingly, removal of the unbound drug by centrifugation significantly reduced the biological activities of the sAuNP conjugates with IC_{50} values > 50 µg/mL. These results suggest that sAuNPs bind part of the natural product irreversibly (under the assay conditions) and prevent its interaction with its biological target. This is partly supported by the fact that removal of the 'unbound' portion significantly reduces the biological activity. The pH at the biological target may be an important factor ensuring the release of the natural product due to the ionic interaction between the negatively charged polysaccharide capping on the AuNP and the basic natural products. In a more acidic environment like cancerous cells, a greater payload release is expected, as shown previously by Wang et al. and Manivasagan et al. [7,26].

To determine whether the decrease in activity (Table 3) was due to poor cell penetration (as shown in Section 3.8) or poor release of the drug payload (Section 3.7), the ability of the sAuNPs to penetrate the cell membrane was carried out. The doxorubicin-bound sAuNP sample's reduced cytotoxic activity (in comparison to 3 alone, Figure 8) is thought to be due to the poor payload release, since the conjugate is taken up by the MCF-7 cell (Figure 7) and doxorubicin alone is highly cytotoxic (Table 3). This is further supported by the release studies carried out (Table 2) where only a 17% release was achieved at neutral pHs. Manivasagan et al. loaded doxorubicin onto similarly synthesized AuNPs using Fucoidan as the capping agent [26]. It was therefore suitable to compare the sAuNPs' performance in the doxorubicin release/desorption studies. At pH 7.4, only 17% was released from the sAuNPs compared to 10% doxorubicin from the fucoidan capped AuNPs after 72 h (Table 2). Thus, the finding was consistent, but there are still some advantages associated with a slow release of drugs in a neutral environment. At pH 4.5 however, Wang et al. and Manivasagan et al. reported a release of 97% [7,26]. Additional studies are required to investigate the effect of pH on drug release under the assay conditions. However, the uptake of the sAuNP also suggest that these conjugates could still find additional scope as multimodal agents, e.g., photothermal ablation agents in cancer therapy [26].

4. Conclusions

In summary, biocompatible, stable, and water soluble AuNPs (sAuNPs) were produced at room temperature and pressure conditions using an aqueous extract of the brown alga *S. incisifolium*. The aqueous extracts contained polysaccharides (Fucoidans) and polyphenols as the major and minor components, respectively, as revealed by NMR and assay data. The stability of the sAuNPs were found to be superior to that of the citrate capped AuNPs in the temperature, pH, ultracentrifugation, freeze-drying, NaCl and HSA concentration studies. The sAuNPs alone do not exert significant cytotoxic activity on their own (where ICP-OES showed that the AuNPs without a drug payload were not able to penetrate the MCF-7 cell membrane), and thus any bioactivity observed is due to the drug/metabolite payload. The known chemotherapeutic agent doxorubicin (3) was used as a control in the cytotoxicity studies, successfully loaded onto the sAuNP carrier and confirmed to adsorb onto the AuNP surface via the anthracene moiety, as described by Curry et al. (2015), using ^{1}H NMR [32]. Doxorubicin and the pyrroloiminoquinone metabolites were successfully adsorbed onto the sAuNPs and were shown to be able to penetrate the cell membrane using ICP-OES, thus they can exert their cytotoxic activity. The activity of the pyrroloiminoquinone metabolites on their own were slightly higher than that for their nanoconjugates, however, the activity was still regarded as significant with cell viabilities at ~60–70% at 12.5 µg/mL. ICP-OES analyses revealed that the decrease in activity may be due to poor payload release rather than MCF-7 cell membrane penetration. It was also clear based on the doxorubicin observations, that the pyrroloiminoquinones are released from the sAuNP carriers/platforms since their activities is still significant. Loading the alkaloids onto a AuNP platform offers additional, more selective treatment modalities in conjunction with the bioactivity such photothermal therapy or theranostics. This work thus shows that (a) the *S. incisifolium* aqueous extract capped AuNPs are stable under physiological conditions and are biocompatible, (b) that conjugation of the sAuNPs with compounds of interest was successful, (c) that the payload is released, and (d) that the NPs bound to the compounds of interest can traverse the MCF-7 cell membrane.

Supplementary Materials: The following supporting information can be downloaded at: https://www.mdpi.com/article/10.3390/ma16031319/s1, The following materials are available: Figure S1. ^{1}H NMR data (DMSO-d_6, 400 MHz) of Tsitsikammamine A (6, TA) obtained at 298 K. Figure S2. ^{13}C NMR data (DMSO-d_6, 100 MHz) of Tsitsikammamine A (6, TA) obtained at 298 K. Figure S3. ^{1}H NMR data (DMSO-d_6, 400 MHz) of Tsitsikammamine B (7, TB) obtained at 298 K. Figure S4. ^{13}C

NMR data (DMSO-d_6, 100 MHz) of Tsitsikammamine B (**7**, TB) obtained at 298 K. Figure S5. ^{1}H NMR data (DMSO-d_6, 400 MHz) of 14-bromodiscorhabdin C (**5**, BDC) obtained at 298 K. Figure S6. ^{13}C NMR data (DMSO-d_6, 100 MHz) of 14-bromodiscorhabdin C (**5**, BDC) obtained at 298 K. Figure S7. ^{1}H-^{13}C HSQC NMR data (400 MHz, D$_2$O) obtained from the *S. incisifolium* aqueous extract. Figure S8. 1H DOSY spectra (400 MHz) of the *S. incisifolium* aqueous extract in D$_2$O at 25 °C. The values obtained in log(m^2/s) for the polyphenol, polysaccharides and water signals are given on the spectrum. Figure S9. Typical UV/Visible spectra obtained for the AuNPs (A) synthesized using the *S. incisifolium* extract (1.0 mg and 20 mg of HAuCl$_4$ salt), with absorbance spectra collected every 5 min for 4.5 h in water. As the amount of extract increased (B), the AuNPs were produced more efficiently, as revealed by the SPR band at 524 nm. Figure S10. Energy Dispersive Spectroscopy (EDS) spectra for cAuNPs (left) and sAuNPs (right). Figure S11. UV/Visible absorption spectra of the sAuNPs in a variety of NaCl concentrations. Figure S12. HRTEM images obtained for the sAuNPs following incubation with HSA. Figure S13. Reconstituted AuNPs following ultrafiltration-centrifugation. The cAuNPs lost the characteristic pink/purple color, becoming blue on conjugation, centrifugation (through the PES membrane), and freeze-drying, while the sAuNPs retained their color and morphology (please see Figure 2G, main text). Figure S14. IC$_{50}$ value determinations for doxorubicin (**3**), 14-bromodiscorhabdin C (**5**), Tsitsikammamine A (**6**) and Tsitsikammamine B (**7**) together with the AuNP conjugates (right). Table S1: NMR data of **6** obtained in DMSO-d_6 at 400 MHz at 298 K in comparison to literature values. Table S2: NMR data of **7** obtained in DMSO-d_6 at 400 MHz at 298 K in comparison to literature values. Table S3: NMR spectroscopic of **5** obtained data in DMSO-d_6 at 298 K in comparison to literature values.

Author Contributions: Conceptualization: D.R.B. and E.M.A.; methodology: B.M., M.S.L. and N.R.S.S.; validation: D.R.B., E.M.A., N.R.S.S. and M.S.L.; formal analysis: D.R.B., E.M.A. and B.M.; investigation: B.M.; resources: D.R.B., T.S., J.J.B., M.M. and E.M.A.; data curation: E.M.A., M.S.L. and D.R.B.; writing—original draft preparation: E.M.A.; writing—review and editing: E.M.A. and D.R.B.; supervision: D.R.B. and E.M.A.; project administration: D.R.B. and E.M.A.; funding acquisition: E.M.A. and D.R.B.; specimen identification T.S. and J.J.B.; sampling: T.S. and J.J.B. All authors have read and agreed to the published version of the manuscript.

Funding: This research was funded by the National Research Foundation (NRF), grant numbers: 93474 (E.M.A.) and 93639 (D.R.B.). The NRF, grant number 93474 (E.M.A.) for financial support in the form of a bursary (B.M.), the University of the Western Cape.

Institutional Review Board Statement: Not applicable.

Data Availability Statement: Not applicable.

Acknowledgments: Use of the NMR and tissue culture facility facilities at the Department of Chemistry and DST/Mintek NIC (University of the Western Cape) is gratefully acknowledged.

Conflicts of Interest: The authors declare no conflict of interest.

References

1. Kim, K.-W.; Roh, J.K.; Wee, H.-J.; Kim, C. *Cancer Drug Discovery*; Springer: Berlin/Heidelberg, Germany, 2016; pp. 113–134.
2. Blagosklonny, M.V. Analysis of FDA approved anticancer drugs reveals the future of cancer therapy. *Cell Cycle* **2004**, *3*, 1035–1042. [CrossRef] [PubMed]
3. Sun, J.; Wei, Q.; Zhou, Y.; Wang, J.; Liu, Q.; Xu, H. A systematic analysis of FDA-approved anticancer drugs. *BMC Syst. Biol.* **2017**, *11*, 87. [CrossRef] [PubMed]
4. Paul, A.T.; Jindal, A. *Anticancer Plants: Clinical Trials and Nanotechnology*; Springer: Singapore, 2017; pp. 27–50.
5. Sztandera, K.; Gorzkiewicz, M.; Klajnert-Maculewicz, M. Gold nanoparticles in cancer treatment. *Mol. Pharm.* **2019**, *16*, 1–23. [CrossRef] [PubMed]
6. Thambiraj, S.; Vijayalakshmi, R.; Shankaran, D.R. An effective strategy for development of docetaxel encapsulated gold nanoformulations for treatment of prostate cancer. *Sci. Rep.* **2021**, *11*, 2808. [CrossRef]
7. Wang, F.; Wang, Y.C.; Dou, S.; Xiong, M.H.; Sun, T.M.; Wang, J. Doxorubicin-tethered responsive gold nanoparticles facilitate intracellular drug delivery for overcoming multidrug resistance in cancer cells. *ACS Nano* **2011**, *5*, 3679–3692. [CrossRef]
8. Hale, S.J.M.; Perrins, R.D.; Garcı, A.C.E.; Pace, A.; Peral, U.; Patel, K.R.; Robinson, A.; Williams, P.; Ding, Y.; Saito, G.; et al. DM1 Loaded Ultrasmall Gold Nanoparticles Display Significant Efficacy and Improved Tolerability in Murine Models of Hepatocellular Carcinoma. *Bioconjugate Chem.* **2019**, *30*, 703–713. [CrossRef]
9. Du, Y.; Xia, L.; Jo, A.; Davis, R.M.; Bissel, P.; Ehrich, M.; Kingston, D.G.I. Synthesis and Evaluation of Doxorubicin-Loaded Gold Nanoparticles for Tumor-Targeted Drug Delivery. *Bioconjugate Chem.* **2018**, *29*, 420–430. [CrossRef]

10. Javed, R.; Ghonaim, R.; Shathili, A.; Khalifa, S.A.; El-Seedi, H.R. Phytonanotechnology: A greener approach for biomedical applications. In *Biogenic Nanoparticles for Cancer Theranostics*; Elsevier: Amsterdam, The Netherlands, 2021; pp. 43–86.
11. Manivasagan, P.; Bharathiraja, S.; Bui, N.Q.; Jang, B.; Oh, Y.O.; Lim, I.G.; Oh, J. Doxorubicin-loaded fucoidan capped gold nanoparticles for drug delivery and photoacoustic imaging. International journal of biological macromolecules. *Int. J. Biol. Macromol.* **2016**, *91*, 578–588. [CrossRef]
12. Alle, M.; G, B.R.; Kim, T.H.; Park, S.H.; Lee, S.H.; Kim, J.C. Doxorubicin-carboxymethyl xanthan gum capped gold nanoparticles: Microwave synthesis, characterization, and anti-cancer activity. *Carbohydr. Polym.* **2020**, *229*, 115511. [CrossRef]
13. Chen, X.; Han, W.; Zhao, X.; Tang, W.; Wang, F. Epirubicin-loaded marine carrageenan oligosaccharide capped gold nanoparticle system for pH-triggered anticancer drug release. *Sci. Rep.* **2019**, *9*, 6754. [CrossRef]
14. Borker, S.; Pokharkar, V. Engineering of pectin-capped gold nanoparticles for delivery of doxorubicin to hepatocarcinoma cells: An insight into mechanism of cellular uptake. *Artif. Cells Nanomed. Biotechnol.* **2018**, *46*, 826–835. [CrossRef]
15. Ragubeer, N.; Beukes, D.; Limson, J.L. Critical assessment of voltammetry for rapid screening of antioxidants in marine algae. *J. Food Chem.* **2010**, *121*, 227–232. [CrossRef]
16. Mmola, M.; Roes-Hill, M.L.; Durrell, K.; Bolton, J.J.; Sibuyi, N.; Meyer, M.R.; Beukes, D.R.; Antunes, E. Enhanced antimicrobial and anticancer activity of silver and gold nanoparticles synthesised using *Sargassum incisifolium* aqueous extracts. *Molecules* **2016**, *21*, 1633. [CrossRef]
17. Seaweeds of the South African Coast. Available online: http://southafrseaweeds.uct.ac.za/descriptions/brown/sargassum_incisifolium.php (accessed on 26 December 2022).
18. Afolayan, A.F.; Bolton, J.J.; Lategan, C.A.; Smith, P.J.; Beukes, D.R. Fucoxanthin, tetraprenylated toluquinone and toluhydroquinone metabolites from Sargassum heterophyllum inhibit the in vitro growth of the malaria parasite Plasmodium falciparum. *Z. Nat. C* **2008**, *63*, 848–852. [CrossRef]
19. Hsu, H.Y.; Hwang, P.A. Clinical applications of fucoidan in translational medicine for adjuvant cancer therapy. *Clin. Transl. Med.* **2019**, *8*, 15. [CrossRef]
20. Zhang, R.; Zhang, X.; Tang, Y.; Mao, J. Composition, isolation, purification and biological activities of Sargassum fusiforme polysaccharides: A review. *J. Carbohydr. Polym.* **2020**, *228*, 115381. [CrossRef]
21. Usoltseva, R.V.; Anastyuk, S.D.; Shevchenko, N.M.; Surits, V.V.; Silchenko, A.S.; Isakov, V.V.; Zvyagintseva, T.N.; Thinh, P.D.; Ermakova, S.P. Polysaccharides from brown algae Sargassum duplicatum: The structure and anticancer activity in vitro. *Carbohydr. Polym.* **2017**, *175*, 547–556. [CrossRef]
22. Antunes, E.M.; Beukes, D.R.; Kelly, M.; Samaai, T.; Barrows, L.R.; Marshall, K.M.; Sincich, C.; Davies-Coleman, M.T. Cytotoxic pyrroloiminoquinones from four new species of South African latrunculid sponges. *J. Nat. Prod.* **2004**, *67*, 1268–1276. [CrossRef]
23. Hooper, G.; Davies-Coleman, M.; Kelly-Borges, M.; Coetzee, P. New alkaloids from a south african Latrunculid sponge. *Tetrahedron Lett.* **1996**, *37*, 7135–7138. [CrossRef]
24. Tobwala, S.; Fan, W.; Hines, C.J.; Folk, W.R.; Ercal, N. Antioxidative potential of Sutherlandia frutescens and its protective effects against oxidative stress in various cell cultures. *BMS Complement. Altern. Med.* **2014**, *14*, 271. [CrossRef]
25. Turkevich, J.; Stevenson, P.; Hillier, J. A study of the nucleation and growth processes in the synthesis of colloidal gold. *Discuss. Faraday Soc.* **1951**, *11*, 55–75. [CrossRef]
26. Manivasagan, P.; Bharathiraja, S.; Moorthy, S.M.; Oh, Y.O.; Song, K.; Seo, H.; Oh, J. Anti-EGFR antibody conjugation of fucoidan-coated gold nanorods as novel photothermal ablation agents for cancer therapy. *ACS Appl. Mater. Interfaces* **2017**, *9*, 14633–14646. [CrossRef] [PubMed]
27. Bhatterjee, S.J. DLS and zeta potential–what they are and what they are not? *J. Control. Release* **2016**, *235*, 337–351. [CrossRef] [PubMed]
28. Shukla, R.; Bansal, V.; Chaudhary, M.; Basu, A.; Bhonde, R.R.; Sastry, M. Biocompatibility of gold nanoparticles and their endocytotic fate inside the cellular compartment: A microscopic overview. *Langmuir* **2005**, *21*, 10644–10654. [CrossRef]
29. Bohnert, T.; Gan, L.J. Plasma protein binding: From discovery to development. *J. Pharm. Sci.* **2013**, *102*, 2953–2994. [CrossRef]
30. Paciotti, G.; Zhao, J.; Cao, S.; Brodie, P.J.; Tamarkin, L.; Huhta, M.; Myer, L.D.; Friedman, J.; Kingston, D.G. Synthesis and evaluation of paclitaxel-loaded gold nanoparticles for tumor-targeted drug delivery. *Bioconjugate Chem.* **2016**, *27*, 2646–2657. [CrossRef]
31. Polizzi, M.A.; Stasko, N.A.; Schoenfisch, M.H. Water-soluble nitric oxide-releasing gold nanoparticles. *Langmuir* **2007**, *23*, 4938–4943. [CrossRef]
32. Curry, D.; Cameron, A.; MacDonald, B.; Nganou, C.; Scheller, H.; Marsh, J.; Beale, S.; Lu, M.; Shan, Z.; Kaliaperumal, R.; et al. Adsorption of doxorubicin on citrate-capped gold nanoparticles: Insights into engineering potent chemotherapeutic delivery systems. *Nanoscale* **2015**, *7*, 19611–19616. [CrossRef]

Disclaimer/Publisher's Note: The statements, opinions and data contained in all publications are solely those of the individual author(s) and contributor(s) and not of MDPI and/or the editor(s). MDPI and/or the editor(s) disclaim responsibility for any injury to people or property resulting from any ideas, methods, instructions or products referred to in the content.

Article

Synthesis and Characterization of Citric Acid-Modified Iron Oxide Nanoparticles Prepared with Electrohydraulic Discharge Treatment

Vladimer Mikelashvili [1,*], Shalva Kekutia [1], Jano Markhulia [1], Liana Saneblidze [1], Nino Maisuradze [1], Manfred Kriechbaum [2] and László Almásy [3,*]

[1] Nanocomposites Laboratory, Vladimer Chavchanidze Institute of Cybernetics of the Georgian Technical University, Z. Anjafaridze Str. 5, 0186 Tbilisi, Georgia; shalva.kekutia@gmail.com (Sh.K.); j.markhulia@gtu.ge (J.M.); liana.saneblidze@gtu.ge (L.S.); nino.maisuradze@ens.tsu.edu.ge (N.M.)

[2] Institute of Inorganic Chemistry, Graz University of Technology, Stremayrgasse 9/5, A-8010 Graz, Austria; manfred.kriechbaum@tugraz.at

[3] Research Institute for Energy Security and Environmental Safety, Centre for Energy Research, Konkoly-Thege Miklós Str. 29-33, 1121 Budapest, Hungary

* Correspondence: vmikelashvili@gtu.ge (V.M.); almasy.laszlo@energia.mta.hu (L.A.)

Citation: Mikelashvili, V.; Kekutia, S.; Markhulia, J.; Saneblidze, L.; Maisuradze, N.; Kriechbaum, M.; Almásy, L. Synthesis and Characterization of Citric Acid-Modified Iron Oxide Nanoparticles Prepared with Electrohydraulic Discharge Treatment. *Materials* 2023, *16*, 746. https://doi.org/10.3390/ma16020746

Academic Editors: Alexandru Mihai Grumezescu and Paul Cătălin Balaure

Received: 6 December 2022
Revised: 28 December 2022
Accepted: 7 January 2023
Published: 12 January 2023

Copyright: © 2023 by the authors. Licensee MDPI, Basel, Switzerland. This article is an open access article distributed under the terms and conditions of the Creative Commons Attribution (CC BY) license (https:// creativecommons.org/licenses/by/ 4.0/).

Abstract: Chemical co-precipitation from ferrous and ferric salts at a 1:1.9 stoichiometric ratio in NH_4OH base with ultrasonication (sonolysis) in a low vacuum environment has been used for obtaining colloidal suspensions of Fe_3O_4 nanoparticles coated with citric acid. Before coating, the nanoparticles were processed by electrohydraulic discharges with a high discharge current (several tens of amperes) in a water medium using a pulsed direct current. Magnetite nanoparticles were obtained with an average crystallite diameter D = 25–28 nm as obtained by XRD and particle sizes of 25 nm as measured by small-angle X-ray scattering. Magnetometry showed that all samples were superparamagnetic. The saturation magnetization for the citric acid covered samples after electrohydraulic processing showed higher value (58 emu/g) than for the directly coated samples (50 emu/g). Ultraviolet-visible spectroscopy and Fourier transform infrared spectroscopy showed the presence and binding of citric acid to the magnetite surface by chemisorption of carboxylate ions. Hydrodynamic sizes obtained from DLS and zeta potentials were 93 and 115 nm, −26 and −32 mV for the citric acid covered nanoparticles and 226 nm and 21 mV for the bare nanoparticles, respectively. The hydraulic discharge treatment resulted in a higher citric acid coverage and better particle dispersion. The developed method can be used in nanoparticle synthesis for biomedical applications.

Keywords: IONPs; SPIONs; biocompatible nanoparticles; citric acid capped iron oxide; SAXS

1. Introduction

Nanomaterials in general, and nanoparticles (NPs) in particular are of great interest to researchers in various fields of science and technology. Because of their small size, high carrier capacity, high stability and the feasibility of incorporating them into both hydrophilic and hydrophobic substances, they are capable of different applications.

Among the various inorganic nanoparticles, iron oxide nanoparticles (IONPs) show unique physicochemical properties such as low Curie transition temperature and superparamagnetic nature; they are biocompatible, present low toxicity and show an antimicrobial activity [1–5]. Below a critical diameter, superparamagnetic iron oxide nanoparticles (SPIONs) possess a large constant magnetic moment which can be crucial for biomedical applications [6–8]. Their small size, which distinguishes them from bulk materials, yields a large surface-to-volume ratio; thus, NPs possess high surface energy which is beneficial for the functionalization of these nanoparticles with bioactive molecules as carriers of drugs, targeting molecules, and in general the creating the possibility of drug concentration by an external magnetic field

Nanosystems consisting of superparamagnetic magnetite (Fe_3O_4) NPs modified and stabilized with different kind of surfactant molecules and suspended in aqueous medium represent a versatile platform for both in vitro and in vivo applications such as drug delivery systems [9–11], hyperthermia [12–14], magnetic resonance imaging (MRI) [2,15,16] and magnetic labels for biosensing [17,18]. Besides biomedical applications, iron-based magnetic nanoparticles (MNPs) can be used in the fields of data storage [19], catalysis [20,21] and environmental remediation [22–26], and as plant protective agents [27,28] or plant growth stimulators [29,30]. Concerning their magnetic properties, iron oxide nanoparticles, in contrast to nanoparticles composed of transition metals such as Co, Ni and Mn, are favoured in biomedical applications because of the high toxicity of the latter compounds [31].

The usage of SPIONs has a great potential in the biomedical field. Their nano-scaled size and biocompatibility make MNPs compatible with cells. They mobilize within the blood stream, and due to their high magnetic moment, in the presence of an external magnetic field they can be targeted to pathologic tissues [32]. Successful applications of SPIONs rely on colloidal stabilization in an aqueous medium and precise control of shape, size, and size distribution that determine the physical and chemical properties of the nanocomposite. In particular, surface modifications are required to avoid agglomeration, as bare nanoparticles have insufficient long-term stability. They require hydrophilic and biocompatible surface coverage before they may be used in medical applications [33–35]. Alternative methods for synthesis and stabilization involve more elaborate procedures, such as a novel approach using microfluidics [36–38].

Surface coverage with citric acid (CA, $C_6H_8O_7$) provides a thermodynamically stable colloidal solution [39]. Citric acid is a widely used organic coating material in the manufacture of nanoparticles since it can change surfaces' charge and hydrophobicity, leaving additional carboxyl groups on the nanoparticle surface. CA shows bactericidal and bacteriostatic effects, and solutions containing citric acid are also used as sterilizing agents [40] and plant growth stimulators [41]. In addition, the biocompatibility and low toxicity of nanosystems consisting of water dispersed SPIONs (Fe_3O_4) modified with citric acid and serving as linker agents with anticancer drugs make them very interesting for biomedical applications, especially as drug delivery systems and MRI agents in modern healthcare [42].

A number of studies have been carried out on citric acid, as a widely accepted stabilizer material in water-based ferrofluids [39,43–46]. Adding CA during the chemical co-precipitation of ferrous salts allows one to control the size of the primary NPs and simultaneously prevent their aggregation [47]. Thus, through adding aqueous CA solution at different stages of synthesis, the core sizes of CA-capped IONPs could be adjusted in the range from 6 nm to 13 nm [48].

MNP synthesis and CA-functionalization in aqueous CA solution using co-participation at lower temperatures and a shortened time compared with conventional methods has been reported [49]. In a two-step process for synthesis, the addition of citric acid at decreasing coating temperatures resulted in increased hydrodynamic sizes of final product which also affect superparamagnetic feature of obtained material [44]. CA is believed to adsorb on the surface of nanoparticles in the monolayer, binding by coordinating ≡FeOH sites via one or two carboxylate groups depending on the steric necessity and surface curvature of NPs [50,51].

The application of electrohydraulic discharges (EHD) in water and organic liquids has been studied for many years. Electrical transmission processes initiate different types of chemical reactions and effects on the physical processes (e.g., degassing, decomposition, homogenization, cavitation, shock waves and ultraviolet/visible electromagnetic radiation), and promote a variety of chemical reactions [52]. A pulsed discharge can be advantageous as a pre-treatment stage before capping the CA, since the shock-wave effect on the suspension destroys the large agglomerates without significant destruction of the initial nanostructures. The efficiency of agglomerate dispersion by electrohydraulic discharge is higher than that of dispersion employing ultrasonication [33,53].

The aim of the present work is to explore the influence of high-voltage pulsed discharge (HVPD) on the stabilization quality of citric acid-coated SPIONs. To date, very few studies report investigations on the usage of electrohydraulic processing during the synthesis of magnetic nanofluids. In this work, we applied HVPD processing in the intermediate phase, before coating the iron oxide nanoparticle with the stabilizing CA molecule and revealing the effect of this treatment on the quality and properties of the resulting magnetic fluid.

2. Materials and Methods

2.1. Materials and Characterization Techniques

The chemicals used for the synthesis of magnetite nanoparticles were of analytical grade without further purification. Ferric chloride hexahydrate ($FeCl_3 \cdot 6H_2O$) ($\geq$98%), ferrous sulfate heptahydrate ($FeSO_4 \cdot 7H_2O$), ammonium hydroxide solution (NH_4OH, 25% of NH_3 basis) and citric acid monohydrate ($HOC(COOH)(CH_2COOH)_2 \cdot H_2O$)) $\geq$99.0% were purchased from Sigma-Aldrich Co. LLC (Darmstadt, Germany).

X-ray powder diffraction (XRD) analysis was performed using a DRON 3M X-ray diffractometer, operating with Cu Kα radiation (λ = 0.1541 nm) filtered by a nickel foil (voltage 40 kV, current 20 mA, and scanning speed 2$°$/min).

Magnetic measurements were performed on a vibrating sample magnetometer (VSM) (7300 Series VSM System, Lake Shore Cryotronics, Inc., Westerville, OH, USA) at room temperature under an applied field up to 1.5 Tesla.

Fourier transform infrared spectroscopy (FTIR) was performed on the Agilent Cary 630 FTIR spectrometer with 320-Cary FTIR Diamond ATR (spectral range: 6300–350 cm^{-1}).

UV-Vis spectroscopy was performed using an AvaSpec-HS2048XL instrument with AvaLight-DHc light source allowing measurements in the 200–1160 nm spectral range.

The hydrodynamic size distribution profile and zeta potential (ζ) of the aqueous suspensions were measured using Anton Paar Litesizer™ 500 equipped with a 658 nm laser, in backscattering geometry, thermostated at 25 °C, with a scattering angle of 173° and adjusted voltage 200 V.

Small-angle X-ray scattering (SAXS) was performed on a SAXSpoint 2.0 instrument (Anton Paar GmbH, Graz, Austria) equipped with a MicroSource Primux 100 copper X-ray generator (λ = 0.154 nm) and an Eiger R 1M position sensitive detector. Liquid samples were injected into quartz capillary and measured at 25 °C.

2.2. Synthesis Methods

2.2.1. Synthesis of Bare Iron Oxide Nanoparticles

Bare (uncovered) iron oxide (Fe_3O_4) nanoparticles were prepared by sonochemical co-precipitation with ultrasound processing using an iron salt ratio Fe^{3+}/Fe^{2+} of 1.9. First, $FeCl_3 \cdot 6H_2O$ (9 g) + 333 mL distilled water (DW) (0.1 M solution) was prepared in the jacketed reactor with mechanical stirring (temperature 45 °C, mixing duration 20 min, vacuum environment), and $FeSO_4 \cdot 7H_2O$ (4.87 g) +175 mL DW (0.1 M solution) in the jacketed ultrasonic reactor (temperature 45 °C, duration 20 min, ultrasonication 30% of 900 W homogenizer). After separate dissolution, the iron salt solutions were collected in an ultrasonic reactor and treated by ultrasonication and vacuum degassing for an additional 15 min. During this time the temperature was raised up to 55 °C, and previously prepared 19 mL NH_4OH (25%) + 35 mL DW (4 M solution) was added dropwise over the course of 16 min by a peristaltic pump in the middle area of the reactor. After the completion of the supply of NH_4OH solution, the sonication continued for additional 120 min without temperature control. The obtained black precipitate was cooled down to room temperature under ultrasonication.

In order to remove residues of the chemical synthesis and reduce the pH (initial pH 10 after synthesis) to the physiological value (pH 7.3), the particles were washed several times with an abundant quantity of DW with magnetic separation using a permanent magnet. After final washing, the vessel was filled up to 500 mL of DW and ultrasonicated with 30% of a 900 W power homogenizer for 30 min. A resulted suspension consisting of bare/uncovered MNPs was labelled bare-SPIONs. A total of 100 mL of this suspension was prepared, with a calculated maximum possible mass of the solid phase of 0.77 g (concentration 0.77 weight/volume percent).

2.2.2. Electrohydraulic (EHD) Processing and Modification with Citric Acid

As in our previous study with folic acid conjugated IONPs, HVPD (electrohydraulic discharge) was applied before capping with CA in an aqueous medium as a surface activation and homogenization technique [33]. To modify the surface of MNPs with a carboxyl group, an aqueous solution of CA was added directly using a peristaltic pump, at atmospheric pressure and room temperature. Briefly, the 100 mL suspension was ultrasonicated with 30% of a 900 W power homogenizer over the course of 10 min, and previously prepared 0.19 g of CA (about 25% of magnetite) + 10 mL DW was dropwise added to the MNPs suspension over 10 min under ultrasonication (14% of 900 W power), followed by an additional 10 min of ultrasonication.

The resulting sample's (SPIONs-CA) pH was regulated using NH_4OH aqueous solution until the pH reached 6 (initial pH 4.3), and the solution was stored for one night. On the following day, the SPIONs-CA solution was washed by DW to remove the excess of CA by decantation on permanent magnet, before being ultrasonicated again for 5 min.

The SPIONs-EHD-CA sample was prepared in a similar way, with the difference that electrohydraulic treatment was applied on the bare-SPIONs suspension by pulsed arc discharges before capping the particles with CA (Figure 1).

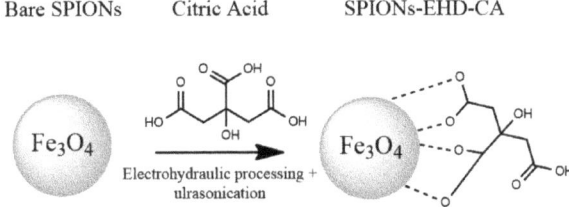

Figure 1. Scheme for preparation of magnetic nanoparticles by electrohydraulic discharges followed by modification with CA.

The electrohydraulic processing was performed in the high current mode, as described in a previous work [33]. This mode allows low voltage and high current discharges inside the closed 300 mL volume reactor in a low vacuum environment (1 kPa). The distance between the electrode rods was d = 0.7 mm, the discharge peak current I_{max} = 30 A, voltage V = 1.2 kV, discharge frequency f = 2 Hz and the maximum impulse duration t_{max} = 20 ms. The experimental setup is described in [33].

3. Results and Discussion

3.1. X-ray Diffraction (XRD)

From the diffraction data of obtained samples, the diffraction peaks at 2θ values were assigned to the crystal planes (220), (311), (400), (422), (511) and (440), respectively (Figure 2).

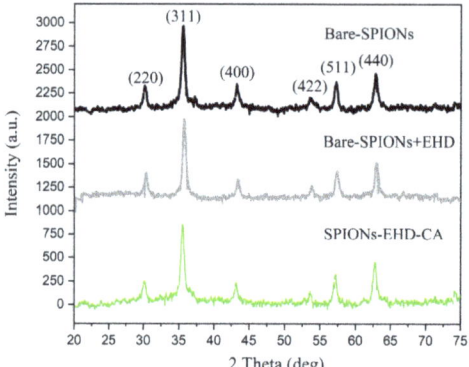

Figure 2. XRD patterns of synthesized bare-SPIONs, electrohydraulically processed bare-SPIONs and the CA-modified sample after EHD processing (SPIONs-EHD-CA). The data are shifted vertically for better visibility.

All peaks match well with characteristic peaks of magnetite (Fe_3O_4) (JCPDS file no. 19-0629). There was no sign of any phase transition between electrohydraulically processed samples and the bare-SPIONs. Additionally, no crystalline impurity phases were observed. The average crystallite diameter (D = 28 ± 2 nm) was calculated using the Scherrer equation from the FWHM (full width at half maximum) of the (311) peak at $2\theta = 35.86°$. The average value of the lattice parameter was found to be a = 0.837 ± 0.001 nm.

3.2. Vibrating Sample Magnetometry (VSM)

The magnetic properties of the obtained nanoparticles were investigated by VSM at room temperature (Figure 3). The hysteresis loops show the superparamagnetic behaviour of all samples.

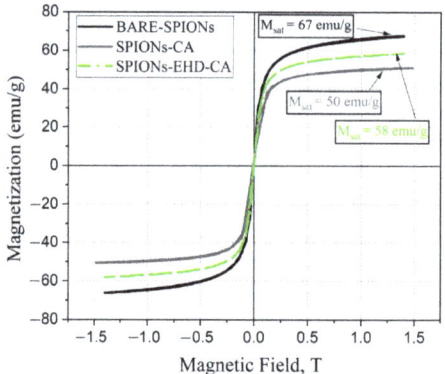

Figure 3. VSM results of synthesized IONPs with no coating (BARE-SPIONs), CA coated after electrohydraulic processing and directly coated with CA.

The samples show no magnetic hysteresis; the magnetization and demagnetization on the curves pass through the origin, implying their superparamagnetic nature, which is an essential property of such nanoparticles in many applications. The bare-SPIONs exhibit higher magnetic saturation (M_{sat} = 67 emu/g) while the nonmagnetic CA reduces M_{sat} to 50 and 58 emu/g for the coated samples. This is the signature of the core–shell structure of the coated magnetite nanoparticles SPIONs-CA and SPIONs-EHD-CA, in which the weight percentage of Fe_3O_4 is lower.

The destruction of the aggregated bare nanoparticles formed after synthesis by the additional electrohydraulic discharge treatment resulted in a more efficient modification (shown by FTIR data next) and higher magnetization (58 emu/g).

3.3. Fourier-Transform Infrared Spectroscopy (FTIR)

The synthesized samples have been characterized by FTIR spectroscopy (Figure 4).

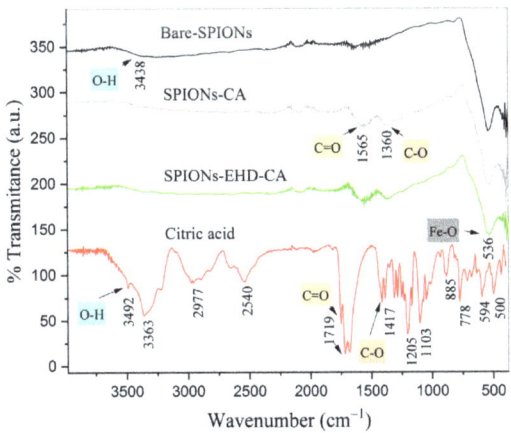

Figure 4. ATR-FTIR spectra of bare magnetite nanoparticles (Fe_3O_4), CA-capped SPIONs and pure citric acid. The data are shifted vertically for better visibility.

The characteristic absorption band of bare magnetic nanoparticles located at 536 cm^{-1} is associated with the stretching vibration mode of Fe–O which is characteristic of iron oxide [54], while the absorption bands at 3438 cm^{-1} of the O–H stretching vibrations indicate the presence of OH groups in the MNPs' surface. For the CA spectrum, an intense band at 3492 cm^{-1} shows the presence of non-dissociated OH groups. The peak at 1719 cm^{-1} can be assigned to the stretching vibration of C=O group [55,56].

Comparing the spectra of bare and CA-capped samples, it can be seen that several peaks appeared in the 1565 and 1360 cm^{-1} for CA-capped samples (Figure 4) due to the binding of a citric acid radical to the magnetite surface by chemisorption of the carboxylate ions [56].

3.4. Ultraviolet-Visible (UV-Vis) Spectroscopy

UV-VIS spectroscopy was used to measure the extinction (scatter plus absorption) of light passing through a sample (Figure 5). Measurements were performed on liquid samples in a quartz cuvette over a spectral range of 200–1100 nm at room temperature.

The citric acid dissolved in distilled water has a strong absorbance band in the UV-region (range 200–260 nm). In this region, bare-SPIONs have also a main adsorption peak at 260 nm and an additional peak in the visible region near 380 nm; this is in agreement with previous reports [57,58]. The influence of CA on the spectra of CA-capped SPIONS is apparent in the steep slopes on both sides of the 240 nm peak, which is uncharacteristic for bare-SPIONs. An apparent shift in the low wavelength peak position of the CA-capped SPIONS can be noticed, which might be related to their less agglomerated morphology compared with that of the bare-SPIONs.

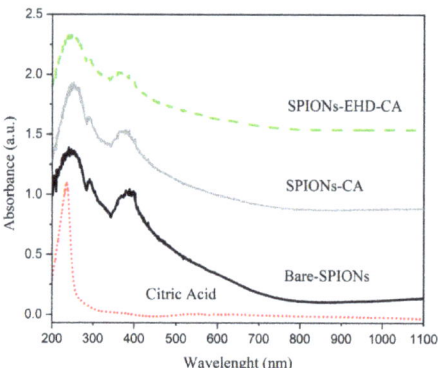

Figure 5. UV-VIS spectra of citric acid solution, bare-SPIONs and CA-capped SPION dispersions. Data are shifted vertically for better visibility.

3.5. Hydrodynamic Sizes and Zeta Potential Measurements

The characteristic hydrodynamic sizes of the particles and zeta potential (ζ) of the aqueous suspensions are shown in Figure 6.

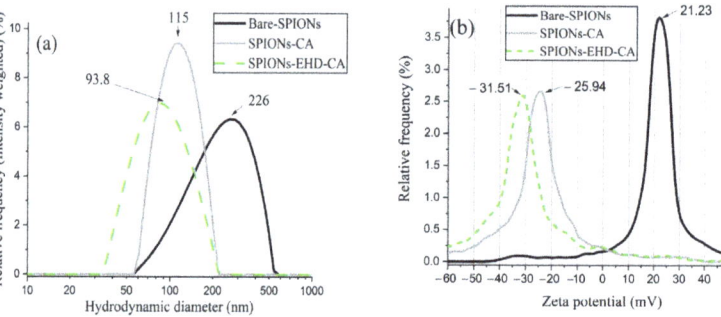

Figure 6. Hydrodynamic size obtained from DLS (**a**) and zeta potentials (**b**).

Dynamic light scattering (Figure 6a) indicates a broad size distribution for the bare-SPIONs with a mean diameter of 226 nm, as the uncoated magnetite nanoparticles tend to aggregate and form clusters with large hydrodynamic sizes. Before CA coating, all the samples were ultrasonicated, which resulted in a narrow size distribution (a mean diameter of 115 nm for sample CA-SPIONs) and smaller diameter. The sample prepared with electrohydraulic discharges displayed an even smaller average cluster size (93.8 nm).

For the assessment of the stability of the colloidal dispersions, and thus the strength of electrostatic repulsion between similarly charged particles, the zeta potential (ζ) of the aqueous suspensions was measured (Figure 6b). CA forms a negative charge around magnetite nanoparticle surfaces, while bare-SPIONs are positively charged. The larger negative value of zeta potential (-31.51 mV) of the electrohydraulically processed samples implies that these nanofluids are more stable than the directly CA-capped samples (-25.94 mV). This is also proven by visual observation of the samples after 7 months storage time (Figure 7). According to our previous studies, electrostatic stabilization employing CA coating provides stability for the magnetic fluids in the range of 8–12 month [34].

Figure 7. Visual observation of the synthesized samples without centrifugation. The picture is taken 7 months after the synthesis.

3.6. Small Angle X-ray Scattering (SAXS)

The angular distribution of X-rays scattered by the samples is displayed in Figure 8. in the form of scattering intensity in the function of the scattering vector magnitude q.

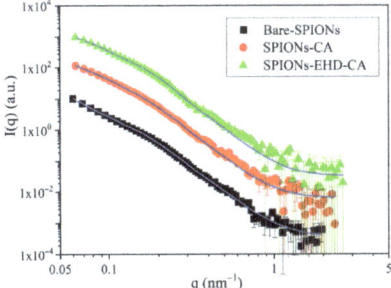

Figure 8. SAXS profiles and fitting by a model of polydisperse spheres using software McSAS. The solid lines are the model fits.

After data treatment that included subtraction of scattering of the carrier liquid by software ATSAS v3.2.0 [59], the scattering data were modelled as originating from a broad size distribution of spherical particles using a Monte Carlo fitting routine implemented in software McSAS v1.3.1 [60,61]. The resulting size distributions (Figure 9) include both single and agglomerated SPIONs. The first maximum of the distribution corresponds to the single nanoparticles, with a mean diameter of 25 nm which is in good agreement with the average crystallite size obtained by XRD. It can be seen that for bare-SPIONs (Figure 9a), the size distribution has a prominent second maximum around sizes 35–40 nm which corresponds to a population of large agglomerates. This feature is reduced in the sample with CA-capped particles (Figure 9b), and it is the weakest, as seen in Figure 9c, for the electrohydraulically processed CA-capped particle system.

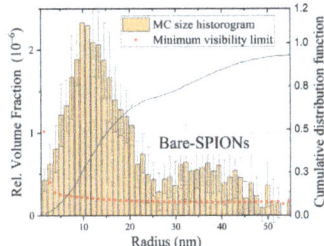 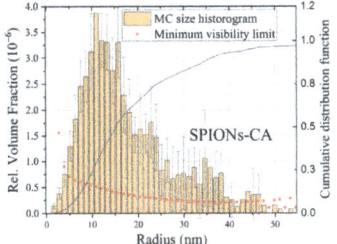 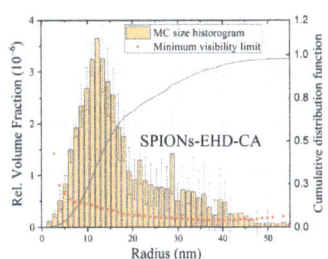

Figure 9. Particle size distributions obtained using McSAS.

4. Conclusions

Biocompatible IONPs were synthesized using a sonochemical co-precipitation method and then capped with citric acid to render particles with reactive carboxyl groups on their surface and to provide additional functionality for biomedical application. The state of dispersion and aggregation has been influenced by electrohydraulic pretreatment of the magnetite particles before the citric acid capping step. XRD, FTIR and zeta potential measurements revealed the magnetite (Fe_3O_4) phase for bare and citric acid functionalized SPIONs and the successful attachment of the functional groups on the particle surface. VSM measurements of the powders revealed higher magnetization of the electrohydraulically processed particles. The saturation magnetization of 61 emu/g and average crystallite and particle diameter ~25–28 nm, as obtained by XRD and SAXS, confers their superparamagnetic nature. The effect of electrohydraulic treatment could be seen as a ~10% decrease of the particle agglomeration and a substantial improvement of the long-term stability of the colloidal dispersion, lasting over 7 months for the presently prepared materials. These properties make the prepared SPIONs suitable candidates for biomedical applications such as cancer therapies, MRI contrast agents, drug transporters and immunotherapy and hyperthermia tools.

Author Contributions: Conceptualization, Sh.K., V.M. and J.M.; methodology, Sh.K., V.M., J.M., N.M. and L.S.; software, V.M.; validation, Sh.K., V.M. and J.M.; formal analysis, Sh.K., V.M., J.M., L.A. and L.S.; investigation, V.M., J.M. and M.K.; resources, V.M., J.M., N.M. and Sh.K.; data curation, V.M., J.M. and L.A.; writing—original draft preparation, V.M.; writing—review and editing, Sh.K., J.M., L.S, N.M. and L.A.; visualization, V.M.; supervision, Sh.K. All authors have read and agreed to the published version of the manuscript.

Funding: This work was supported by Shota Rustaveli National Science Foundation of Georgia (SRNSFG) [grant number: AR-19-1211, project title: "Synthesis of the multifunctional magnetic nanosystem with innovative technology for medical application"] and Central European Research Infrastructure Consortium (CERIC), grant number: 20192124.

Institutional Review Board Statement: Not applicable.

Informed Consent Statement: Not applicable.

Data Availability Statement: Not applicable.

Acknowledgments: The research was carried out in the Nanocomposites Laboratory, Vladimer Chavchanidze Institute of Cybernetics of the Georgian Technical University, Georgia. The authors would like to acknowledge use of the SOMAPP Lab (the laboratory where the SAXS and DLS measurements were carried out), a core facility supported by the Austrian Federal Ministry of Education, Science and Research, the Graz University of Technology, the University of Graz and the Anton Paar GmbH.

Conflicts of Interest: The authors declare no conflict of interest.

References

1. Farag, R.K.; Labena, A.; Fakhry, S.H.; Safwat, G.; Diab, A.; Atta, A.M. Antimicrobial activity of hybrids terpolymers based on magnetite hydrogel nanocomposites. *Materials* **2019**, *12*, 3604. [CrossRef] [PubMed]
2. Uwaya, G.E.; Fayemi, O.E.; Sherif, E.S.M.; Junaedi, H.; Ebenso, E.E. Synthesis, electrochemical studies, and antimicrobial properties of Fe_3O_4 nanoparticles from *Callistemon viminalis* plant extracts. *Materials* **2020**, *13*, 4894. [CrossRef] [PubMed]
3. Otari, S.V.; Kalia, V.C.; Bisht, A.; Kim, I.W.; Lee, J.K. Green synthesis of silver-decorated magnetic particles for efficient and reusable antimicrobial activity. *Materials* **2021**, *14*, 7893. [CrossRef]
4. Wojciechowska, A.; Markowska-Szczupak, A.; Lendzion-Bieluń, Z. TiO_2-modified magnetic nanoparticles (Fe_3O_4) with antibacterial properties. *Materials* **2022**, *15*, 1863. [CrossRef] [PubMed]
5. Khan, S.; Shah, Z.H.; Riaz, S.; Ahmad, N.; Islam, S.; Akram Raza, M.; Naseem, S. Antimicrobial activity of citric acid functionalized iron oxide nanoparticles—Superparamagnetic effect. *Ceram. Int.* **2020**, *46*, 10942–10951. [CrossRef]
6. Tóth, I.Y.; Illés, E.; Szekeres, M.; Zupkó, I.; Turcu, R.; Tombácz, E. Chondroitin-Sulfate-A-Coated Magnetite Nanoparticles: Synthesis, Characterization and Testing to Predict Their Colloidal Behavior in Biological Milieu. *Int. J. Mol. Sci.* **2019**, *20*, 4096. [CrossRef]

7. Lu, A.H.; Salabas, E.L.; Schüth, F. Magnetic nanoparticles: Synthesis, protection, functionalization, and application. *Angew. Chem. Int. Ed.* **2007**, *46*, 1222–1244. [CrossRef]
8. Socoliuc, V.; Peddis, D.; Petrenko, V.I.; Avdeev, M.V.; Susan-Resiga, D.; Szabó, T.; Turcu, R.; Tombácz, E.; Vékás, L. Magnetic nanoparticle systems for nanomedicine—A materials science perspective. *Magnetochemistry* **2020**, *6*, 2. [CrossRef]
9. Veiseh, O.; Gunn, J.W.; Zhang, M. Design and fabrication of magnetic nanoparticles for targeted drug delivery and imaging. *Adv. Drug Deliv. Rev.* **2010**, *62*, 284–304. [CrossRef]
10. Lachowicz, D.; Kaczyńska, A.; Wirecka, R.; Kmita, A.; Szczerba, W.; Bodzoń-Kulakowska, A.; Sikora, M.; Karewicz, A.; Zapotoczny, S. A hybrid system for magnetic hyperthermia and drug delivery: SPION functionalized by curcumin conjugate. *Materials* **2018**, *11*, 2388. [CrossRef]
11. Yang, W.J.; Lee, J.H.; Hong, S.C.; Lee, J.; Lee, J.; Han, D.W. Difference between toxicities of iron oxide magnetic nanoparticles with various surface-functional groups against human normal fibroblasts and fibrosarcoma cells. *Materials* **2013**, *6*, 4689–4706. [CrossRef] [PubMed]
12. Revia, R.A.; Zhang, M. Magnetite nanoparticles for cancer diagnosis, treatment, and treatment monitoring: Recent advances. *Mater. Today* **2016**, *19*, 157–168. [CrossRef] [PubMed]
13. Lemine, O.M.; Madkhali, N.; Alshammari, M.; Algessair, S.; Gismelseed, A.; El Mir, L.; Hjiri, M.; Yousif, A.A.; El-Boubbou, K. Maghemite (γ-Fe$_2$O$_3$) and γ-Fe$_2$O$_3$-TiO$_2$ nanoparticles for magnetic hyperthermia applications: Synthesis, characterization and heating efficiency. *Materials* **2021**, *14*, 5691. [CrossRef] [PubMed]
14. Chung, R.J.; Shih, H.T. Preparation of multifunctional Fe@Au core-shell nanoparticles with surface grafting as a potential treatment for magnetic hyperthermia. *Materials* **2014**, *7*, 653–661. [CrossRef]
15. Chertok, B.; Moffat, B.A.; David, A.E.; Yu, F.; Bergemann, C.; Ross, B.D.; Yang, V.C. Iron oxide nanoparticles as a drug delivery vehicle for MRI monitored magnetic targeting of brain tumors. *Biomaterials* **2008**, *29*, 487–496. [CrossRef]
16. Lee, Y.T.; Woo, K.; Choi, K.S. Preparation of water-dispersible and biocompatible iron oxide nanoparticles for MRI agent. *IEEE Trans. Nanotechnol.* **2008**, *7*, 111–117.
17. Wang, Z.; Cuschieri, A. Tumour cell labelling by magnetic nanoparticles with determination of intracellular iron content and spatial distribution of the intracellular iron. *Int. J. Mol. Sci.* **2013**, *14*, 9111–9125. [CrossRef] [PubMed]
18. Wang, Y.-X.J.; Quercy-Jouvet, T.; Wang, H.-H.; Li, A.-W.; Chak, C.-P.; Xuan, S.; Shi, L.; Wang, D.F.; Lee, S.F.; Leung, P.-C.; et al. Efficacy and durability in direct labeling of mesenchymal stem cells using ultrasmall superparamagnetic iron oxide nanoparticles with organosilica, dextran, and peg coatings. *Materials* **2011**, *4*, 703–715. [CrossRef]
19. Zhang, X.X.; Wen, G.H.; Huang, S.; Dai, L.; Gao, R.; Wang, Z.L. Magnetic properties of Fe nanoparticles trapped at the tips of the aligned carbon nanotubes. *J. Magn. Magn. Mater.* **2001**, *231*, 9–12. [CrossRef]
20. Vengsarkar, P.S.; Xu, R.; Roberts, C.B. Deposition of iron oxide nanoparticles onto an oxidic support using a novel gas-expanded liquid process to produce functional Fischer-Tropsch synthesis catalysts. *Ind. Eng. Chem. Res.* **2015**, *54*, 11814–11824. [CrossRef]
21. Rajabi, F.; Abdollahi, M.; Luque, R. Solvent-free esterification of carboxylic acids using supported iron oxide nanoparticles as an efficient and recoverable catalyst. *Materials* **2016**, *9*, 557. [CrossRef]
22. Zhang, W.X. Nanoscale iron particles for environmental remediation: An overview. *J. Nanoparticle Res.* **2003**, *5*, 323–332. [CrossRef]
23. Nicola, R.; Costişor, O.; Ciopec, M.; Negrea, A.; Lazău, R.; Ianăşi, C.; Piciorus, E.-M.; Len, A.; Almásy, L.; Szerb, E.I.; et al. Silica-coated magnetic nanocomposites for Pb^{2+} removal from aqueous solution. *Appl. Sci.* **2020**, *10*, 2726. [CrossRef]
24. Nicola, R.; Costişor, O.; Muntean, S.G.; Nistor, M.-A.; Putz, A.-M.; Ianăşi, C.; Lazău, R.; Almásy, L.; Săcărescu, L. Mesoporous magnetic nanocomposites: A promising adsorbent for the removal of dyes from aqueous solutions. *J. Porous Mater.* **2019**, *27*, 413–428. [CrossRef]
25. Zhu, S.; Leng, Y.; Yan, M.; Tuo, X.; Yang, J.; Almásy, L.; Tian, Q.; Sun, G.; Zou, L.; Li, Q.; et al. Bare and polymer coated iron oxide superparamagnetic nanoparticles for effective removal of U (VI) from acidic and neutral aqueous medium. *Appl. Surf. Sci.* **2018**, *447*, 381–387. [CrossRef]
26. Erdem, B.; İşcan, K.B. Multifunctional magnetic mesoporous nanocomposites towards multiple applications in dye and oil adsorption. *J. Sol-Gel Sci. Technol.* **2021**, *98*, 528–540. [CrossRef]
27. Saqib, S.; Zaman, W.; Ayaz, A.; Habib, S.; Bahadur, S.; Hussain, S.; Muhammad, S.; Ullah, F. Postharvest disease inhibition in fruit by synthesis and characterization of chitosan iron oxide nanoparticles. *Biocatal. Agric. Biotechnol.* **2020**, *28*, 101729. [CrossRef]
28. Saqib, S.; Zaman, W.; Ullah, F.; Majeed, I.; Ayaz, A.; Hussain Munis, M.F. Organometallic assembling of chitosan-Iron oxide nanoparticles with their antifungal evaluation against Rhizopus oryzae. *Appl. Organomet. Chem.* **2019**, *33*, e5190. [CrossRef]
29. Răcuciu, M.; Tecucianu, A.; Oancea, S. Impact of magnetite nanoparticles coated with aspartic acid on the growth, antioxidant enzymes activity and chlorophyll content of maize. *Antioxidants* **2022**, *11*, 1193. [CrossRef]
30. Răcuciu, M.; Creangă, D. Magnetite/Tartaric acid nanosystems for experimental study of bioeffects on Zea mays growth. *Rom. J. Phys.* **2017**, *62*, 804.
31. Mittal, A.; Roy, I.; Gandhi, S. Magnetic nanoparticles: An overview for biomedical applications. *Magnetochemistry* **2022**, *8*, 107. [CrossRef]
32. Ganapathe, L.S.; Mohamed, M.A.; Mohamad Yunus, R.; Berhanuddin, D.D. Magnetite (Fe$_3$O$_4$) Nanoparticles in Biomedical Application: From Synthesis to Surface Functionalisation. *Magnetochemistry* **2020**, *6*, 68. [CrossRef]
33. Mikelashvili, V.; Kekutia, Sh.; Markhulia, J.; Saneblidze, L.; Jabua, Z.; Almásy, L.; Kriechbaum, M. Folic acid conjugation of magnetite nanoparticles using pulsed electrohydraulic discharges. *J. Serb. Chem. Soc.* **2021**, *86*, 181–194. [CrossRef]

34. Markhulia, J.; Kekutia, Sh.; Mikelashvili, V.; Almásy, L.; Saneblidze, L.; Tsertsvadze, T.; Maisuradze, N.; Leladze, N.; Kriechbaum, M. Stable aqueous dispersions of bare and double layer functionalized superparamagnetic iron oxide nanoparticles for biomedical applications. *Mater. Sci. Pol.* **2021**, *39*, 331–345. [CrossRef]
35. Serantes, D.; Baldomir, D. Nanoparticle size threshold for magnetic agglomeration and associated hyperthermia performance. *Nanomaterials* **2021**, *11*, 2786. [CrossRef]
36. Niculescu, A.-G.; Chircov, C.; Grumezescu, A.M. Magnetite nanoparticles: Synthesis methods—A comparative review. *Methods* **2022**, *199*, 16–27. [CrossRef]
37. Chircov, C.; Bîrcă, A.C.; Vasile, B.S.; Oprea, O.-C.; Huang, K.-S.; Grumezescu, A.M. Microfluidic synthesis of -NH_2- and -COOH-functionalized magnetite nanoparticles. *Nanomaterials* **2022**, *12*, 3160. [CrossRef]
38. Chircov, C.; Bîrcă, A.C.; Grumezescu, A.M.; Vasile, B.S.; Oprea, O.; Nicoară, A.I.; Yang, C.-H.; Huang, K.-S.; Andronescu, E. Synthesis of magnetite nanoparticles through a lab-on-chip device. *Materials* **2021**, *14*, 5906. [CrossRef]
39. Răcuciu, M.; Creangă, D.E.; Airinei, A. Citric-acid–coated magnetite nanoparticles for biological applications. *Eur. Phys. J. E* **2006**, *21*, 117–121. [CrossRef]
40. Kirimura, K.; Yoshioka, I. Citric acid. In *Comprehensive Biotechnology*, 3rd ed.; Moo-Young, M., Ed.; Elsevier: Amsterdam, The Netherlands, 2019; Volume 3, pp. 158–165.
41. Iannone, M.F.; Groppa, M.D.; Zawoznik, M.S.; Coral, D.F.; Fernández van Raap, M.B.; Benavides, M.P. Magnetite nanoparticles coated with citric acid are not phytotoxic and stimulate soybean and alfalfa growth. *Ecotoxicol. Environ. Saf.* **2021**, *211*, 111942. [CrossRef]
42. Sousa, M.E.; Raap, M.B.F.; Rivas, P.C.; Zélis, P.M.; Girardin, P.; Pasquevich, G.A.; Alessandrini, J.L.; Muraca, D.; Sánchez, F.H. Stability and relaxation mechanisms of citric acid coated magnetite nanoparticles for magnetic hyperthermia. *J. Phys. Chem. C* **2013**, *117*, 5436–5445. [CrossRef]
43. Ferreira, L.P.; Reis, C.P.; Robalo, T.T.; Jorge, M.E.M.; Ferreira, P.; Gonçalves, J.; Hajalilou, A.; Cruz, M.M. Assisted synthesis of coated iron oxide nanoparticles for magnetic hyperthermia. *Nanomaterials* **2022**, *12*, 1870. [CrossRef] [PubMed]
44. Liu, J.; Dai, C.; Hu, Y. Aqueous aggregation behavior of citric acid coated magnetite nanoparticles: Effects of pH, cations, anions, and humic acid. *Environ. Res.* **2018**, *161*, 49–60. [CrossRef] [PubMed]
45. Goodarzi, A.; Sahoo, Y.; Swihart, M.T.; Prasad, P.N. Aqueous ferrofluid of citric acid coated magnetic particles. *Mat. Res. Soc. Symp. Proc.* **2003**, *789*, N6.6. [CrossRef]
46. Răcuciu, M.; Creangă, D.E.; Airinei, A.; Chicea, D.; Bădescu, V. Synthesis and properties of magnetic nanoparticles coated with biocompatible compounds. *Mater. Sci. Pol.* **2010**, *28*, 609–616.
47. Atrei, A.; Mahdizadeh, F.F.; Baratto, M.C.; Scala, A. Effect of citrate on the size and the magnetic properties of primary Fe_3O_4 nanoparticles and their aggregates. *Appl. Sci.* **2021**, *11*, 6974. [CrossRef]
48. Li, L.; Mak, K.Y.; Leung, C.W.; Chan, K.Y.; Chan, W.K.; Zhong, W.; Pong, P.W.T. Effect of synthesis conditions on the properties of citric-acid coated iron oxide nanoparticles. *Microelectron. Eng.* **2013**, *110*, 329–334. [CrossRef]
49. Dheyab, M.A.; Aziz, A.A.; Jameel, M.S.; Abu Noqta, O.; Khaniabadi, P.M.; Mehrdel, B. Simple rapid stabilization method through citric acid modification for magnetite nanoparticles. *Sci. Rep.* **2020**, *10*, 10793. [CrossRef]
50. Campelj, S.; Makovec, D.; Drofenik, M. Preparation and properties of water-based magnetic fluids. *J. Phys. Condens. Matter* **2008**, *20*, 204101. [CrossRef]
51. Hajdú, A.; Tombácz, E.; Illés, E.; Bica, D.; Vékás, L. Magnetite nanoparticles stabilized under physiological conditions for biomedical application. *Progr. Colloid Polym. Sci.* **2008**, *135*, 29–37. [CrossRef]
52. Locke, B.R.; Sato, M.; Sunka, P.; Hoffmann, M.R.; Chang, J.-S. Electrohydraulic discharge and nonthermal plasma for water treatment. *Ind. Eng. Chem. Res.* **2006**, *45*, 882–905. [CrossRef]
53. Lerner, M.I.; Gorbikov, I.A.; Bakina, O.V.; Kasantzev, S.O. Deagglomeration of nanostructured aluminum oxyhydroxide upon shock wave impact of electrohydraulic discharge. *Inorg. Mater. Appl. Res.* **2017**, *8*, 473–478. [CrossRef]
54. Pinheiro, P.C.; Daniel-da-Silva, A.L.; Tavares, D.S.; Calatayud, M.P.; Goya, G.F.; Trindade, T. Fluorescent magnetic bioprobes by surface modification of magnetite nanoparticles. *Materials* **2013**, *6*, 3213–3225. [CrossRef]
55. Tian, Q.; Krakovský, I.; Yan, G.; Bai, L.; Liu, J.; Sun, G.; Rosta, L.; Chen, B.; Almásy, L. Microstructure changes in polyester polyurethane upon thermal and humid aging. *Polymers* **2016**, *8*, 197. [CrossRef]
56. Sahoo, Y.; Goodarzi, A.; Swihart, M.T.; Ohulchanskyy, T.Y.; Kaur, N.; Furlani, E.P.; Prasad, P.N. Aqueous ferrofluid of magnetite nanoparticles: Fluorescence labeling and magnetophoretic control. *J. Phys. Chem. B* **2005**, *109*, 3879–3885. [CrossRef] [PubMed]
57. Saif, S.; Tahir, A.; Asim, T.; Chen, Y.; Adil, S.F. Polymeric nanocomposites of iron–oxide nanoparticles (IONPs) synthesized using *Terminalia chebula* leaf extract for enhanced adsorption of Arsenic(V) from water. *Colloids Interfaces* **2019**, *3*, 17. [CrossRef]
58. Farhanian, D.; De Crescenzo, G.; Tavares, J.R. Large-scale encapsulation of magnetic iron oxide nanoparticles via syngas photo-initiated chemical vapor deposition. *Sci. Rep.* **2018**, *8*, 12223. [CrossRef]
59. Manalastas-Cantos, K.; Konarev, P.V.; Hajizadeh, N.R.; Kikhney, A.G.; Petoukhov, M.V.; Molodenskiy, D.S.; Panjkovich, A.; Mertens, H.D.T.; Gruzinov, A.; Borges, C.; et al. ATSAS 3.0: Expanded functionality and new tools for small-angle scattering data analysis. *J. Appl. Crystallogr.* **2021**, *54*, 343–355. [CrossRef]

60. Pauw, B.R.; Pedersen, J.S.; Tardif, S.; Takata, M.; Iversen, B.B. Improvements and considerations for size distribution retrieval from small-angle scattering data by Monte Carlo methods. *J. Appl. Crystallogr.* **2013**, *46*, 365–371. [CrossRef]
61. Bressler, I.; Pauw, B.R.; Thünemann, A.F. McSAS: Software for the retrieval of model parameter distributions from scattering patterns. *J. Appl. Crystallogr.* **2015**, *48*, 962–969. [CrossRef]

Disclaimer/Publisher's Note: The statements, opinions and data contained in all publications are solely those of the individual author(s) and contributor(s) and not of MDPI and/or the editor(s). MDPI and/or the editor(s) disclaim responsibility for any injury to people or property resulting from any ideas, methods, instructions or products referred to in the content.

Article

One Pot Synthesis of Copper Oxide Nanoparticles for Efficient Antibacterial Activity

Rajaram Rajamohan *, Chaitany Jayprakash Raorane, Seong-Cheol Kim * and Yong Rok Lee *

School of Chemical Engineering, Yeungnam University, Gyeongson 38541, Republic of Korea
* Correspondence: rajmohanau@yu.ac.kr (R.R.); sckim07@ynu.ac.kr (S.-C.K.); yrlee@yu.ac.kr (Y.R.L.)

Abstract: The unique semiconductor and optical properties of copper oxides have attracted researchers for decades. However, using fruit waste materials such as peels to synthesize the nanoparticles of copper oxide (CuO NPs) has been rarely described in literature reviews. The main purpose of this part of the research was to report on the CuO NPs with the help of apple peel extract under microwave irradiation. Metal salts and extracts were irradiated at 540 W for 5 min in a microwave in a 1:2 ratio. The crystallinity of the NPs was confirmed by the XRD patterns and the crystallite size of the NPs was found to be 41.6 nm. Elemental mapping of NPs showed homogeneous distributions of Cu and O. The NPs were found to contain Cu and O by EDX and XPS analysis. In a test involving two human pathogenic microbes, NPs showed antibacterial activity and the results revealed that the zone of inhibition grew significantly with respect to the concentration of CuO NPs. In a biofilm, more specifically, NPs at 25.0 μg/mL reduced mean thickness and biomass values of *S. aureus* and *E. coli* biofilms by >85.0 and 65.0%, respectively, with respect to untreated controls. In addition, environmentally benign materials offer a number of benefits for pharmaceuticals and other biomedical applications as they are eco-friendly and compatible.

Keywords: microwave-assisted synthesis; CuO; XPS analysis; BET; antimicrobial activity; biofilm

Citation: Rajamohan, R.; Raorane, C.J.; Kim, S.-C.; Lee, Y.R. One Pot Synthesis of Copper Oxide Nanoparticles for Efficient Antibacterial Activity. *Materials* **2023**, *16*, 217. https://doi.org/10.3390/ma16010217

Academic Editors: Alexandru Mihai Grumezescu and Paul Cătălin Balaure

Received: 25 November 2022
Revised: 14 December 2022
Accepted: 21 December 2022
Published: 26 December 2022

Copyright: © 2022 by the authors. Licensee MDPI, Basel, Switzerland. This article is an open access article distributed under the terms and conditions of the Creative Commons Attribution (CC BY) license (https://creativecommons.org/licenses/by/4.0/).

1. Introduction

Infections caused by bacteria are a growing public health concern and are the major reason for the spread of serious diseases worldwide, with millions of new cases and deaths per year [1,2]. Fresh products contaminated with bacteria are the most common source of bacterial illnesses. Therefore, individual consumers, industries, and regulatory authorities are concerned about food safety. Most commonly, *Salmonella*, *Escherichia coli*, and *Staphylococcus aureus* cause foodborne illnesses [2,3]. It is possible for people to experience diarrhea, abdominal cramps, and nausea after consuming food that contains these pathogenic bacteria. As a result, it may also cause chronic illnesses such as cancer, brain disorders, kidney failure, and liver failure [4]. Bacterial infections caused by these microorganisms remain a challenge because they often form biofilms on surfaces and have developed an enhanced resistance to commonly used antimicrobial agents [5]. Drying, pickling, thermal processing, and freezing are traditional methods of extending food shelf life. As a result of these procedures, most nutrients in food are denaturalized or destroyed [6].

Metal oxides are widely used in modern technology owing to their excellent electrical, chemical, and optical properties [7,8]. Furthermore, metal oxide nanoparticles are increasingly being investigated for their biological properties. Several researchers have assessed the biologically effective activity of nanoparticles of metal oxide, especially copper oxide (CuO), which have established improved biological activity in comparison with metal NPs [9]. Among the various metal and their oxide-based NPs, copper has gained recognition due to its high redox potential [10]. Copper-based NPs such as CuO, amorphous and crystalline CuS, $CuPO_4$, and CuI is reported to have biological activity [11–15].

A variety of physical and chemical routes have been used to obtain CuO NPs with desired morphologies [16–20]. It is important to note, however, that these methods require a lot of labor, a lot of energy, an intensive route, and hazardous chemicals [21]. It is, therefore, essential that new biocompatible approaches be developed that can help to rectify the above-mentioned limitations [22]. The synthesis of NPs including metal as well as metal oxide is shifting from physical and chemical methods to biological methods termed biosynthesis or green synthesis [23,24]. Recently, fruit peels have been used to synthesize metal or metal oxide nanoparticles [25–27]. Due to its sustainability, cost-effectiveness, and simplicity, the photosynthesis of CuO NPs has gained more attention recently [28]. The purpose of this research work was to provide an environmentally friendly synthetic process for CuO NPs characterization followed by an antibacterial activity.

To the best of our knowledge, there are no other reports on the synthesis of CuO NPs with the help of apple peel extracts via microwave irradiation. The obtained NPs were characterized by analytical techniques which include XRD, FE-SEM, HR-TEM, XPS, and BET surface analysis. Moreover, our aim from the application point of view was to test the antimicrobial and antibiofilm efficacy of synthesized CuO NPs against Gram-positive as well as Gram-negative bacterial pathogens.

2. Materials and Methods

2.1. Materials

The Sigma Aldrich Company, Seoul, Republic of Korea, provided copper nitrate (molecular formula: $CuNO_3 \cdot 2H_2O$, Purity: >99%) for the synthesis of NPs without further purification. The $CuNO_3 \cdot 2H_2O$ was readily soluble in distilled water. After thoroughly washing the glassware with distilled water, they were dried in the oven for 30.0 min to avoid contamination of the glassware by the deposition of impurities.

2.2. Preparation of Apple Peel Extract

From the fresh and delicious apples, apple peel extract has been prepared with the help of a homogenizer. A detailed procedure has been given in the Supplementary Materials.

2.3. Preparation of CuO NPs

AP extract (10.0 mL) and $CuNO_3 \cdot 2H_2O$ (3.146 g in 50.0 mL distilled water) were taken separately with the required amounts. A scheme for synthesizing NPs is shown in Scheme 1. An extract of AP was typically added directly to aqueous solutions of $CuNO_3 \cdot 2H_2O$ at 0.003 mol/L with constant stirring for 10 min at 60.0 °C [29]. A pale blue color developed in the solution after ten minutes. In a microwave oven (Panasonic N-ST342, Seoul, Republic of Korea), the mixture was irradiated for 5 min at 90.0 °C under an N_2 atmosphere. Pale blue turned into light brown in less than a minute. A 15 min centrifuge with three 5 min intervals was used to separate the CuO NPs from the solvent, followed by numerous washes with ethanol and also deionized water. Centrifugation caused the CuO NPs to sediment after washing and was followed by sonication in water for one minute. A refrigerator was used to store the CuO NPs after they were dried for 24 h at 400.0 °C. It was highly probable that CuO nanoparticles would form NPs when exposed to air. AP extract was an efficient reducing agent as well as a stabilizing agent in the formation of NPs.

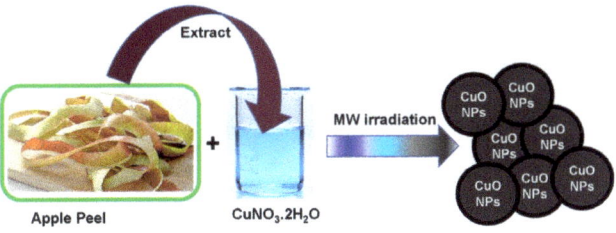

Scheme 1. Synthesis of bio-reduced CuO NPs.

2.4. Antibacterial Activity

2.4.1. In-Vitro Antibacterial Efficacy

The assessment of the antibacterial efficacy of CuO NPs was performed using the agar well diffusion method [30]. For this study, *E. coli* as Gram-negative bacteria (ATCC 43895) and *S. aureus* as Gram-positive bacteria (ATCC 6538) were used. Briefly, on sterile Mueller Hinton agar (MHA) plates, overnight cultures of each bacterial strain at 0.5 McFarland standard were spread, which were pierced with a 7 mm diameter cork borer and loaded up with 50.0 µL of CuO NPs diluted in 1.0% DMSO at different concentrations (10.0 µg/mL, 200.0 µg/mL, and 300.0 µg/mL (w/v)). After the incubation process, the radius of the inhibition zone was measured by the use of a Vernier caliper. Clinical Laboratory Standards Institute (CLSI) bacteria and Cation-adjusted Mueller–Hinton broth media were used in this study. For reliable findings and reproducibility, experiments were carried out using at least two different cultures.

2.4.2. Antibiofilm Potency of CuO NPs against Bacterial Pathogens

A biofilm experiment was carried out on 96-well microtiter plates using the crystal violet staining technique [31]. The initial turbidities of OD 0.05 (~10^6 CFU mL^{-1}) for *S. aureus* and OD 0.1 (~10^6 CFU mL^{-1}) for *E. coli* at 600.0 nm were inoculated into an LB culture media (final volume 300.0 µL) with or without the CuO NPs and incubated for 24 h without shaking at 37.0 °C. The formation of biofilm was confirmed by staining with 0.1% crystal violet for 30 min and washed frequently with distilled water and then 95.0% ethanol was added to each well. The absorbance of each plate well was recorded at 570.0 nm using a spectramax 190 microplate reader equipped with a xenon flash lamp (Molecular device, San Jose, CA, USA). Biofilm assays were carried out with two independent cultures in triplicate.

2.4.3. Antibiofilm Potency of CuO NPs against Bacterial Pathogens

A biofilm observation of the CuO NPs against both Gram-positive and Gram-negative bacterial pathogens have been measured by CLSM, and the detailed procedure has been provided in the Supplementary Materials.

2.5. Instruments Used

The bio-reduced CuO NPs are characterized by analytical instruments, and details are provided in the Supplementary Materials.

3. Results and Discussion

3.1. Analysis of Bio-Reduced CuO NPs by DRS

Figure 1 shows the DRS spectrum of pure CuO NPs. DRS curves were measured by UV-VIS-NIR spectrophotometers. NPs exhibit absorption bands in the range of 365.0 nm [32]. At 560.0 nm, a peak was observed as a shoulder, which implies the existence of CuO on the surface of NPs [32]. Furthermore, there was a weak reflectance in both the UV and visible ranges (200.0–800.0 nm). It also provided information about the greater absorption in the regions, as it gives weak transmittance. The band gap for the NPs was also calculated from the spectra and the value was 1.58 eV. The bandgap energy was higher than the bulk CuO material [33] and very close to the synthesized CuO NPs [34,35]. In the results, it consists only of CuO NPs and not of Cu.

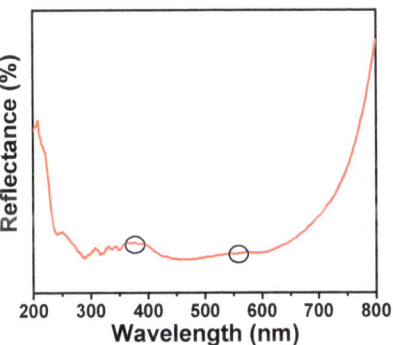

Figure 1. DRS spectra of bio-reduced CuO NPs.

3.2. Analysis of Bio-Reduced CuO NPs by FT-IR Spectrum

The effects of the peel extract used in the synthesis of NPs were analyzed by FT-IR analysis to obtain the structural and chemical properties of the synthesized metal oxides [36,37]. FT-IR spectra were recorded between 400.0 and 4000.0 cm^{-1} (Figure 2). CuO NPs have a peak at 3416.0 cm^{-1}, which was correlated with hydroxyl group stretching. NPs have a peak at 1650 cm^{-1} caused by a bending O–H. Another peak at 1376.0 cm^{-1} was related to C-O asymmetry in NPs. Another peak in the structure was at 1115.0 cm^{-1}, which was related to C-O symmetry. A peak at 533.0 cm^{-1} was associated with Cu-O bonds. There were three characteristic bands of CuO including the A_u mode, and two B_u modes of CuO appeared at 432.3 cm^{-1}, 497.0 cm^{-1}, and 603.3 cm^{-1}, respectively [38]. The high-frequency mode can be observed at 603.3 cm^{-1} and it has been attributed to the Cu-O stretching vibration in the [101] direction. The [101] direction of the Cu-O stretching vibration has been linked to the other peak, which can be seen at 497.0 cm^{-1} [39]. Therefore, the FT-IR analysis indicates that CuO NPs are in their pure phase and have a monoclinic structure.

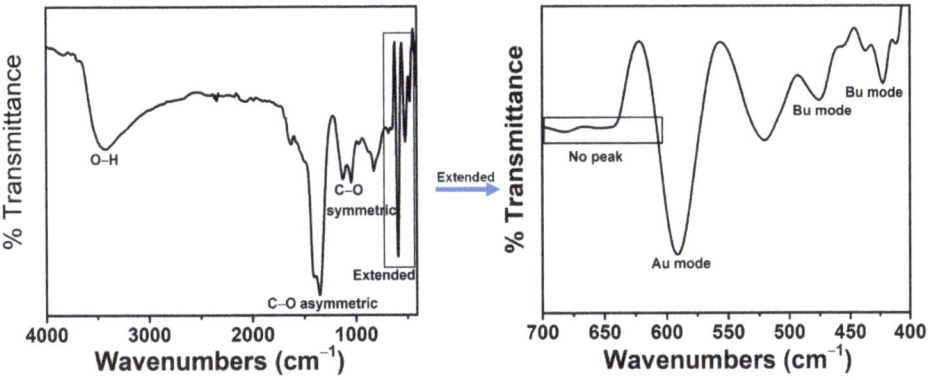

Figure 2. FT-IR spectrum of bio-reduced CuO NPs.

3.3. Analysis of Bio-Reduced CuO NPs by Raman Spectrum

Raman spectral analysis can be used as a major analytical technique to identify the vibrations of metal oxide NPs and local atomic arrangements and analyze their structural features [40,41]. It can also be used to determine how crystalline the NPs are. Figure 3 shows a strong peak at 283.0 cm^{-1}, which is associated with the A_g mode of vibration. The weak peaks that appeared at 312.0 cm^{-1} and 612.0 cm^{-1} corresponded to the B_g modes of vibration [41,42]. Only vertically and with a displacement do oxygen atoms move to the b-axis in Raman modes for both the A_g and B_g bands. Decreasing the size of the NPs altered a Raman shift and bandwidth [42].

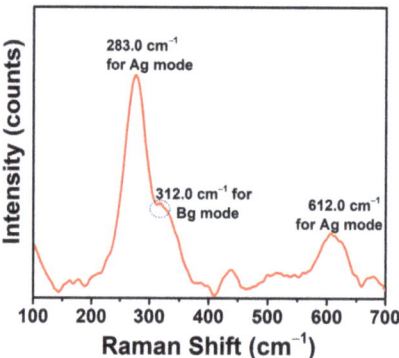

Figure 3. Raman spectra of bio-reduced CuO NPs.

3.4. Analysis of Bio-Reduced CuO NPs by XRD

Analysis of NPs obtained from metal oxides by XRD patterns is a powerful analytical technique to obtain information about the crystalline peaks [43]. Figure 4 shows the XRD patterns of CuO NPs at 2θ ranges from 20 to 80 (in degrees). According to the Joint Committee on Powder Diffraction Standards (JCPDS) database, crystalline phases were recognized. The patterns of NPs exhibited a significant peak at (2θ) 25.41, 32.51, 35.52, 38.32, 40.10, 42.51, 48.62, 53.10, 58.68, 61.52, and 62.64 (JCPDS01-080-1268) which belong to miller indices [021], [110], [002], [200], [130], [131], [202], [020], [002], [113], and [311], respectively [44,45]. Furthermore, there were some low-intensity peaks which may be due to the negligible amount of impurities in the NPs. Slight variations were observed in the obtained peaks at their position with respect to the JCPDS data, which may be due to the slight modifications in terms of phase on the surfaces of NPs. A strong intensity peak at 35.52 and a low-intensity peak at 38.32 appeared which revealed the formation of CuO and existed as the monoclinic phase [46]. Additionally, the sharp XRD patterns were evident for the crystalline nature as well as the monoclinic phase of the CuO on the whole surface of NPs. The lattice parameters a, b, and c were found to be 4.68 Å, 3.41 Å, and 5.08 Å, respectively. The average crystalline size was found to be 41.6 nm for the NPs according to the well-known Scherrer equation as follows [46,47],

$$D = (K\lambda)/(\beta \cdot \cos\theta) \quad (1)$$

where D stands for crystallite size (nm); K stands for Scherrer's constant that is associated with crystallite shape, normally taken as 0.9; β is the full width half maximum (radians), λ is the wavelength of the Cu Kα radiation (1.54 Å); and θ is the Bragg angle (Å).

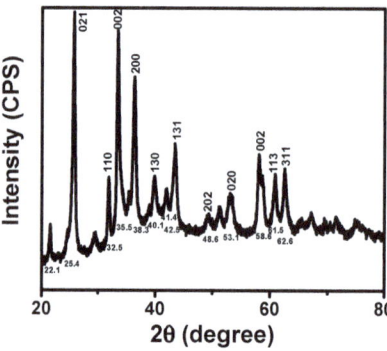

Figure 4. XRD patterns of bio-reduced CuO NPs.

3.5. Analysis of Bio-Reduced CuO NPs by FE-SEM

FE-SEM images were used for visual examination and analysis of the surface morphology of the NPs. An FE-SEM image of CuO NPs at different magnifications can be seen in Figure 5A–D. The CuO NPs were regular in shape with a particle-like structure. CuO NPs have particle sizes ranging from 25.0 to 55.0 nm and uniform distributions. Images show some particles with square shapes and clusters that have agglomerated together. Stabilized NPs can also form clusters of material-like particles relatively close together. The NPs were stabilized and reduced by the peel extract, allowing them to be re-dispersed [48]. A peel extract limits clustering and flocculation in order to control particle size distribution. In order to fabricate nanoparticles within the ranges of small sizes, apple peel extract was found to be an effective stabilizing agent. As a result, the assembly of NPs was found by processes of aggregation, growth of particles, and also impurity adsorption [49]. EDX results of the synthesized NPs confirm their chemical composition and particle distribution of NPs on the whole surface. The existence of Cu and O in the NPs is demonstrated in Figure 5E. Thus, the only two components of synthesized NPs were copper and oxygen. The pattern made it clear that the NPs are crystalline structures made of two elements, such as Cu and O. Solid and strong signals were observed around 0.92 keV, 8.05 keV, and 8.91 keV for Cu with Cu La, Cu Ka, and Cu Kb representations, respectively, and 0.53 keV for O with O Ka representation [50,51]. In Figure 5G,H, the elemental map of CuO NPs shows homogeneous distributions of Cu and O, respectively. It was confirmed by EDX elemental analysis that Cu and O were present in a single particle, with an atomic percent composition of Cu at 64.19% and O at 35.81%. Due to the coating of the carbon with the NPs to measure the SEM analysis, it was not considered for the composition of elements present in the NPs. It was confirmed by these results that a CuO structure can be formed within five minutes through microwave synthesis.

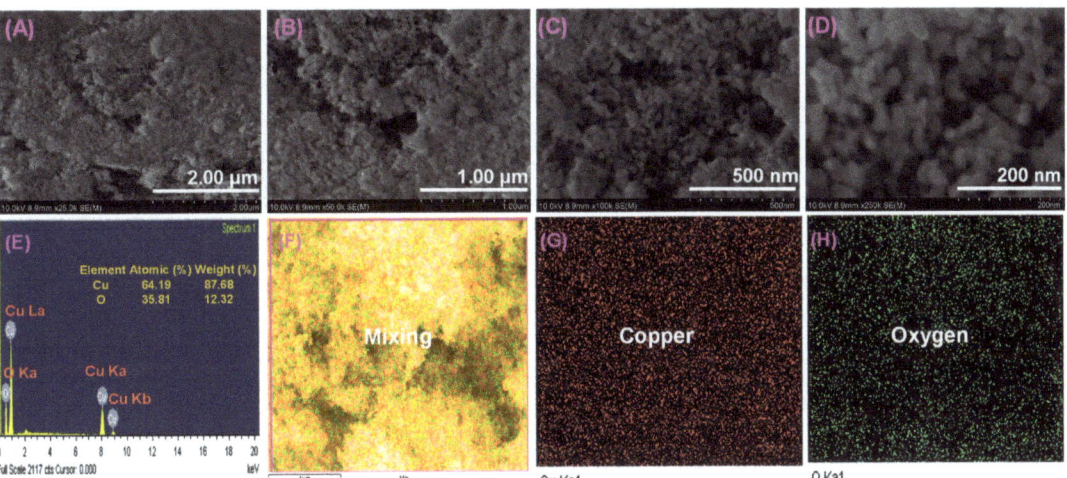

Figure 5. FE-SEM images of bio-reduced CuO NPs (**A–D**), EDX spectrum of bio-reduced CuO NPs (**E**), elemental mapping with mixing (**F**), elemental mapping of Cu (**G**), and elemental mapping of O (**H**).

3.6. Analysis of Bio-Reduced CuO NPs by HR-TEM

Figure 6A–E show HR-TEM images of the obtained CuO NPs with 31 nm scale bars. These TEM images and their SAED patterns indicate the size and crystallinity of the NPs [52,53]. These images show spherical NPs with an average diameter of 40.2 ± 4.0 nm and a narrow distribution of particle sizes. Analyses of particle size distributions were performed using ImageJ software. Due to NP agglomeration, spherical nature, and interconnections, Figure 6 closely matches SEM results. According to Figure 6F, the SAED pattern corresponds to the BCC crystalline structure of CuO, indexed to planes (021), (110), (002), (200), (130), (131), (202), (020), (002), (113), and (311). The spacing of the lattice fringes in one direction was about 0.21 nm. Morphological characterization of the NPs using FE-SEM and HR-TEM revealed that they were agglomerated.

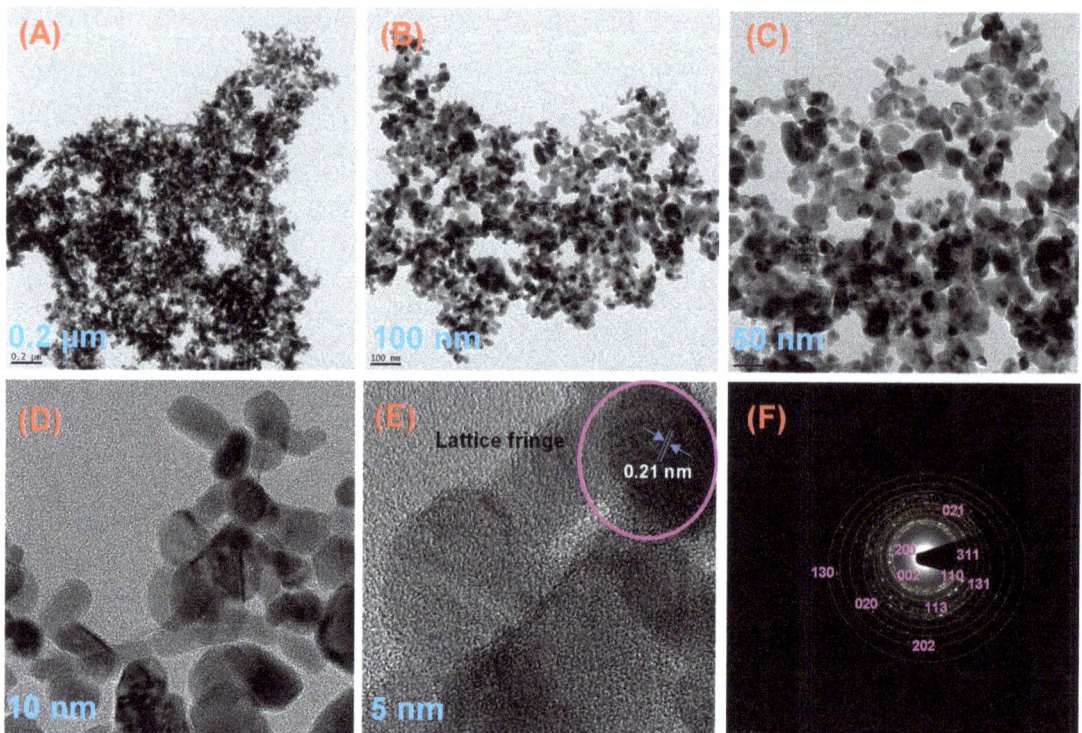

Figure 6. HR-TEM images of bio-reduced CuO NPs (**A–E**); SAED patterns of bio-reduced CuO NPs (**F**).

3.7. Analysis of Bio-Reduced CuO NPs by XPS

CuO NPs were analyzed using XPS analysis—a powerful surface-sensitive technique for the determination of oxidation state as well as chemical composition in the NPs [54]. For the standardization of all binding energies, the C 1s peak that appeared at 284.60 eV was used as a reference. According to Figure 7A, the peaks of the XPS wide scan spectrum were associated with Cu, C, and O elements. According to Figure 7B–D, the XPS spectra of Cu 2p, C 1s, and O 1s were measured with high-resolution (core XPS) spectra. The narrow energy range spectra of Cu 2p demonstrated a predominant peak at the stronger binding side of Cu $2p_{3/2}$ and increased binding energy, indicating an unfilled Cu $3d_9$ shell. The presence of Cu^{2+} in the CuO sample [55] was further confirmed by the presence of an unfilled Cu $3d_9$ shell. Additionally, the peaks at 953.28 eV and 933.38 eV in the core level spectrum of Cu 2p (deconvolution of CuO NPs, Figure 7B) can be attributed to two possibilities, such as Cu $2p_{3/2}$ and Cu $2p_{1/2}$ of CuO NPs, respectively. A high-resolution spectrum of carbon (C 1s) is shown in Figure 7C, which confirms the reference peak at 284.48 eV and other higher energy peaks at 286.08 eV and 288.41 eV. As a charge reference for the XPS spectra on the surface of NPs, the three peaks of the C 1s spectrum were known as contamination of adventitious carbon. The first peak of C 1s had a binding energy of

284.67 eV, indicating adventitious carbon containing the C–C bond; the second peak of C 1s had a binding energy of 286.08 eV, indicating adventitious carbon containing the C-O-C bond; and the third peak of C 1s had a binding energy of 288.08 eV, indicating adventitious carbon containing O-C=O bonds. According to the Gaussian–Lorentzian fit of O1s, two components were present at 529.08 eV and 530.88 eV (Figure 7D). This peak at 529.08 eV was attributed to the binding energy of lattice oxygen $(O_L)^{2-}$ in CuO lattices and agrees with O^{2-} in metal oxides $(Cu^{2+} - O^{2-})$ [56]. It can be determined that the second peak at 530.88 eV reflects the binding energy for the oxygen vacancies or defects within the environment of CuO NPs [57]. According to the XPS spectra of NPs, there was no possibility of residual nitrogen in the precursor. As a result of measuring the XPS spectrum, CuO NPs were verified to be structurally stable.

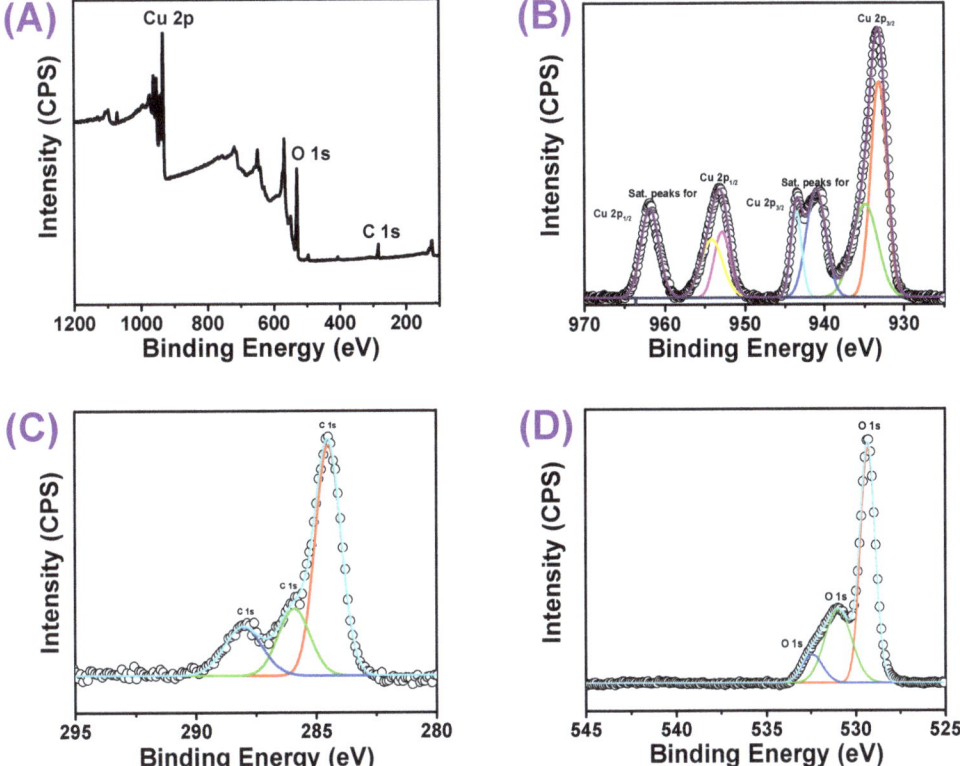

Figure 7. XPS of (A) survey scan spectrum, (B) Cu 2p, (C) C 1s, and (D) O 1s.

3.8. Analysis of Bio-Reduced CuO NPs by TG/DTA

The thermal stability of the NPs was typically evaluated using the TGDTA curves, and a weight loss (in%) can be determined in relation to an increase in temperature [58]. The TGA and DTA curves of CuO NPs are displayed in Figure 8. The exothermic peak in the DTA curves at 201.9 °C indicated the release of energy from the surface of NPs into the surrounding environment. There was not much weight loss observed in the ranges of 30.0–800.0 °C [59]. However, the weight loss percentage was very little at three stages of temperatures, 110.0–210.0 °C, 300.0–410.0 °C, and 670.0–800.0 °C. The above three weight stages were mainly due to the release of moisture content and peel extract from the surface of NPs, as desorption had taken place during the thermal analysis [60]. Thus, the NPs were very stable up to 800.0 °C without decomposition.

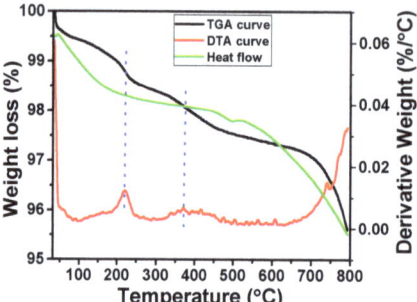

Figure 8. TGDTA curves of bio-reduced CuO NPs.

3.9. Analysis of Bio-Reduced CuO NPs by BET Surface Area

The pore size distribution and surface area of CuO NPs were studied using N_2 gas adsorption with BET surface area analysis. The adsorption–desorption curve follows a type IV isotherm obtained for the CuO NPs and is shown in Figure 9A. The hysteresis loop within the relative pressure (P/P_0) ranges from 0.7 to 0.9 confirmed the presence of the mesoporous nature of the obtained NPs [61]. The surface areas were calculated to be 31.8240 m^2/g by the standard multi-point BET (Figure 9B). The pore-size distribution was investigated for the obtained NPs by the desorption of the BJH method [62,63]. The pore size value was found to be 39.19 nm, which was clearly shown in Figure 9C, D. The results of pore size and surface area are consolidated in Table 1. Thus, the pore size shows that the NPs are mesoporous in nature which matched well with the porosity results obtained from the XRD analysis, FESEM, and also HRTEM images.

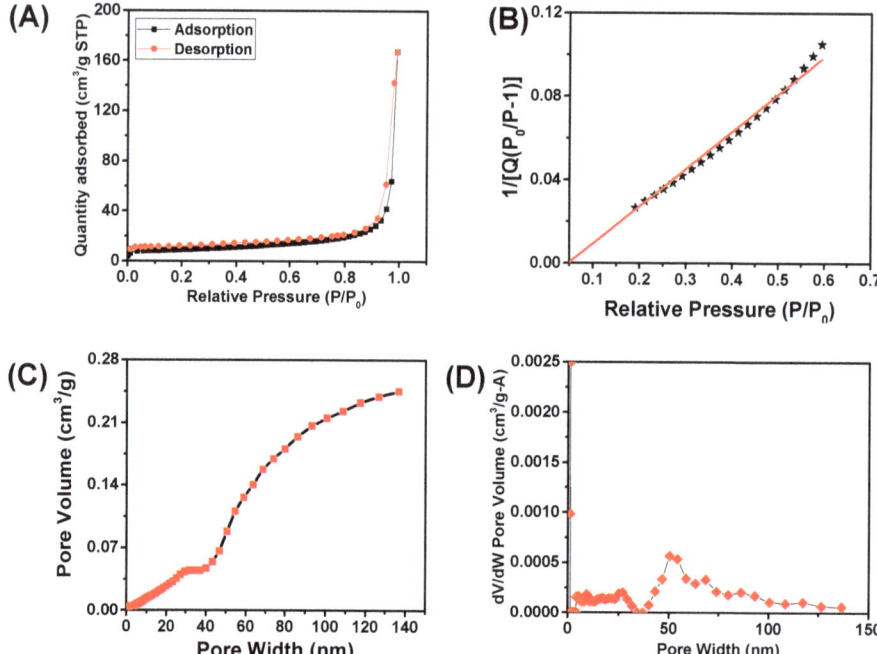

Figure 9. BET surface analysis with N_2 gas adsorption–desorption isotherms of bio-reduced CuO NPs (**A**), surface area plot (**B**), BJH desorption pore size distribution (**C**), and differential pore volume plot (**D**).

Table 1. Surface area, pore volume, and pore size distribution of CuO NPs by BET analysis.

Material	Surface Area (m²/g)	Pore Volume (m²/g)			Pore Size (nm)	
		t-Plot Micropore Volume	BJH Adsorption Cumulative Volume of Pores	BJH Desorption Cumulative Volume of Pores	BJH Adsorption Average Pore Diameter (4V/A)	BJH Desorption Average Pore Diameter (4V/A)
CuO NPs	31.8240	0.005294	0.255223	0.251015	42. 44	39.19

3.10. A Probable Mechanism of Bio-Reduced CuO NPs

An outline of the suggested CuO NPs mechanism is given in Scheme 2. As part of the biosynthetic process, there were usually three processing stages, which include the activation, growth, and termination phases [64,65]. In this process, the extract may act as a reducing as well as a stabilizing agent. In an initial step, a $CuNO_3 \cdot 2H_2O$ salt precursor would be dissolved in distilled water and activated by removing its cations. Copper ions and AP extract undergo an oxidation–reduction reaction during mixing. Cationic copper was reduced and the oxidation state became 2+ into a metallic form. The nucleation process (Cu^0) involves the direct conversion of electron-rich natural constituents (AP extract as a bio-reducing agent) directly into CuO NPs due to the greater chemical reactivity on their surfaces. The hydroxyl group of the AP extract effectively participated in the reducing process, and as evident in Figure 2, a broad peak is visible at 3416.0 cm^{-1} for the hydroxyl groups in the AP extract. During the oxidation process, the Cu^0 was progressively combined and the growth of CuO NPs commenced [66]. In the third stage, CuO NPs were stabilized. A strong protective shield layer formed and it surrounded the entire surface of the nucleated NPs; hence, limiting the growth of NPs. Additionally, these extracts use steric pressures to keep the capped NPs apart [67]. Similarly, food waste materials contain a variety of bioactive components that could aid in lowering metal ions or metal oxides as well as stabilizing metallic NPs [68].

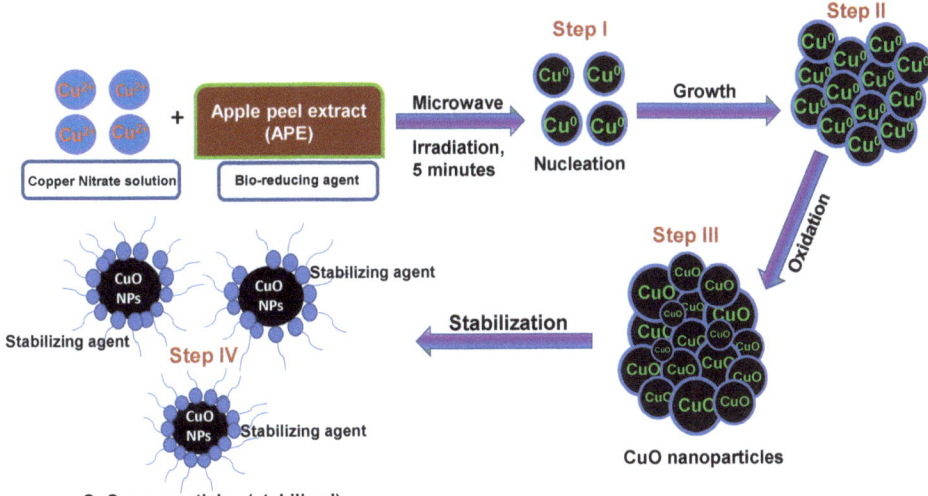

Scheme 2. Schematic procedure for the bio-reduced CuO NPs [62].

3.11. Antibacterial Efficacy of Bio-Reduced CuO NPs

Both Gram-positive and Gram-negative bacteria tested were susceptible to CuO NPs. E. coli and S. aureus were susceptible to NPs with MIC values of 50.0 µg/mL and MBC 100 µg/mL, respectively. Additionally, the agar diffusion test was carried out for the assessment of the antibacterial activity of CuO NPs and showed a clear zone for the activity against both bacteria. The diameters of inhibition zones after the treatment are consolidated in Table 2 and representative images are shown in Figure 10. The results revealed that the zone of inhibition grew significantly with respect to the concentration of CuO NPs. There was a direct relationship between the zone of inhibition and the concentrations of NPs. In *E. coli* and *S. aureus*, the zone of inhibition was determined to be 29.0 ± 2.3 mm and 26.0 ± 1.1 mm, respectively. CuO NPs have been shown to be particularly effective against various types of bacterial strains [69]. The highly unique surface characteristics of CuO NPs to volume ratio permit them to interact across the surface of the cell membrane of the bacterial pathogen, finally killing the bacterial pathogen [70]. Particularly, electronic interactions, produced by CuO NPs with a smaller size and a larger surface area, were helpful to enhance the surface responsiveness of NPs. Additionally, the improved surface area percent instantly acts together with the bacterium, causing enhanced bacteria interaction during the process. These two crucial elements (Cu and O in NPs) play a significant role in enhancing the antibacterial efficacy of NPs with a large surface area [71].

Table 2. Antibacterial efficacy of bio-reduced CuO NPs at different concentrations against *S. aureus* and *E. coli* by a zone of inhibition.

Name of the Bacterial Strains	Zone of Inhibition (mm)			
	Conc. of CuO NPs (µg/mL)			
	300.0	200.0	100.0	0
S. aureus	26.0 ± 1.1	23.0 ± 1.3	19.0 ± 0.9	ND
E. coli	29.0 ± 2.3	26.0 ± 0.8	20.0 ± 0.5	ND

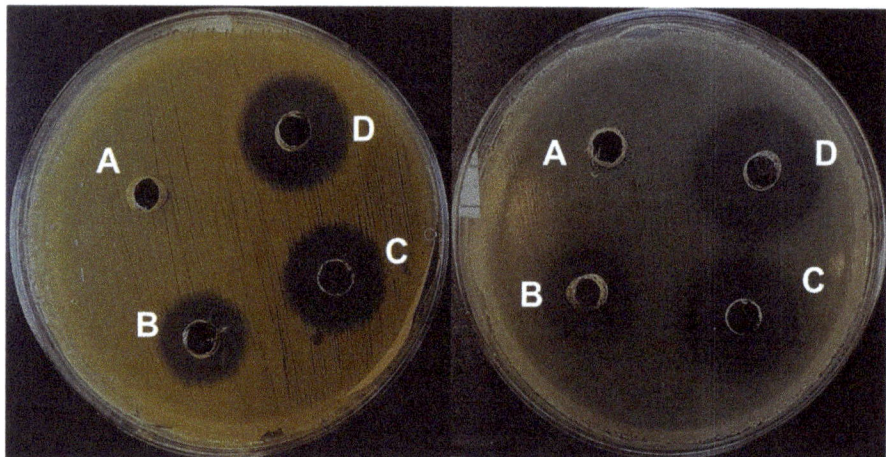

Figure 10. Antibacterial efficacy of bio-reduced CuO NPs with different concentrations (**A**, 0 µg/mL; **B**, 100 µg/mL; **C**, 200 µg/mL; and **D**, 300 µg/mL).

3.12. Antibiofilm Potency of Bio-Reduced CuO NPs against E. coli and S. aureus

A biofilm potency of NPs was performed against *E. coli* and *S. aureus*. Figure 11 shows the dose-dependent antibiofilm inhibition after the treatment with NPs at doses of 5.0 µg/mL, 10.0 µg/mL, and 25.0 µg/mL. As a result, NPs at a lower dose of 10.0 µg/mL inhibited 12.0 ± 10.7 and 31.0 ± 7.2% biofilm formation by *E. coli* and *S. aureus* after 24 h of incubation, respectively. Furthermore, when the dose of NPs was increased to 25.0 µg/mL, significant inhibition of *E. coli* and *S. aureus* biofilm formation was observed >60.0 ± 2.3

and 80.0 ± 0.3%, respectively. The CuO-containing polymeric or non-polymeric NPs were reported to be antibacterial as well as antibiofilm (28.0 to 69.0%) agents against *E. coli* and *S. aureus* [69], *Bacillus subtilis*, and *Pseudomonas aeruginosa* [72]. In the current study, the synthesized NPs inhibited selective *E. coli* and *S. aureus* biofilm at lower concentrations, whereas as concentration increased the NPs showed potential antibacterial efficacy against both tested bacterial strains. Additionally, reduced biofilm thickness was confirmed by confocal laser microscopy (Figure 11) and COMSTAT biofilm analysis (Table 3). NPs at 25.0 µg/mL reduced biomass and mean thickness of both pathogens, *E. coli* and *S. aureus*, by >65.0 and 85.0%, respectively, according to untreated controls (Table 3). From an applications point of view, CuO NPs are used in various industry sectors; moreover, there is a medical application of CuO NPs as antibacterial material [73]. Concerns are raised by their toxicity, which includes toxicity to the blood and immune system, but knowledge of their immunotoxicity is still relatively restricted. CuO NPs or material decorated with NPs prevents the growth of biofilm or adherent populations of microorganisms on the surface of materials [74]. Recently, Boliang et al. reported in 2022 that biosynthesized CuO NPs enhance antibiofilm activity against *K. pneumonia* and *S. aureus* [75].

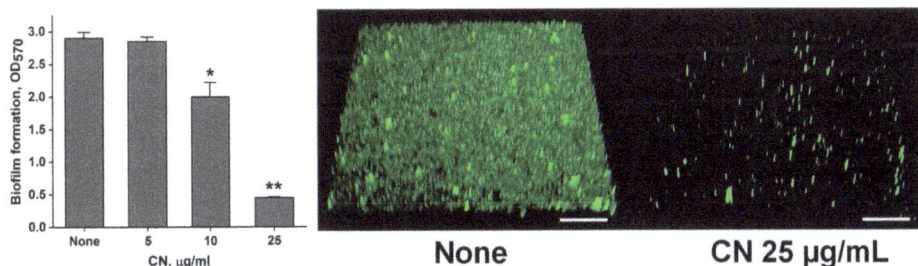

Figure 11. Effects of bio-reduced CuO NPs on *S. aureus* (**A**) and *E. coli* (**B**) biofilm formation. ** $p < 0.01$ * $p < 0.05$ Vs. non-treated controls. Scale bars represent 100.0 µm.

Table 3. COMSTAT analysis of biofilm biomass, mean thickness, and substratum coverage spatial characteristics.

Bacterial Strains	Biofilm Biomass ($\mu m^3 \, \mu m^{-2}$)		Mean Thickness (μm)		Substratum Coverages (%)	
	None	25.0 µg/mL	None	25.0 µg/mL	None	25.0 µg/mL
S. aureus	43.02 ± 2.4	6.11 ± 0.6	46.38 ± 3.1	11.36 ± 1.1	100 ± 0.8	8.01 ± 1.2
E. coli	40.80 ± 1.9	9.01 ± 1.0	39.42 ± 1.1	16.40 ± 0.4	100 ± 0.4	12.4 ± 2.1

4. Conclusions

For the quick one-pot production of CuO NPs, pressure control or a higher temperature was not necessary to maintain. The XRD pattern revealed that all of the NPs had a monoclinic crystalline structure. According to XPS and EDX studies, Cu and O make up the NPS. The pore-size distribution was investigated for the obtained NPs by the desorption of the BJH method, which revealed that the NPs are mesoporous in nature. The pore size for the NPs was found to be 39.19 nm. An in vitro antibacterial efficacy of CuO NPs was tested against the bacterial pathogens *E. coli* and *S. aureus*. The results revealed a clear zone of inhibition against both bacterial strains. The highly unique surface characteristics of CuO NPs to volume ratio enable them to interact across the surface of the bacterial cell membrane eventually leads to killing them. A biofilm assay was also performed to examine the antibiofilm potency of NPs against both pathogens. NPs at 25.0 µg/mL reduced the biomass and mean thickness of both the bacterial pathogens by >85.0% and 65.0%, with respect to the untreated controls. In the near future, disease detection may be the primary focus of CuO NPs' biological uses; however, they may also have potential applications in a wide range of other fields.

Supplementary Materials: The following supporting information can be downloaded at: https://www.mdpi.com/article/10.3390/ma16010217/s1; the preparation of extract, biofilm observations by confocal laser scanning microscopy, and instruments used for the characterization of CuO NPs are provided in the Supplementary Materials.

Author Contributions: Conceptualization, methodology, interpretation, and writing—original draft, R.R.: methodology, interpretation, and writing—original draft, C.J.R.: review and editing, S.-C.K.: review, editing, and supervision, Y.R.L. All authors have read and agreed to the published version of the manuscript.

Funding: This research received no external funding.

Institutional Review Board Statement: Not applicable.

Informed Consent Statement: Not applicable.

Data Availability Statement: The datasets used or analyzed during the current study are available from the corresponding author upon reasonable request.

Conflicts of Interest: The authors declare no conflict of interest.

References

1. Myrna, R.-B.; Nadja, M.M.-L.; Carla, M.R.-Q.; Ana, L.V.-A.; Felix, R.R.-V.; Oscar, J.P.-P. Single Step Microwave Assisted Synthesis and Antimicrobial Activity of Silver, Copper and Silver-Copper Nanoparticles. *J. Mater. Sci. Chem. Eng.* **2020**, *8*, 13–29.
2. CDC. Annual Summaries of Foodborne Outbreaks | Foodborne Outbreak Surveillance System. Food Safety. 2017. Available online: https://www.cdc.gov/fdoss/annual-reports/index.html (accessed on 31 December 2017).
3. CDC. Foodborne Germs and Illness. 2020. Available online: https://www.cdc.gov/foodsafety/foodborne-germs.html (accessed on 31 December 2020).
4. Beyth, N.; Houri-Haddad, Y.; Domb, A.; Khan, W.; Hazan, R. Alternative Antimicrobial Approach: Nano-Antimicrobial Materials. *Evid.-Based Complement. Altern. Med.* **2015**, *2015*, 246012. [CrossRef] [PubMed]
5. Das, B.; Dash, S.K.; Mandal, D.; Ghosh, T.; Chattopadhyay, S.; Tripathy, S. Green Synthesized Silver Nanoparticles Destroy Multidrug Resistant Bacteria via Reactive Oxygen Species Mediated Membrane Damage. *Arab. J. Chem.* **2017**, *10*, 862–876. [CrossRef]
6. Sharif, Z.; Mustapha, F.J.Y. Revisión de métodos de preservación y conservantes naturales para extender la longevidad de los alimentos. *Ing. Quim.* **2017**, *19*, 145–153.
7. Molkenova, A.; Sarsenov, S.; Atabaev, S.; Khamkhash, L.; Atabaev, T.S. Hierarchically-structured hollow CuO microparticles for efficient photo-degradation of a model pollutant dye under the solar light illumination. *Environ. Nanotechnol. Monit. Manag.* **2021**, *16*, 100507. [CrossRef]
8. Ambardekar, V.; Sahoo, S.; Srivastava, D.K.; Majumder, S.B.; Bandyopadhyay, P.P. Plasma sprayed CuO coatings for gas sensing and catalytic conversion applications. *Sens. Actuators B* **2021**, *331*, 129404. [CrossRef]
9. Vasantharaj, S.; Sathiyavimal, S.; Senthilkumar, P.; Oscar, F.L.; Pugazhendhi, A. Biosynthesis of iron oxide nanoparticles using leaf extract of Ruellia tuberosa: Antimicrobial properties and their applications in photocatalytic degradation. *J. Photochem. Photobiol. B* **2019**, *192*, 74–82. [CrossRef]

10. Pramanika, A.; Lahaa, D.; Chattopadhyay, S.; Dash, S.; Roy, S.; Pramanik, P.; Karmakar, P. Targeted delivery of chitosan-Folic acid coated "copper carbonate" nanoparticle to cancer cells in-vivo. *Toxicol. Res.* **2015**, *4*, 1604–1612. [CrossRef]
11. Laha, D.; Pramanik, A.; Maity, J.; Mukherjee, A.; Pramanik, P.; Karmakar, P. Interplay between autophagy and apoptosis mediated by copper oxide nanoparticles in human breast cancer cells MCF7. *Biochim. Biophys. Acta* **2014**, *1840*, 1–9. [CrossRef]
12. Maqbool, Q.; Iftikhar, S.; Nazar, M.; Abbas, F.; Saleem, A.; Hussain, T.; Kausar, R.; Anwaar, S.; Jabeen, N. Green fabricated CuO nanobullets via Olea europaea leaf extract shows auspicious antimicrobial potential. *IET Nanobiotechnol.* **2017**, *11*, 463–468. [CrossRef]
13. Mohammed, W.M.; Mubark, T.H.; Al-Haddad, R.M.S. Effect of CuO nanoparticles on antimicrobial activity prepared by sol-gel method. *Int. J. Appl. Eng. Res. Dev.* **2018**, *13*, 10559–10562.
14. Akhavan, O.; Ghaderi, E. Cu and CuO nanoparticles immobilized by silica thin films as antibacterial materials and photocatalysts. *Surf. Coat. Technol.* **2010**, *205*, 207–213. [CrossRef]
15. Heinlaan, M.; Ivask, A.; Bilnova, I.; Dubourguier, H.C.; Kahru, A. Toxicity of nanosized and bulk ZnO, CuO and TiO_2 to bacteria Vibrio fischeri and crustaceans Daphnia magna and Thamnocephalusplatyurus. *Chemosphere* **2008**, *71*, 1308–1316. [CrossRef] [PubMed]
16. Akintelu, S.A.; Folorunso, A.S.; Folorunso, F.A.; Oyebamiji, A.K. Green synthesis of copper oxide nanoparticles for biomedical application and environmental remediation. *Heliyon* **2020**, *6*, e04508. [CrossRef] [PubMed]
17. Rania, D.; Rabah, A.; Mamadou, T.; Christine, M.; Andrei, K. Antibacterial activity of ZnO and CuO nanoparticles against gram positive and gram negative strains. *Mater. Sci. Eng. C* **2019**, *104*, 109968.
18. Ameer, A.; Arham, S.A.; Oves, M.; Khan, M.S.; Adnan, M. Size-dependent antimicrobial properties of CuO nanoparticles against Gram-positive and -negative bacterial strains. *Int. J. Nanomed.* **2012**, *7*, 3527–3535.
19. Hina, Q.; Sumbul, R.; Dushyant Kumar, C.; Ashok Kumar, T.; Vikramaditya, U. Green Synthesis, Characterization and Antimicrobial Activity of Copper Oxide Nanomaterial Derived from Momordica charantia. *Int. J. Nanomed.* **2020**, *15*, 2541–2553.
20. Jeronsia, J.E.; Joseph, L.A.; Vinosha, P.A.; Mary, A.J.; Das, S.J. Camellia sinensis leaf extract mediated synthesis of copper oxide nanostructures for potential biomedical applications. *Mater. Today Proc.* **2019**, *8*, 214–222. [CrossRef]
21. Koupaei, M.H.; Shareghi, B.; Saboury, A.A.; Davar, F.; Semnani, A.; Evini, M. Green synthesis of zinc oxide nanoparticles and their effect on the stability and activity of proteinase K. *RSC Adv.* **2016**, *6*, 42313–42323. [CrossRef]
22. Awwad, A.M.; Albiss, B.A.; Salem, N.M. Antibacterial activity of synthesized copper oxide nanoparticles using malva sylvestris leaf extract. *SMU Med. J.* **2015**, *2*, 91–101.
23. Pugazhendhi, A.; Kumar, S.S.; Manikandan, M.; Saravanan, M. Photocatalytic properties and antimicrobial efficacy of Fe doped CuO nanoparticles against the pathogenic bacteria and fungi. *Microb. Pathog.* **2018**, *122*, 84–89. [CrossRef]
24. Vasantharaj, S.; Sathiyavimal, S.; Saravanan, M.; Senthilkumar, P.; Gnanasekaran, K.; Shanmugavel, M.; Manikandan, E.; Pugazhendhi, A. Synthesis of ecofriendly copper oxide nanoparticles for fabrication over textile fabrics: Characterization of antibacterial activity and dye degradation potential. *J. Photochem. Photobiol. B Biol.* **2019**, *191*, 143–149. [CrossRef]
25. Yang, N.; Weihong, L.; Hao, L. Biosynthesis of Au nanoparticles using agricultural waste mango peel extract and its in vitro cytotoxic effect on two normal cells. *Mater. Lett.* **2014**, *134*, 67–70. [CrossRef]
26. Yang, B.; Qi, F.; Tan, J.; Yu, T.; Qu, C. Study of green synthesis of ultrasmall gold nanoparticles using Citrus sinensis peel. *Appl. Sci.* **2019**, *9*, 2423. [CrossRef]
27. Gangapuram, B.R.; Bandi, R.; Alle, M.; Dadigala, R.; Kotu, G.M.; Guttena, V. Microwave assisted rapid green synthesis of gold nanoparticles using Annona squamosa L peel extract for the efficient catalytic reduction of organic pollutants. *J. Mol. Struct.* **2018**, *1167*, 305–315. [CrossRef]
28. Mahmoud, N.; Fatemeh, G.; Zahra, I.; Mohammad, S. Recent developments in the biosynthesis of Cu-based recyclable nanocatalysts using plant extracts and their application in the chemical reactions. *Chem. Rec.* **2019**, *19*, 601–643.
29. He, X.; Liu, R.H. Phytochemicals of Apple Peels: Isolation, Structure Elucidation, and Their Antiproliferative and Antioxidant Activities. *J. Agric. Food Chem.* **2008**, *56*, 9905–9910. [CrossRef]
30. Mohamed, A.A.; Abu-Elghait, M.; Ahmed, N.E.; Salem, S.S. Eco-friendly mycogenic synthesis of ZnO and CuO nanoparticles for in-vitro antibacterial, antibiofilm, and antifungal applications. *Biol. Trace Elem. Res.* **2021**, *199*, 2788–2799. [CrossRef]
31. Kim, Y.-G.; Lee, J.-H.; Raorane, C.J.; Oh, S.T.; Park, J.G.; Lee, J. Herring Oil and Omega Fatty Acids Inhibit Staphylococcus aureus Biofilm Formation and Virulence. *Front. Microbiol.* **2018**, *9*, 1241. [CrossRef]
32. Junfei, F.; Yimin, X. Investigation of optical absorption and photothermal conversion characteristics of binary CuO/ZnO nanofluids. *RSC Adv.* **2022**, *7*, 56023–56033.
33. Mohammad, S.D.; Mostafa, Z.M. Experimental study of water-based CuO nanofluid flow in heat pipe solar collector. *J. Therm. Anal. Calorim.* **2019**, *137*, 2061–2072.
34. Maji, S.K.; Mukherjee, N.; Mondal, A.; Adhikary, B.; Karmakar, B. Chemical synthesis of mesoporous CuO from a single precursor: Structural, optical and electrical properties. *J. Solid State Chem.* **2010**, *183*, 1900–1904. [CrossRef]
35. Lin, H.-H.; Wang, C.-Y.; Shih, H.C.; Chen, J.-M.; Hsieh, C.-T. Characterizing well-ordered CuO nanofibrils synthesized through gas-solid reactions. *J. Appl. Phys.* **2004**, *95*, 5889–5895. [CrossRef]
36. Kaur, M.; Muthe, K.P.; Despande, S.K.; Shipra, C.; Singh, J.B.; Neeika, V.; Gupta, S.K.; Yakh-mi, J.V. Growth and branching of CuO nanowires by thermal oxidation of copper. *J. Cryst. Growth* **2006**, *289*, 670–675. [CrossRef]

37. Anita, S.E.; Dae, J.K. Synthesis and characterization of CuO nanowires by a simple wet chemical method. *Nanoscale Res. Lett.* **2012**, *7*, 70.
38. Nyquist, R.A.; Kagel, R.O. *Infrared Spectra of Inorganic Compounds*; Academic Press: New York, NY, USA; London, UK, 1997; Volume 220.
39. Kliche, K.; Popovic, Z.V. Far-infrared spectroscopic investigations on CuO. *Phys. Rev. B* **1990**, *42*, 10060–10066. [CrossRef]
40. Benhammada, A.; Trache, D.; Chelouche, S.; Mezroua, A. Catalytic Effect of green Cu_2O nanoparticles on the thermal decomposition kinetics of ammonium perchlorate. *Z. Anorg. Allg. Chem.* **2021**, *647*, 312–325. [CrossRef]
41. Karthikeyan, B. Raman spectral probed electron–phonon coupling and phonon lifetime properties of Ni-doped CuO nanoparticles. *Appl. Phys. A* **2021**, *127*, 205. [CrossRef]
42. Angeline Mary, A.P.; Thaminum Ansari, A.; Subramanian, R. Sugarcane juice mediated synthesis of copper oxide nanoparticles, characterization and their antibacterial activity. *J. King Saud Univ. Sci.* **2019**, *31*, 1103–1114. [CrossRef]
43. Saad, N.A.; Dar, M.H.; Ramya, E.; Naraharisetty, S.R.G.; Narayana Rao, D. Saturable and reverse saturable absorption of a Cu_2O-Ag nanoheterostructure. *J. Mater. Sci.* **2019**, *54*, 188–199. [CrossRef]
44. Nagajyothi, P.; Muthuraman, P.; Sreekanth, T.; Kim, D.H.; Shim, J. Green synthesis: In-vitro anticancer activity of copper oxide nanoparticles against human cervical carcinoma cells. *Arab. J. Chem.* **2017**, *10*, 215–225. [CrossRef]
45. Chandan, T.; Indranirekha, S.; Moushumi, H.; Manash, R.D. Reduction of aromatic nitro compounds catalyzed by biogenic CuO nanoparticles. *RSC Adv.* **2014**, *4*, 53229–53236.
46. Ehsan, A.; Mohammad, H.; Reza, S.; Maryam, S. Copper plasmon-induced Cu-doped ZnO-CuO double-nanoheterojunction: In-situ combustion synthesis and pho-to-decontamination of textile effluents. *Mater. Res. Bull.* **2020**, *129*, 110880.
47. Masoud, M.; Abdullah, I.; Reza, I. A performance study on the electrocoating process with CuZnAl nanocatalyst for a methanol steam reformer: The effect of time and voltage. *RSC Adv.* **2016**, *6*, 25934–25942.
48. Qiuli, Z.; Zhimao, Y.; Bingjun, D.; Xinzhe, L.; Yingjuan, G. Preparation of copper nanoparticles by chemical reduction method using potassium borohydride. *Trans. Nonferrous Met. Soc. China* **2010**, *20*, s240–s244.
49. Cornell, R.M.; Schwertmann, U. *The Iron Oxides Structure, Properties, Reactions Occurrences and Uses*; Wiley-VCH: Weinheim, Germany, 1996.
50. Ismail, M.; Khan, M.I.; Khan, S.B.; Khan, M.A.; Akhtar, K.; Asiri, A.M. Green synthesis of plant supported Cu single bond Ag and Cu single bond Ni bimetallic nanoparticles in the reduction of nitrophenols and organic dyes for water treatment. *J. Mol. Liqs.* **2018**, *260*, 78–91. [CrossRef]
51. Issaabadi, Z.; Nasrollahzadeh, M.; Sajadi, S.M. Green synthesis of the copper nanoparticles supported on bentonite and investigation of its catalytic activity. *J. Clean. Prod.* **2017**, *142*, 3584–3591. [CrossRef]
52. Karthik, K.V.; Raghu, A.V.; Reddy, K.R.; Ravishankar, R.; Sangeeta, M.; Nagaraj, P.S.; Reddy, C.V. Green synthesis of Cu-doped ZnO nanoparticles and its application for the photocatalytic degradation of hazardous organic pollutants. *Chemosphere* **2022**, *287*, 132081. [CrossRef]
53. Yadav, D.; Subodh, S.K. Awasthi, Recent advances in the design, synthesis and catalytic applications of triazine-based covalent organic polymers. *Mater. Chem. Front.* **2022**, *6*, 1574–1605. [CrossRef]
54. Arif Khan, M.; Nafarizal, N.S.; Mohd, K.A.; Chin, F.S. Surface Study of CuO Nanopetals by Advanced Nanocharacterization Techniques with Enhanced Optical and Catalytic Properties. *Nanomat* **2020**, *10*, 1298. [CrossRef]
55. Shinde, S.K.; Dubal, D.P.; Ghodake, G.S.; Fulari, V.J. Hierarchical 3D-flower-like CuO nanostructure on copper foil for supercapacitors. *RSC Adv.* **2014**, *5*, 4443–4447. [CrossRef]
56. Molazemhosseini, A.; Magagnin, L.; Vena, P.; Liu, C.C. Single-Use Nonenzymatic Glucose Biosensor Based on CuO Nanoparticles Ink Printed on Thin Film Gold Electrode by Micro-Plotter Technology. *J. Electroanal. Chem.* **2017**, *789*, 50–57. [CrossRef]
57. Vaseem, M.; Hong, A.R.; Kim, R.T.; Hahn, Y.B. Copper Oxide Quantum Dot Ink for Inkjet-Driven Digitally Controlled High Mobility Field Effect Transistors. *J. Mater. Chem. C* **2013**, *1*, 2112–2120. [CrossRef]
58. Manjari, G.; Saran, S.; Arun, T.; Vijaya Bhaskara Rao, A.; Suja, P.D. Catalytic and recyclability properties of phytogenic copper oxide nanoparticles derived from *Aglaia elaeagnoidea* flower extract. *J. Saudi. Chem. Soc.* **2017**, *21*, 610–618. [CrossRef]
59. Lakshmi, K.; Revathi, S. A facile route to synthesize CuO sphere-like nanostructures for supercapacitor electrode application. *J. Mater. Sci. Mater. Electron.* **2020**, *31*, 21528–21539.
60. Manoj, D.; Saravanan, R.; Santhanalakshmi, J.; Agarwal, S.; Gupta, V.K.; Boukherroub, R. Towards green synthesis of monodisperse Cu nanoparticles: An efficient and high sensitive electrochemical nitrite sensor. *Sens. Actuators B* **2018**, *266*, 873–882. [CrossRef]
61. Mugundan, A.; Rajamannan, B.; Viruthagiri, G.; Shanmugam, N.; Gobi, R.; Praveen, P. Synthesis and characterization of undoped and cobalt-doped TiO_2 nanoparticles via sol–gel technique. *Appl. Nanosci.* **2015**, *5*, 449–456. [CrossRef]
62. Govindaraj, R.; Pandian, S.M.; Ramasamy, P.; Mukhopadhyay, S. Sol–gel synthesized mesoporous anatase titanium dioxide nanoparticles for dye sensitized solar cell (DSSC) applications. *Bull. Mater. Sci.* **2015**, *38*, 291–296. [CrossRef]
63. Manikandan, B.; Rita, J. Impact of Ni metal ion concentration in TiO_2 nanoparticles for enhanced photovoltaic performance of dye sensitized solar cell. *Mater. Sci. Mater. Electron.* **2021**, *32*, 5295–5308.
64. Shamaila, S.; Sajjad, A.K.L.; Farooqi, S.A.; Jabeen, N.; Majeed, S.; Farooq, I. Advancements in nanoparticle fabrication by hazard free eco-friendly green routes. *Appl. Mater. Today* **2016**, *5*, 150–199. [CrossRef]
65. Foad, B.; Sajjad, S.; Mohammad, B.; Feisal, K. Biofabrication of highly pure copper oxide nanoparticles using wheat seed extract and their catalytic activity: A mechanistic approach. *Green Process. Synth.* **2019**, *8*, 691–702.

66. Sutradhar, P.; Debnath, N.; Saha, M. Microwave-assisted rapid synthesis of alumina nanoparticles using tea, coffee and triphala extracts. *Adv. Manuf.* **2013**, *1*, 357–361. [CrossRef]
67. Raveendran, P.; Fu, J.; Wallen, S.L. Completely "green" synthesis and stabilization of metal nanoparticles. *J. Am. Chem. Soc.* **2003**, *125*, 13940–13941. [CrossRef] [PubMed]
68. Koopi, H.; Buazar, F. A novel one-pot biosynthesis of pure alpha aluminum oxide nanoparticles using the macroalgae *Sargassum ilicifolium* A green marine approach. *Ceram. Int.* **2018**, *44*, 8940–8945. [CrossRef]
69. Sathya, S.; Murthy, P.S.; Devi, V.G.; Das, A.; Anandkumar, B.; Sathyaseelan, V.S.; Doble, M.; Venugopalan, V.P. Antibacterial and cytotoxic assessment of poly (methyl methacrylate) based hybrid nanocomposites. *Mater. Sci. Eng. C* **2019**, *100*, 886–896. [CrossRef]
70. Nabila, M.I.; Kannabiran, K. Biosynthesis, characterization and antibacterial activity of copper oxide nanoparticles (CuO NPs) from actinomycetes. *Biocatal. Agric. Biotechnol.* **2018**, *15*, 56–62. [CrossRef]
71. Rudramurthy, G.R.; Swamy, M.K.; Sinniah, U.R.; Ghasemzadeh, A. Nanoparticles: Alternatives against drug-resistant pathogenic microbes. *Molecules* **2016**, *21*, 836. [CrossRef]
72. Logpriya, S.; Bhuvaneshwari, V.; Vaidehi, D.; SenthilKumar, R.P.; Nithya Malar, R.S.; Pavithra Sheetal, B.; Amsaveni, R.; Kalaiselvi, M. Preparation and characterization of ascorbic acid-mediated chitosan–copper oxide nanocomposite for anti-microbial, sporicidal and biofilm-inhibitory activity. *J. Nanostruct. Chem.* **2018**, *8*, 301–309. [CrossRef]
73. Tulinska, J.; Mikusova, M.L.; Liskova, A.; Busova, M.; Masanova, V.; Uhnakova, I.; Rollerova, E.; Alacova, R.; Krivosikova, Z.; Wsolova, L.; et al. Copper Oxide Nanoparticles Stimulate the Immune Response and Decrease Antioxidant Defense in Mice After Six-Week Inhalation. *Front. Immunol.* **2022**, *13*, 874253. [CrossRef]
74. Grigore, M.E.; Biscu, E.R.; Holban, A.M.; Gestal, M.C.; Grumezescu, A.M. Methods of Synthesis, Properties and Biomedical Applications of CuO Nanoparticles. *Pharmaceuticals* **2016**, *9*, 75. [CrossRef]
75. Boliang, B.; Sivakumar, S.; Vaitheeswaran, D.; Muniasamy, S.; Naiyf, S.A.; Barathi, S.; Vinod, S.U.; Balasubramanian, M.G. Biosynthesized copper oxide nanoparticles (CuO NPs) enhances the anti-biofilm efficacy against K. pneumoniae and S. aureus. *J. King Saud Uni. Sci.* **2022**, *34*, 102120.

Disclaimer/Publisher's Note: The statements, opinions and data contained in all publications are solely those of the individual author(s) and contributor(s) and not of MDPI and/or the editor(s). MDPI and/or the editor(s) disclaim responsibility for any injury to people or property resulting from any ideas, methods, instructions or products referred to in the content.

Article

Macroporous Mannitol Granules Produced by Spray Drying and Sacrificial Templating

Morgane Valentin [1,*], Damien Coibion [1], Bénédicte Vertruyen [1,*], Cédric Malherbe [2], Rudi Cloots [1] and Frédéric Boschini [1]

[1] GREEnMat, CESAM Research Unit, University of Liège, 4000 Liège, Belgium
[2] Mass Spectrometry Laboratory, MolSys Research Unit, University of Liège, 4000 Liège, Belgium
* Correspondence: morgane.valentin@uliege.be (M.V.); b.vertruyen@uliege.be (B.V.)

Abstract: In pharmaceutical applications, the porous particles of organic compounds can improve the efficiency of drug delivery, for example into the pulmonary system. We report on the successful preparation of macroporous spherical granules of mannitol using a spray-drying process using polystyrene (PS) beads of ~340 nm diameter as a sacrificial templating agent. An FDA-approved solvent (ethyl acetate) was used to dissolve the PS beads. A combination of infrared spectroscopy and thermogravimetry analysis proved the efficiency of the etching process, provided that enough PS beads were exposed at the granule surface and formed an interconnected network. Using a lab-scale spray dryer and a constant concentration of PS beads, we observed similar granule sizes (~1–3 microns) and different porosity distributions for the mannitol/PS mass ratio ranging from 10:1 to 1:2. When transferred to a pilot-scale spray dryer, the 1:1 mannitol/PS composition resulted in different distributions of granule size and porosity depending on the atomization configuration (two-fluid or rotary nozzle). In all cases, the presence of PS beads in the spray-drying feedstock was found to favor the formation of the α mannitol polymorph and to lead to a small decrease in the mannitol decomposition temperature when heating in an inert atmosphere.

Keywords: mannitol; spray drying; polystyrene; porosity; templating

Citation: Valentin, M.; Coibion, D.; Vertruyen, B.; Malherbe, C.; Cloots, R.; Boschini, F. Macroporous Mannitol Granules Produced by Spray Drying and Sacrificial Templating. *Materials* **2023**, *16*, 25. https://doi.org/10.3390/ma16010025

Academic Editors: Alexandru Mihai Grumezescu and Paul Cătălin Balaure

Received: 22 November 2022
Revised: 16 December 2022
Accepted: 17 December 2022
Published: 21 December 2022

Copyright: © 2022 by the authors. Licensee MDPI, Basel, Switzerland. This article is an open access article distributed under the terms and conditions of the Creative Commons Attribution (CC BY) license (https://creativecommons.org/licenses/by/4.0/).

1. Introduction

Nowadays, powder engineering is increasingly used in the pharmaceutical field to improve the efficiency of drugs by enhancing their physicochemical, micromeritic and/or pharmacokinetic properties. In this context, porous particles are studied because their low density and/or their high specific surface area can help to control drug delivery [1–3] and improve bioavailability [4]. Some porous particles also display improved compaction ability during tablet manufacturing [4]. Porous particles have been reported in the literature for both active pharmaceutical ingredients (API) [5–7] and excipients [2–4], in addition to many other fields in materials science, such as electrocatalysis [8–11], gas sensing [12] and other applications [9,13–15].

Several of the methods used to prepare low-density porous excipient and/or APIs particles are based on spray drying, a technique where droplets of solution/suspension are converted into a dry powder by the evaporation of the solvent/liquid [16–19].

The different spray-drying strategies to create porous dried particles differ by the nature of the precursors that will be transformed into pores. In the PulmoSpheres™ technology [5], small emulsion droplets of a less volatile liquid are dispersed in the liquid containing the pharmaceutical substance; the evaporation of these emulsion sub-droplets during the last stage of the spray-drying process creates pores [6,7,17]. Another approach relies on ammonium bicarbonate as a pore-forming agent whose decomposition takes place after the vaporization of the solvent [17,20,21]. The strategy used in the present work involves spherical beads as a sacrificial template that is removed after the spray-drying step. This sacrificial templating approach has been much studied for the preparation of metal oxides or silica with controlled porosity using a polymer template removed by heat treatment [22–28]. Polystyrene beads are one of the most common sacrificial templates because their diameter can be controlled during the synthesis allowing a broad range of pore sizes [13–15,27–29].

However, the sacrificial template strategy has been much less studied in the case of porous organic particles [29,30], where the low thermal stability of the host requires that the template be removed by dissolution/etching instead of a heat treatment. In this context, the best-documented system is that of hyaluronic acid combined with polystyrene beads, later dissolved in toluene [29,30].

Here, we focus on mannitol, an excipient that is one of the most popular alternatives to lactose because it offers improved chemical stability and lower intolerance/allergy issues [31,32]. In the context of the Dry Powder Inhaler (DPI) technology, mannitol can act as an API to improve mucus clearance in people living with cystic fibrosis [33], but is more commonly used as a carrier of the API, for example in the treatment of asthma or chronical obstructive pulmonary disease (COPD) [34,35]. The most common DPI delivery strategy relies on fine (<10 µm) and/or coarse (20–100 µm) mannitol particles mixed together and then mixed with the API [36]. As a result, the API is located at the surface of the excipient particles. Co-spraying the API with the excipient would allow for process intensification but corresponds to a different drug delivery profile [16,37–39]. Using polystyrene beads as the sacrificial template, a 1:1 mannitol/polystyrene mass ratio would correspond to a maximum apparent density of 0.62 g/cm^3 for the porous mannitol particles, compared to 1.514 g/cm^3 for dense mannitol [40]. Porous mannitol granules are therefore expected to travel more efficiently into the pulmonary system.

Because toluene [29,30] is a Class 2 organic solvent suspected of causing irreversible and reversible toxicity [41], we selected ethyl acetate as the solvent to dissolve the polystyrene beads. Indeed, the Hildebrand solubility parameters of ethyl acetate and polystyrene are similar (18.2 and 18.3 MPa$^{1/2}$, respectively) [42], and ethyl acetate is approved by the Food and Drug Administration (FDA) as a Class 3 solvent [43,44], and by the European regulatory committee for pharmaceutical applications [40,45].

In the present work, our main objective was to investigate the influence of the mass ratio between mannitol and polystyrene beads on the porosity observed after etching in ethyl acetate. The efficiency of the etching was examined through infrared spectroscopy and thermogravimetric analyses. We also report the influence of the mass ratio between mannitol and polystyrene beads on the polymorphism of mannitol. Then, we transferred the synthesis from lab-scale to pilot-scale conditions and considered how this affects the homogeneity of the porosity distribution. These final sections also include preliminary tests on the possibility of preparing formulations of the spray-dried porous mannitol with an API (budesonide) for inhalation therapy.

2. Materials and Methods

2.1. Materials

Mannitol Pearlitol® was obtained from Rocquette Frères (Lestrem, France). Potassium persulfate and styrene (≥99% purity) were purchased from Sigma-Aldrich. Ethyl acetate (≥99.8% purity) was purchased from VWR Chemicals. MilliQ (Millipore Milli-Q Plus) water was used for all syntheses. Budesonide was supplied by Sicor Societa Italiana Corticosteroidi Srl (Milano, Italy).

2.2. Aqueous Suspension of PS Beads

Aqueous suspensions of polystyrene beads were freshly prepared before each series of experiments by surfactant-free emulsion polymerization as described in the literature [46,47]. In a typical synthesis (Figure 1a), 2.72 g of styrene were added to 100 g of milliQ water in a round-bottom flask with a magnetic stirrer rotating at 450 rpm. After N$_2$ flushing and heating at 70 °C in an oil bath, 175 mg of potassium persulfate (KPS) initiator dissolved in 25 g of milliQ water was introduced via a dropping funnel. The reaction mixture was maintained under stirring at 450 rpm for 24 h. The as-obtained PS latex with a concentration of 20.18 g/L was used without dilution for the syntheses in the pilot-scale GEA Niro Mobile Minor spray dryer. The latex was diluted by a factor of 6 for the syntheses in the lab-scale Büchi Mini B-191 spray dryer.

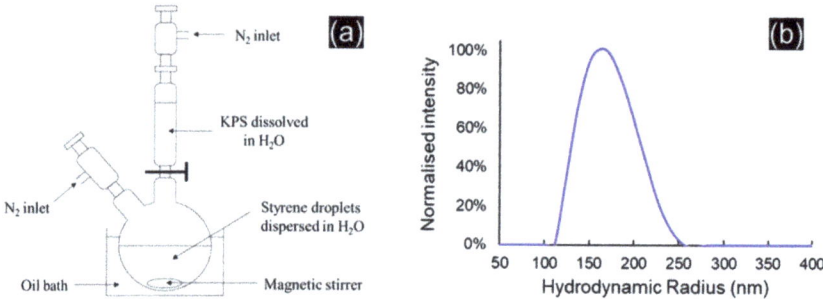

Figure 1. (a) Schematic illustration of the apparatus used to produce the PS beads by surfactant-free emulsion polymerization and (b) distribution of hydrodynamic radius obtained with dynamic light scattering.

2.3. Spray Drying of Mannitol/PS Granules

Mannitol/PS granules were obtained by dissolving mannitol Pearlitol® powder in the water-based PS latex and spray drying the suspension. Table 1 provides an overview of the samples: mannitol/PS granules with different mannitol:PS ratios were prepared in the Büchi lab-scale spray dryer while upscaling experiments for the 1:1 ratio were carried out in the pilot-scale Niro spray dryer. The spray-drying parameters are listed in Table 2.

Table 1. Overview of the M/PS granules obtained with different sets of synthesis conditions (M = mannitol, PS = polystyrene beads).

Label	M/PS Ratio (w/w)	Spray Dryer	(Mannitol) (g/L)	(PS Beads) (g/L)	Expected PS Content in Spray-Dried Granules	TGA-Measured PS Content in Spray-Dried Granules	$T_{melting}$ for Mannitol
B-1/0	1/0	Büchi, Two-fluid nozzle	36	0	0 wt%	Not applicable	165
B-10/1	10/1	Büchi, Two-fluid nozzle	33.6	3.36	9 wt%	7.9 wt%	165
B-10/6	10/6	Büchi, Two-fluid nozzle	5.60	3.36	37.5 wt%	34.9 wt%	162
B-1/1	1/1	Büchi, Two-fluid nozzle	3.36	3.36	50 wt%	48.7 wt%	160
B-1/2	1/2	Büchi, Two-fluid nozzle	1.18	3.36	66.7 wt%	-	-
N-1/1-TF	1/1	Niro, Two-fluid nozzle	20.18	20.18	50 wt%	43.0 wt%	161
N-1/1-R	1/1	Niro, Rotary nozzle	20.18	20.18	50 wt%	-	162

Table 2. Experimental parameters for the syntheses in the lab- and pilot-scale spray dryers.

Spray-Drying Parameters	Büchi (Lab-Scale)	Niro (Pilot-Scale)	
	Two-Fluid Nozzle	Two-Fluid Nozzle	Rotary Nozzle
Inlet temperature (°C)	120	120	120
Outlet temperature (°C)	43	72	75
Atomization gas flow rate (m^3/h)	0.6	6.7	6.7
Drying gas flow rate (m^3/h)	29.4	67	67
Liquid flow rate (mL/min)	5	25	25
Injected volume (mL)	25	250	250
Typical experimental yield	25%	50%	20%

2.4. Etching of the PS Beads to Form Porous Mannitol Granules

Etching of the PS beads to produce porous mannitol granules was performed by dispersing 200 mg of mannitol/PS granules in 20 mL of ethyl acetate and stirring for 60 min. The etched granules were then recovered by filtration on a 0.45 µm membrane nylon filter, washing with small volumes of ethyl acetate and drying in an oven at 50 °C for 24 h.

2.5. Mannitol-Budesonide Formulations

Formulations containing 0.8% w/w of API were prepared in a Turbula® mixer (45 rpm for 30 min, Eskens Benelux, Mechelen, Belgium) by mixing about 1 g of porous mannitol granules (excipient) with micronized budesonide (active pharmaceutical ingredient). Cylindrical aluminum containers (3 cm diameter, 3 cm height) were used in order to minimize electrostatic charging.

2.6. Characterization Techniques

The hydrodynamic radius distribution of the PS beads was measured using dynamic light scattering (DLS-Viscotek model 802, DLS Viscotek Europe, London, UK) after 100× dilution of the latex. The stability of the latex over 24 h was studied using a Turbiscan®. The zeta potential was measured under constant stirring with a DT-1200 zetameter.

The shape and the surface of the granules were examined with a scanning electron microscope (FEI XL30 ESEM-FEG, FEI Eindhoven The Netherlands) operating at an acceleration voltage of 4 keV for mannitol-containing samples and at 15 keV for PS beads. Prior to imaging, a few milligrams of samples were dispersed onto carbon sticky tabs and coated with a gold layer of approximately 20 nm.

Thermal analysis of the granules before and after etching was performed with a TGA-DTA-DSC analyzer (Setaram labsys Evo instrument), previously calibrated using indium, tin, lead, zinc and aluminium fusion points. A total of 5 mg of each sample was heated at 5 °C/min in alumina crucibles under helium atmosphere. The temperatures for the fusion and the decomposition of mannitol reported later in the text were taken as the extremum in the DSC peak (fusion) and the middle of the mass loss in the TG curve (decomposition).

X-ray diffractograms were collected in Bragg–Brentano geometry using a Bruker D8 Twin-Twin diffractometer with a Lynxeye-XET 1D detector, Bruker, Karlsruhe, Germany (Cu-Kα radiation, 2θ = 5–60°, 0.02° step size and 0.1 s/channel step time). The percentages of the α and β mannitol phases were estimated by Rietveld refinement with the TOPAS software, Version 4.2 using the fundamental parameters approach to model the instrumental contribution [48]. The structural models were taken from the PDF4+ database (PDF 00-022-1793 for the α polymorph and PDF 00-022-1797 for the β polymorph).

Attenuated total reflectance infrared (ATR-IR) spectra were recorded on a few milligrams of powders using a Nicolet IS5 with an ID7 ATR accessory (Thermo Scientific) equipped with a diamond crystal.

The homogeneity of the budesonide–mannitol mixing was determined on two aliquots of 150 mg of the mixed powder dissolved in a 40/60 (*v/v*) MeOH/water solvent mixture (100 mL). The dosage of budesonide was carried out by an ultra-high performance liquid chromatography UHPLC (Agilent Technologies 1290 infinity) using a reversed-phase column (Acquity CSH C18 1.7 µm, Waters, Milford, USA) coupled to a UV detector for the detection and the quantification of budesonide. The amount of budesonide in the aliquots was determined using a calibration curve with a reference standard.

3. Results and Discussion

3.1. Preparation of the Aqueous Suspension of PS Beads

Figure 1b shows the radius size distribution of the PS beads measured by dynamic light scattering, revealing a narrow distribution with a diameter of 335 ± 40 µm (mean PS bead diameter ± 3 standard deviations). The SEM micrograph in Figure 2a shows the smooth spherical shape of the PS beads. The zeta potential of −96 mV ensures good dispersion and avoids the aggregation of the beads in the suspension. The migration rate inside the suspension was 0.2 mm/h, confirming the good stability of the suspension of the PS beads.

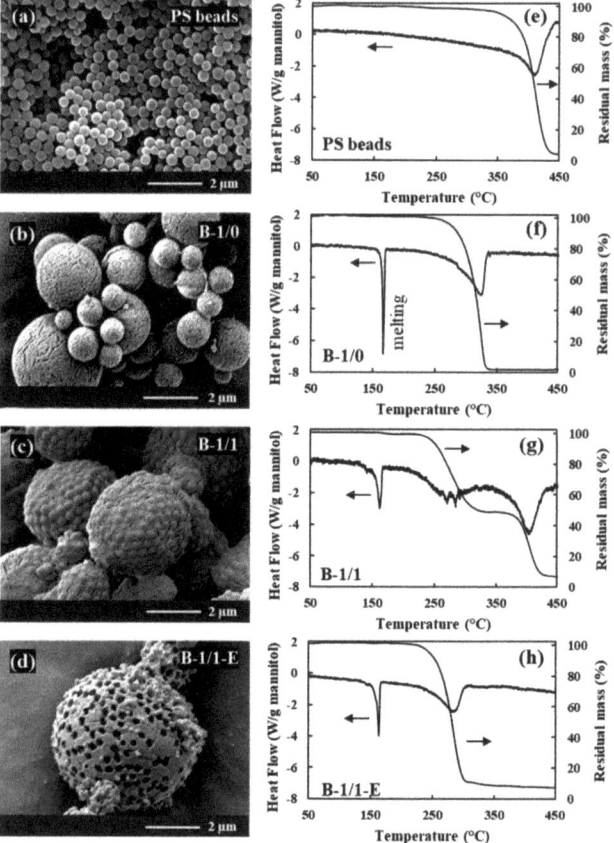

Figure 2. The SEM micrographs, TG and DSC curves (5 K/min in helium) for (**a,e**) PS beads, (**b,f**) spray-dried mannitol (B-1/0), (**c,g**) spray-dried granules with a mass ratio of mannitol to PS beads of 1:1 (B-1/1) and (**d,h**) the same granules after etching in ethyl acetate (B-1/1-E). The spray-dried samples shown in this figure were obtained with the lab-scale Büchi spray dryer.

3.2. Preparation of Porous Mannitol Granules by Spray Drying and PS Etching

Figure 2 shows SEM micrographs and thermal analysis curves for (a, e) the PS beads, (b, f) a mannitol-only spray-dried sample and (c, d, g, h) 1:1 mannitol/PS spray-dried granules before and after etching in ethyl acetate. As can be seen in Figure 2c, the 1:1 mannitol/PS granules retain the spherical microstructure observed in the mannitol-only sample (Figure 2b). The possibility of dissolving the polystyrene beads in ethyl acetate without destroying the mannitol network is evidenced by the porosity clearly visible in the etched granules shown in Figure 2d. The dissolution of the PS beads is further confirmed by the absence of any mass loss in the temperature range of the thermal decomposition of the PS (~380 °C–420 °C) in the thermogravimetric curve of the etched granules (Figure 2h, to be compared to Figure 2e,g for the PS beads and the as-sprayed granules, respectively). The thermal events observed at lower temperatures correspond to the fusion of mannitol (160–165 °C, endothermic peak, no mass loss) and the decomposition of mannitol (endothermic peak + mass loss between 250 °C and 330 °C).

3.3. Influence of the Mannitol/PS Ratio on the Etching Efficiency and on the Mannitol Polymorphism

The left panel of Figure 3 compares the thermal analysis curves of the granules obtained for different mannitol/PS ratios. The data for the 10:6 mannitol/PS ratio (Figure 3b) are similar to those already described in the case of the 1:1 mannitol/PS ratio (shown again in Figure 3c). On the contrary, in the case of the lowest PS content (10:1 mannitol/PS ratio- Figure 3a), the mass loss curves before and after etching reveal that a significant fraction of the PS beads was not dissolved during the etching procedure. This can be explained by the SEM micrographs of sample B-10/1 in Figure 4a,e, where only a few PS beads can be observed at the surface of the as-sprayed granules, suggesting that most of the PS beads are trapped inside the mannitol matrix and cannot be reached by the ethyl acetate solvent.

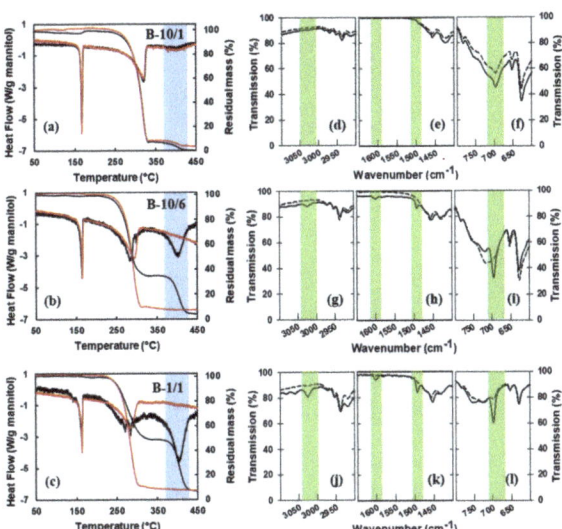

Figure 3. Influence of the M/PS mass ratio on the etching efficiency: (**a**–**c**) TG–DSC curves (5 K/min in helium) of the spray dried granules before etching (black curves) and after etching (red curves); (**d**–**l**) zooms in three regions of the IR spectra of the spray-dried granules before etching (plain curves) and after etching (dashed curves), with the green-shaded areas highlighting characteristic PS vibrations: aromatic C–H stretching at 3025 cm^{-1}, aromatic C=C stretching at 1500 cm^{-1} and 1600 cm^{-1} and aromatic C–H bending at 700 cm^{-1} [49]. The spray-dried samples in this figure were obtained with the lab-scale Büchi spray dryer.

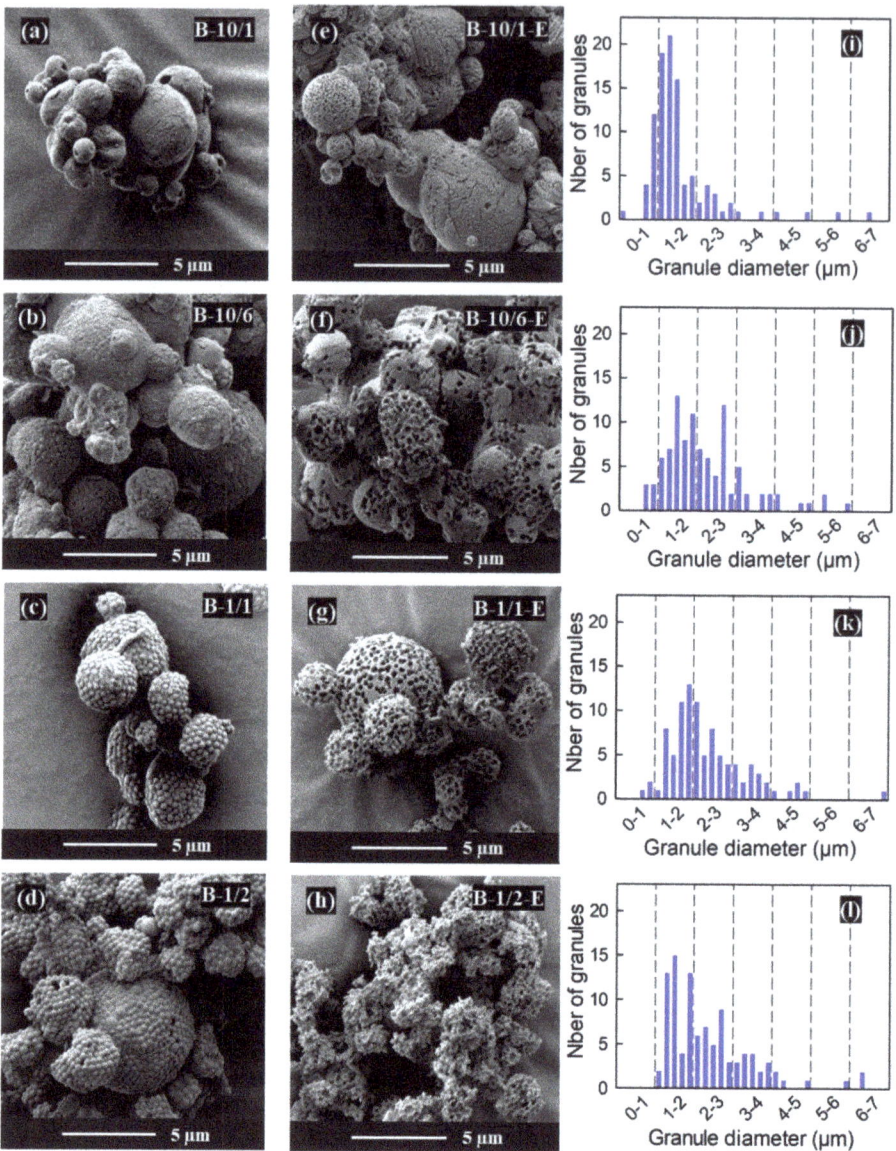

Figure 4. Influence of the M/PS mass ratio on the granule microstructures: (**a–d**) SEM micrographs of spray-dried granules before etching; (**e–h**) SEM micrographs of spray dried granules after etching; (**i–l**) histograms of granule diameters before etching. The spray-dried samples shown in this figure were obtained with the lab-scale Büchi spray dryer.

As a complement to thermal analysis, the same set of samples were characterized by IR spectroscopy, following the approach used by Iskandar et al. (2009) [30] and Nandiyanto and Okuyama (2017) [29] for the hyaluronic acid/PS system. Selected wavenumber ranges of the attenuated total reflectance infrared (ATR-IR) spectra before and after etching are shown in the right-hand part of Figure 3 (see Figure S1 for typical complete spectra). In the ATR configuration, the evanescent IR beam probes only the top ~2 μm of the powder in contact with the diamond crystal and the PS content in the sample with 10:1 mannitol/PS ratio turned out to be too small to be detected, both in the as-sprayed state and after etching. This indicates that the absence of the characteristic PS peaks in the spectra of the etched 10:6 and 1:1 samples should not be considered sufficient proof of a successful etching and needs to be confirmed, as was done here and in Nandiyanto and Okuyama (2017) [29], by thermal analysis results.

The thermal analysis curves in Figures 2 and 3 also provide information about the mannitol component of the granules: Figure 5 (left) plots the temperatures of the fusion and the decomposition of mannitol in the samples prepared with different mannitol/PS ratios. While the fusion temperature is hardly affected, the decomposition temperature decreases by about 40 °C when going from the mannitol-only spray-dried sample to the one prepared with a 1:1 mannitol/PS ratio. The heating rate and the sample masses used for the thermal analysis experiments were the same; therefore, the difference in decomposition temperature can be attributed to actual differences between the samples. The polymorphism was characterized by X-ray diffraction (Figure 5 DRX) and only α and β polymorphs were observed. The percentage of the thermodynamically favored β polymorph is plotted in Figure 5 (right): starting from 100% β polymorph in the mannitol-only spray-dried sample, the β content drops sharply as soon as PS beads are present in the spray-drying feedstock, suggesting that the surface of the PS beads could act as heterogeneous nucleation sites for the kinetically favored α polymorph. Comparing the evolution of the decomposition temperature in Figure 5 (left) and the percentage of β polymorph in Figure 5 (right) shows that polymorphism could be a factor contributing to the evolution of the decomposition temperature but that other effects must play a role. This is confirmed by the data for the commercial Pearlitol mannitol sample, which do not follow the trend. Finally, Figure 5 also shows that neither the decomposition temperature of mannitol nor its polymorphism are much affected by the etching procedure in ethyl acetate.

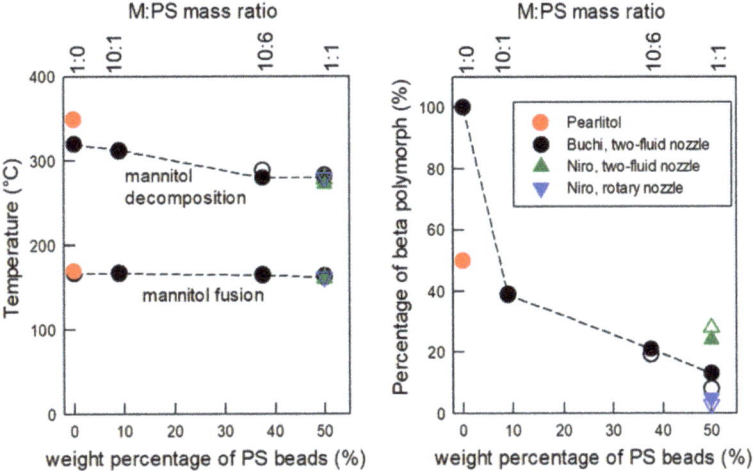

Figure 5. Influence of the M/PS ratio on (**left**) the temperatures of fusion and decomposition of mannitol determined from TGA/DSC curves and (**right**) the percentage of the β polymorph, determined from XRD patterns (see Figure 6) for samples prepared with the lab-scale (Büchi) or the pilot-scale (Niro) spray-driers. The data for the commercial mannitol Pearlitol® is shown for comparison. Empty symbols correspond to the etched samples. The dashed lines are guides for the eye.

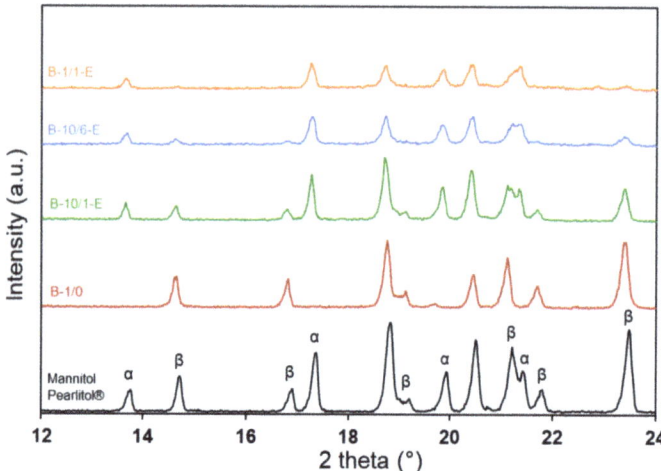

Figure 6. Influence of the M/PS ratio on the polymorphism of mannitol: XRD patterns of the spray-dried granules after etching (lab-scale Büchi spray dryer samples).

3.4. Influence of the Mannitol/PS Ratio on the Granule Microstructure

The SEM micrographs in Figure 4 show the evolution of the microstructure as a function of the mannitol/PS ratio. Micrographs of a sample with 1:2 mannitol/PS ratio prepared during preliminary tests are included here although this composition was not further investigated due to the irregular and fragile granule surface structure obtained after etching (Figure 4h). All the other samples are made up from partly agglomerated granules of roughly spherical shape. Table 1 lists the TGA-measured PS contents of these samples and finds them only slightly lower than the nominal ones, indicating that only a minor fraction of the PS beads are lost by attrition in the spray dryer (ending up in the back filter).

The granule size distributions, plotted as histograms of granule diameters measured on SEM micrographs (Figure 4i–l), are similar, with a modest broadening of the distribution towards larger sizes when the PS content increases. This suggests that the granule size is mostly controlled by the constant atomization parameters and the constant concentration of PS beads in this series of experiments (see Table 1). An explanation can be proposed by considering the transport phenomena taking place in the droplets during spray drying [16,50]: the granule size is established when the slowly moving PS beads come into contact with each other because the diffusion of liquid from the center to the surface of the shrinking droplets can no longer compensate for the solvent evaporation from the surface.

While the granule size seems to be dictated by the PS beads, the appearance of the granule surfaces in the SEM micrographs can be related to the mannitol concentration in the droplets, which decreases when the mannitol/PS ratio changes from 10:1 to 1:2 (see Table 1). For a mannitol/PS ratio of 10:1, the surface of the granules is mostly constituted of mannitol, with only a few visible PS beads (Figure 4a). This suggests that the precipitation of mannitol took place when the surface of the granules was still strongly impregnated by liquid. For the other samples, the lower mannitol concentration means that mannitol saturation (and hence precipitation) took place at a later stage of the droplet drying, so that the granules display exposed PS beads on their surfaces. The granules of the sample with an intermediate mannitol/PS ratio of 10:6 display surfaces where both behaviors co-exist (Figure 4b); the homogeneity of the PS bead distribution is much better for the mannitol/PS ratio of 1:1 (Figure 4c), while the highest mannitol/PS of 1:2 leads to such a low content of mannitol at the surface (Figure 4d) that the granules do not retain their spherical shape after etching (Figure 4h).

The micrographs of the etched granules (Figure 4e–h) confirm that ethyl acetate dissolves the PS beads in the samples where they are accessible from the surface through an interconnected network (see above, discussion of the thermal analysis results).

The results in this section cannot be compared in detail with the work of Nandiyanto et al. [29] and Iskandar et al. [30] on hyaluronic acid (HA)/PS beads because these authors used smaller beads (100 nm, 200 nm, 300 nm), lower concentrations (hence smaller granules with a diameter typically below 1 μm) and different ways of varying the HA/PS ratios. When keeping the HA concentration constant and decreasing the PS bead content, Nandiyanto et al. observed a decrease in the number of surface pores [29]. This is coherent with our results, where we kept the PS concentration constant and increased the mannitol concentration. More generally, our results on the mannitol/PS system confirm their conclusion that the "spray-drying + etching of PS beads" strategy is not limited to inorganic matrix materials and can be successfully applied to the preparation of the porous particles of organic compounds.

3.5. Upscaled Production for the 1:1 Mannitol/PS Ratio

Based on the results discussed in the previous section, the 1:1 mannitol/PS ratio was considered as the most suitable to obtain spherical granules with a good homogeneity of porosity. Therefore, granules with a 1:1 mannitol/PS ratio were produced in larger quantity in a pilot-scale spray dryer equipped with a two-fluid nozzle. In order to maintain a granule size similar to the granules prepared with the lab-scale spray dryer, the experimental parameters had to be adapted (see Tables 1 and 2, N-1/1-TF sample): a higher concentration of the feedstock solution was required to achieve efficient drying at the higher liquid flow rate and was compensated by the smaller droplet size resulting from the higher atomization flow rate. As expected, due to the larger drying chamber, the yield (50%) was improved with respect to the values obtained with the Büchi lab-scale spray dryer (25%). A higher yield could probably be obtained for larger feedstock volumes.

The histogram of granule diameters in Figure 7a shows that the size range of the distribution is similar to the sample prepared in the lab-scale spray dryer (Figure 4k), although a larger proportion of granules seems to be found in the highest diameter zone of the range. This may explain why the distribution of the pores at the surface of the granules is less homogeneous (Figure 7b), while sufficient to ensure effective etching as seen in the thermal analysis curve (Figure 7c). Figure 5 shows that the shift to the pilot-scale spray dryer did not affect the fusion and decomposition temperatures of mannitol, or the percentage of β polymorph.

The pore distribution could have been optimized by fine-tuning the mannitol/PS ratio, possibly through a DoE (Design of Experiments) procedure. However, the preliminary tests of mixing the porous mannitol granules with the budesonide API revealed a severe agglomeration (Figure 7d), which would need to be solved before the porous mannitol granules could be considered for application as a fine porous excipient for DPI technology.

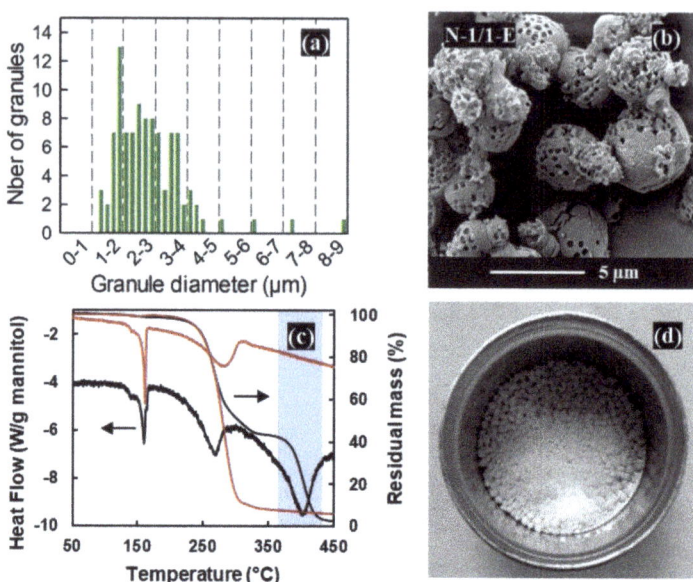

Figure 7. Upscaling of the synthesis for the M:PS ratio = 1:1 composition in the Niro pilot-scale spray dryer with the two-fluid nozzle: (**a**) histogram of the granule diameters, (**b**) SEM micrograph of the granules after etching, (**c**) TGA/DSC curves before (in black) and after (in red) etching, (**d**) photograph of the agglomerated formulation after mixing with the budesonide API.

3.6. Larger Porous Granules Prepared with Rotary Nozzle Atomisation

As a complement to the experiments reported above for spray-drying conditions yielding granules with diameters < 5 µm, a solution with a 1:1 mannitol/PS ratio was also sprayed in the pilot-scale spray dryer equipped with a rotary nozzle. The larger droplet sizes and different trajectories in the drying chamber [51] offered by this atomization mode expand the panorama of possible microstructures for the 1:1 mannitol/PS ratio.

The histograms and SEM micrographs in Figure 8 reveal a broad distribution of granule diameters, especially for the granules collected at the bottom of the drying chamber (between 4 and 25 µm — Figure 8d) when compared to the granules collected at the bottom of the cyclone (between 3 and 12 µm — Figure 8a). The overall yield was about 20%, from which about 70 wt% was recovered at the bottom of the cyclone.

The micrographs of the etched granules in Figure 9 show that these spray-drying conditions result in a variety of microstructures and pore distributions. For the majority of the granules, only a fraction of their surfaces display open pores resulting from the etching of exposed PS spheres. As discussed in Section 3.4, the non-porous fraction of the surfaces probably corresponds to areas where the PS beads near the droplet surface were still impregnated by the solution when mannitol saturation was reached during spray drying [16,50]. A few broken spheres (marked with yellow arrows in Figure 9) reveal a thick, porous layer below the surface; the center of the largest amongst these spheres is occupied by a more disordered core including some large mannitol crystallites. A minority of granules, amongst the smallest ones, display a much more homogeneous distribution of surface pores (marked with light blue arrows in Figure 9) in agreement with the similar-sized granules obtained in the previous sections. Finally, a few large hollow granules, such as the one shown in Figure 9f, display open pores on their whole surface.

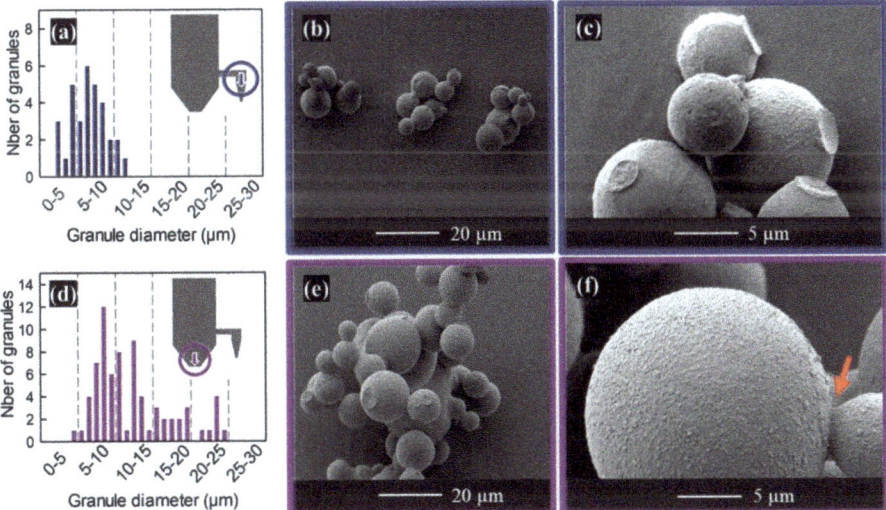

Figure 8. Histograms of granule diameters and SEM micrographs of as-sprayed granules prepared for the M:PS ratio = 1:1 composition in the Niro pilot-scale spray dryer with the rotary nozzle: (**a–c**) granules collected at the bottom of the cyclone (blue arrow on 8a); (**d–f**) granules collected at the bottom of the drying chamber (purple arrow on 8d). The red arrow highlights a neck between two granules.

The SEM micrographs in Figures 8 and 9 also reveal the presence of necks connecting the granules. The red arrow in Figure 8f points to one such neck, while several of the granules in Figure 8c,e display circular shallow craters that probably result from the breaking of such necks. Most of these observations suggest that these necks were created by the sticking together of still-wet granules and that the necks break easily to free the individual granules. However, in a few cases the degree of coalescence is more advanced, with the radius of the neck approaching that of the granules; an extreme case is that of the non-spherical granule in Figure 9e.

This microstructural inspection of the granules prepared with the rotary nozzle configuration confirms that larger droplet/granule sizes and the resultant change in surface/volume ratio require a modification of the mannitol/PS ratio to reach a good homogeneity of the distribution of the surface pores created by the etching of the exposed PS spheres. In its present state, the range of granule sizes prepared with the rotary nozzle configuration (3–25 µm) overlaps the ranges of fine (<10 µm) and coarse (20–100 µm) excipients. However, the fraction with the smaller granule diameters and narrower size distribution (i.e., the fraction collected at the bottom of the cyclone) was tested by mixing with 0.8 wt% budesonide API to check whether the severe agglomeration observed for the granules prepared in two-fluid mode would also be observed. As can be seen in Figure 9d, there was almost no agglomeration. Therefore, the budesonide content was measured by UHPLC in two aliquots taken from the formulation and values corresponding to 93.1% and 99.7% of the amount expected in the case of a perfectly homogeneous mixing were obtained. These preliminary tests are promising indications that the agglomeration problem could be solved and that the spray-dried porous granules could perform well as a porous fine excipient.

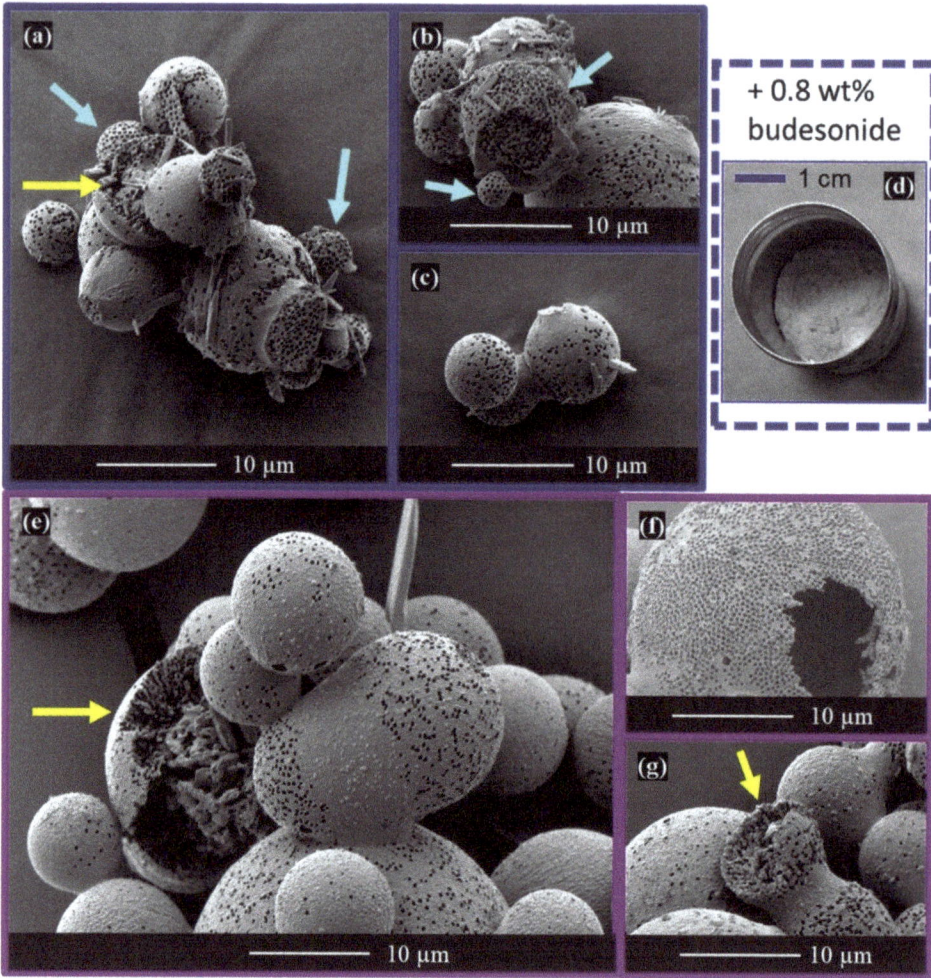

Figure 9. (a–c,e–g) The SEM micrographs of etched granules prepared for the M:PS ratio = 1:1 composition in the Niro pilot-scale spray dryer with the rotary nozzle; (d) photograph of the formulation after mixing with the budesonide API. Figures corresponding to granules collected at the bottom of the cyclone or at the bottom of the drying chamber are outlined in dark blue and dark pink, respectively. The arrows highlight examples of granule microstructures discussed in the text.

4. Conclusions

Porous mannitol granules with a diameter of about 2–3 microns were successfully produced by etching spray-dried composite granules of mannitol and polystyrene beads with ethyl acetate, an FDA approved solvent, in order to remove the polymer template. The lowest PS content (10:1 M:PS ratio) led to the encapsulation of most of the PS beads inside the mannitol matrix. Higher PS contents resulted in an open porous network after etching. In lab-scale conditions, the 1:1 M:PS ratio was found to provide the best homogeneity of porosity distribution, but the impact of other parameters (such as the droplet size) on the porosity distribution was evidenced when transferring to the pilot scale. The presence of the polystyrene beads in the spray-drying feedstock affected the polymorphism of mannitol (favoring the α polymorph) and its decomposition temperature in helium (decreasing by about 40 °C). The spray-drying configuration with a two-fluid nozzle yielded spherical porous mannitol granules with a diameter of about 2–3 microns (slightly larger for the pilot-scale system), which is in the suitable range for application as a porous fine excipient.

Supplementary Materials: The following supporting information can be downloaded at: https://www.mdpi.com/article/10.3390/ma16010025/s1, Figure S1: IR spectra before and after etching of PS beads (B-1/1 sample).

Author Contributions: Conceptualization, F.B.; Formal analysis, M.V., D.C. and B.V.; Investigation, M.V. and D.C.; Supervision, F.B.; Validation, M.V. and D.C.; Visualization, M.V., D.C. and B.V.; Writing—original draft, M.V. and D.C.; Writing—review and editing, M.V., D.C., B.V., C.M., R.C. and F.B. All authors have read and agreed to the published version of the manuscript.

Funding: This research received no external funding.

Data Availability Statement: The data presented in this study are available from the corresponding author, M.V., upon reasonable request.

Acknowledgments: M.V. and D.C. are grateful to the University of Liège for PhD grants.

Conflicts of Interest: The authors declare no conflict of interest.

References

1. Zhou, M.; Shen, L.; Lin, X.; Hong, Y.; Feng, Y. Design and Pharmaceutical Applications of Porous Particles. *RSC Adv.* **2017**, *7*, 39490–39501. [CrossRef]
2. Ahuja, G.; Pathak, K. Porous Carriers for Controlled/Modulated Drug Delivery. *Indian J. Pharm. Sci.* **2009**, *71*, 599–607. [CrossRef] [PubMed]
3. Thananukul, K.; Kaewsaneha, C.; Opaprakasit, P.; Lebaz, N.; Errachid, A.; Elaissari, A. Smart Gating Porous Particles as New Carriers for Drug Delivery. *Adv. Drug Deliv. Rev.* **2021**, *174*, 425–446. [CrossRef] [PubMed]
4. Saffari, M.; Ebrahimi, A.; Langrish, T. A Novel Formulation for Solubility and Content Uniformity Enhancement of Poorly Water-Soluble Drugs Using Highly-Porous Mannitol. *Eur. J. Pharm. Sci.* **2016**, *83*, 52–61. [CrossRef]
5. Duddu, S.P.; Sisk, S.A.; Walter, Y.H.; Tarara, T.E.; Trimble, K.R.; Clark, A.R.; Eldon, M.A.; Elton, R.C.; Pickford, M.; Hirst, P.H.; et al. Improved Lung Delivery from a Passive Dry Powder Inhaler Using an Engineered PulmoSphere? *Powder. Pharm. Res.* **2002**, *19*, 689–695. [CrossRef]
6. Geller, D.E.; Weers, J.; Heuerding, S. Development of an Inhaled Dry-Powder Formulation of Tobramycin Using Pulmosphere™ Technology. *J. Aerosol Med. Pulm. Drug Deliv.* **2011**, *24*, 175–182. [CrossRef]
7. Dellamary, L.A.; Tarara, T.E.; Smith, D.J.; Woelk, C.H.; Adractas, A.; Costello, M.L.; Gill, H.; Weers, J.G. Hollow Porous Particles in Metered Dose Inhalers. *Pharm. Res.* **2000**, *17*, 168–174. [CrossRef]
8. Xu, S.; Shi, Y.; Wen, Z.; Liu, X.; Zhu, Y.; Liu, G.; Gao, H. Applied Catalysis B: Environmental Polystyrene Spheres-Templated Mesoporous Carbonous Frameworks Implanted with Cobalt Nanoparticles for Highly Efficient Electrochemical Nitrate Reduction to Ammonia. *Appl. Catal. B Environ.* **2023**, *323*, 122192. [CrossRef]
9. Liu, X.; Verma, G.; Chen, Z.; Hu, B.; Huang, Q.; Yang, H.; Ma, S.; Wang, X.; Liu, X.; Verma, G.; et al. Metal-Organic Framework Nanocrystal-Derived Hollow Porous Materials: Synthetic Strategies and Emerging Applications Metal-Organic Framework Nanocrystal-Derived Hollow Porous Materials: Synthetic Strategies and Emerging Applications. *Innovation* **2022**, *3*, 100281. [CrossRef]
10. Chen, X.; Zhu, Z.; Vargun, E.; Li, Y.; Saha, P.; Cheng, Q. Atomic Fe on Hierarchically Ordered Porous Carbon towards High-Performance Lithium-Sulfur Batteries. *J. Electroanal. Chem.* **2023**, *928*, 117046. [CrossRef]
11. Chen, Y.; Zhou, M.; Huang, Y.; Ma, Y.; Yan, L.; Zhou, X.; Ma, X.; Zhao, X.; Chen, C.; Bai, J.; et al. Enhanced Ethanol Oxidation over Pd Nanoparticles Supported Porous Graphene-Doped MXene Using Polystyrene Particles as Sacrificial Templates. *Rare Met.* **2022**, *41*, 3170–3179. [CrossRef]
12. Shakeel, A.; Rizwan, K.; Farooq, U.; Iqbal, S.; Ali, A. Chemosphere Advanced Polymeric/Inorganic Nanohybrids: An Integrated Platform for Gas Sensing Applications. *Chemosphere* **2022**, *294*, 133772. [CrossRef] [PubMed]
13. Gurung, S.; Gucci, F.; Cairns, G.; Chianella, I.; Leighton, G.J.T. Hollow Silica Nano and Micro Spheres with Polystyrene Templating: A Mini-Review. *Materials* **2022**, *15*, 8578. [CrossRef] [PubMed]
14. Daem, N.; Mayer, A.; Spronck, G.; Colson, P.; Loicq, J.; Henrist, C.; Cloots, R.; Maho, A.; Dewalque, J. Inverse Opal Photonic Nanostructures for Enhanced Light Harvesting in CH3NH3PbI3 Perovskite Solar Cells. *Appl. Nano Mater.* **2022**, *5*, 13583–13593. [CrossRef]
15. Lv, K.; Zhang, J.; Zhao, X.; Kong, N.; Tao, J.; Zhou, J. Understanding the Effect of Pore Size on Electrochemical Capacitive Performance of MXene Foams. *Small* **2022**, *18*, 1–11. [CrossRef]
16. Alhajj, N.; O'Reilly, N.J.; Cathcart, H. Designing Enhanced Spray Dried Particles for Inhalation: A Review of the Impact of Excipients and Processing Parameters on Particle Properties. *Powder Technol.* **2021**, *384*, 313–331. [CrossRef]
17. Vehring, R. Pharmaceutical Particle Engineering via Spray Drying. *Pharm. Res.* **2008**, *25*, 999–1022. [CrossRef]

18. Nandiyanto, A.B.D.; Okuyama, K. Progress in Developing Spray-Drying Methods for the Production of Controlled Morphology Particles: From the Nanometer to Submicrometer Size Ranges. *Adv. Powder Technol.* **2011**, *22*, 1–19. [CrossRef]
19. Nandiyanto, A.B.D.; Ogi, T.; Wang, W.N.; Gradon, L.; Okuyama, K. Template-Assisted Spray-Drying Method for the Fabrication of Porous Particles with Tunable Structures. *Adv. Powder Technol.* **2019**, *30*, 2908–2924. [CrossRef]
20. Gervelas, C.; Serandour, A.L.; Geiger, S.; Grillon, G.; Fritsch, P.; Taulelle, C.; Le Gall, B.; Benech, H.; Deverre, J.R.; Fattal, E.; et al. Direct Lung Delivery of a Dry Powder Formulation of DTPA with Improved Aerosolization Properties: Effect on Lung and Systemic Decorporation of Plutonium. *J. Control Release* **2007**, *118*, 78–86. [CrossRef]
21. Straub, J.A.; Chickering, D.E.; Lovely, J.C.; Zhang, H.; Shah, B.; Waud, W.R.; Bernstein, H. Intravenous Hydrophobic Drug Delivery: A Porous Particle Formulation of Paclitaxel (AI-850). *Pharm. Res.* **2005**, *22*, 347–355. [CrossRef] [PubMed]
22. Kaplin, I.Y.; Lokteva, E.S.; Golubina, E.V.; Lunin, V.V. Template Synthesis of Porous Ceria-Based Catalysts for Environmental Application. *Molecules* **2020**, *25*, 4242. [CrossRef] [PubMed]
23. Shchukin, D.G.; Caruso, R.A. Template Synthesis and Photocatalytic Properties of Porous Metal Oxide Spheres Formed by Nanoparticle Infiltration. *Chem. Mater.* **2004**, *16*, 2287–2292. [CrossRef]
24. Zhang, H.; Hardy, G.C.; Khimyak, Y.Z.; Rosseinsky, M.J.; Cooper, A.I. Synthesis of Hierarchically Porous Silica and Metal Oxide Beads Using Emulsion-Templated Polymer Scaffolds. *Chem. Mater.* **2004**, *16*, 4245–4256. [CrossRef]
25. Gradoń, L.; Janeczko, S.; Abdullah, M.; Iskandar, F.; Okuyama, K. Self-Organization Kinetics of Mesoporous Nanostructured Particles. *AIChE J.* **2004**, *50*, 2583–2593. [CrossRef]
26. Iskandar, F.; Mikrajuddin; Okuyama, K. In Situ Production of Spherical Silica Particles Containing Self-Organized Mesopores. *Nano Lett.* **2001**, *1*, 231–234. [CrossRef]
27. Balgis, R.; Ogi, T.; Arif, A.F.; Anilkumar, G.M. Morphology Control of Hierarchical Porous Carbon Particles from Phenolic Resin and Polystyrene Latex Template via Aerosol Process. *Carbon N. Y.* **2014**, *84*, 281–289. [CrossRef]
28. Nandiyanto, A.B.D.; Hagura, N.; Iskandar, F.; Okuyama, K. Design of a Highly Ordered and Uniform Porous Structure with Multisized Pores in Film and Particle Forms Using a Template-Driven Self-Assembly Technique. *Acta Mater.* **2010**, *58*, 282–289. [CrossRef]
29. Nandiyanto, A.B.D.; Okuyama, K. Influences of Size and Amount of Colloidal Template and Droplet Diameter on the Formation of Porous-Structured Hyaluronic Acid Particles. *Indones. J. Sci. Technol.* **2017**, *2*, 152–165. [CrossRef]
30. Iskandar, F.; Nandiyanto, A.B.D.; Widiyastuti, W.; Young, L.S.; Okuyama, K.; Gradon, L. Production of Morphology-Controllable Porous Hyaluronic Acid Particles Using a Spray-Drying Method. *Acta Biomater.* **2009**, *5*, 1027–1034. [CrossRef]
31. Rahimpour, Y.; Kouhsoltani, M.; Hamishehkar, H. Alternative Carriers in Dry Powder Inhaler Formulations. *Drug Discov. Today* **2014**, *19*, 618–626. [CrossRef] [PubMed]
32. Maas, S.G.; Schaldach, G.; Littringer, E.M.; Mescher, A.; Griesser, U.J.; Braun, D.E.; Walzel, P.E.; Urbanetz, N.A. The Impact of Spray Drying Outlet Temperature on the Particle Morphology of Mannitol. *Powder Technol.* **2011**, *213*, 27–35. [CrossRef]
33. De Boeck, K.; Haarman, E.; Hull, J.; Lands, L.C.; Moeller, A.; Munck, A.; Riethmüller, J. Inhaled Dry Powder Mannitol in Children with Cystic Fi Brosis: A Randomised Ef Fi Cacy and Safety Trial. *J. Cyst. Fibros.* **2017**, *16*, 380–387. [CrossRef] [PubMed]
34. Mönckedieck, M.; Kamplade, J.; Fakner, P.; Urbanetz, N.A.; Walzel, P.; Steckel, H.; Scherließ, R. Dry Powder Inhaler Performance of Spray Dried Mannitol with Tailored Surface Morphologies as Carrier and Salbutamol Sulphate. *Int. J. Pharm.* **2017**, *524*, 351–363. [CrossRef]
35. Littringer, E.M.; Mescher, A.; Schroettner, H.; Achelis, L.; Walzel, P.; Urbanetz, N.A. Spray Dried Mannitol Carrier Particles with Tailored Surface Properties—The Influence of Carrier Surface Roughness and Shape. *Eur. J. Pharm. Biopharm.* **2012**, *82*, 194–204. [CrossRef] [PubMed]
36. Jones, M.D.; Price, R. The Influence of Fine Excipient Particles on the Performance of Carrier-Based Dry Powder Inhalation Formulations. *Pharm. Res.* **2006**, *23*, 1665–1674. [CrossRef] [PubMed]
37. Chow, M.Y.T.; Qiu, Y.; Lo, F.F.K.; Lin, H.H.S.; Chan, H.K.; Kwok, P.C.L.; Lam, J.K.W. Inhaled Powder Formulation of Naked SiRNA Using Spray Drying Technology with L-Leucine as Dispersion Enhancer. *Int. J. Pharm.* **2017**, *530*, 40–52. [CrossRef]
38. Arzi, R.S.; Sosnik, A. Electrohydrodynamic Atomization and Spray-Drying for the Production of Pure Drug Nanocrystals and Co-Crystals ☆. *Adv. Drug Deliv. Rev.* **2018**, *131*, 79–100. [CrossRef]
39. Walsh, D.; Serrano, D.R.; Marie, A.; Norris, B.A. Production of Cocrystals in an Excipient Matrix by Spray Drying. *Int. J. Pharm.* **2018**, *536*, 467–477. [CrossRef]
40. Rowe, R.C.; Sheskey, P.J.; Quinn, M.E. *Handbook of Pharmaceutical Excipients*, 6th ed.; Rowe, R.C., Sheskey, P.J., Quinn, M.E., Eds.; The Pharmaceutical Press: London, UK; The American Pharmacists Association: Washington, DC, USA, 2009.
41. Grodowska, K.; Parczewski, A. Organic Solvents in the Pharmaceutical Industry. *Acta Pol. Pharm. Drug Res.* **2010**, *67*, 3–12.
42. Dimian, A.C.; Bildea, C.S.; Kiss, A.A. *Chemical Product Design*; Elsevier Science: Amsterdam, The Netherlands, 2014; Volume 35, ISBN 9780444627001.
43. Food and Drug Administration, Center for Drug Evaluation and Research. Q3C—Tables and List Guidance for Industry Q3C—Tables and List Guidance for Industry. Available online: https://www.fda.gov/media/71737/download (accessed on 15 December 2022).
44. Food and Drug Administration, Center for Drug Evaluation and Research. Appendix 6. Toxicological Data For Class 3 Solvents. Available online: https://www.fda.gov/regulatory-information/search-fda-guidance-documents/q3c-appendix-6 (accessed on 15 December 2022).

45. Hartwig, A.M.C. Ethyl Acetate Ethyl Acetate. *MAK Collect. Occup. Heal. Saf.* **2019**, *4*, 2027–2044. [CrossRef]
46. Yohanala, P.T.F.; Mulya Dewa, R.; Quarta, K.; Widiyastuti, W.; Winardi, S. Preparation of Polystyrene Spheres Using Surfactant-Free Emulsion Polymerization. *Mod. Appl. Sci.* **2015**, *9*, 121. [CrossRef]
47. Cai, Z.; Teng, J.; Yan, Q.; Zhao, X.S. Solvent Effect on the Self-Assembly of Colloidal Microspheres via a Horizontal Deposition Method. *Colloids Surfaces A Physicochem. Eng. Asp.* **2012**, *402*, 37–44. [CrossRef]
48. Cheary, B.Y.R.W.; Coelho, A. A Fundamental Parameters Approach to X-Ray Line-Profile Fitting. *J. Appl. Crystallogr.* **1992**, *25*, 109–121. [CrossRef]
49. Fang, J.; Xuan, Y.; Li, Q. Preparation of Polystyrene Spheres in Different Particle Sizes and Assembly of the PS Colloidal Crystals. *Sci. China Technol. Sci.* **2010**, *53*, 3088–3093. [CrossRef]
50. Handscomb, C.S.; Kraft, M.; Bayly, A.E. A New Model for the Drying of Droplets Containing Suspended Solids after Shell Formation. *Chem. Eng. Sci.* **2009**, *64*, 228–246. [CrossRef]
51. Zafiryadis, F.L. Numerical Modeling of Droplet Trajectories in Pilot Plant Spray Dryer Fitted with Rotary Atomizer. Masters' Thesis, Technical University of Denmark (Department of Mechanical Engineering), Lyngby, Denmark, 2019. Available online: http://s3-eu-west-1.amazonaws.com/foreninglet-wordpress-offload-s3/wp-content/uploads/sites/53/2018/01/07112723/FrederikZafiryadis_Spray_Dryer.pdf (accessed on 5 October 2022).

Disclaimer/Publisher's Note: The statements, opinions and data contained in all publications are solely those of the individual author(s) and contributor(s) and not of MDPI and/or the editor(s). MDPI and/or the editor(s) disclaim responsibility for any injury to people or property resulting from any ideas, methods, instructions or products referred to in the content.

Article

Leonotis nepetifolia Flower Bud Extract Mediated Green Synthesis of Silver Nanoparticles, Their Characterization, and *In Vitro* Evaluation of Biological Applications

Shashiraj Kariyellappa Nagaraja [1],*, Shaik Kalimulla Niazi [2], Asmatanzeem Bepari [3], Rasha Assad Assiri [3] and Sreenivasa Nayaka [1],*

[1] P.G. Department of Studies in Botany, Karnatak University, Dharwad 580003, Karnataka, India
[2] Department of Preparatory Health Sciences, Riyadh Elm University, Riyadh 12611, Saudi Arabia
[3] Department of Basic Medical Sciences, College of Medicine, Princess Nourah bint Abdulrahman University, Riyadh 11671, Saudi Arabia
* Correspondence: rajscbz@gmail.com (S.K.N.); sreenivasankud@gmail.com (S.N.)

Abstract: Biosynthesis of silver nanoparticles (AgNPs) using the green matrix is an emerging trend and is considered green nanotechnology because it involves a simple, low-cost, and environmentally friendly process. The present research aimed to synthesize silver nanoparticles from a *Leonotis nepetifolia* (L.) R.Br. flower bud aqueous extract, characterize these nanoparticles, and perform *in vitro* determination of their biological applications. UV-Vis spectra were used to study the characterization of biosynthesized *L. nepetifolia*-flower-bud-mediated AgNPs (LnFb-AgNPs); an SPR absorption maximum at 418 nm confirmed the formation of LnFb-AgNPs. The presumed phytoconstituents subjected to reduction in the silver ions were revealed by FTIR analysis. XRD, TEM, EDS, TGA, and zeta potential with DLS analysis revealed the crystalline nature, particle size, elemental details, surface charge, thermal stability, and spherical shape, with an average size of 24.50 nm. In addition, the LnFb-AgNPs were also tested for antimicrobial activity and exhibited a moderate zone of inhibition against the selected pathogens. Concentration-dependent antioxidant activity was observed in the DPPH assay. Further, the cytotoxicity increased proportionate to the increasing concentration of the biosynthesized LnFb-AgNPs with a maximum effect at 200 μg/mL by showing the inhibition cell viability percentages and an IC_{50} of 35.84 μg/mL. Subsequently, the apoptotic/necrotic potential was determined using Annexin V/Propidium Iodide staining by the flow cytometry method. Significant early and late apoptosis cell populations were observed in response to the pancreatic ductal adenocarcinoma (PANC-1) cell line, as demonstrated by the obtained results. In conclusion, the study's findings suggest that the LnFb-AgNPs could serve as remedial agents in a wide range of biomedical applications.

Keywords: *Leonotis nepetifolia* flower buds; green synthesis; cytotoxicity; PANC-1 cell line; apoptosis; flow cytometry

Citation: Nagaraja, S.K.; Niazi, S.K.; Bepari, A.; Assiri, R.A.; Nayaka, S. *Leonotis nepetifolia* Flower Bud Extract Mediated Green Synthesis of Silver Nanoparticles, Their Characterization, and *In Vitro* Evaluation of Biological Applications. *Materials* **2022**, *15*, 8990. https://doi.org/10.3390/ma15248990

Academic Editor: Alexandru Mihai Grumezescu

Received: 9 November 2022
Accepted: 13 December 2022
Published: 16 December 2022

Publisher's Note: MDPI stays neutral with regard to jurisdictional claims in published maps and institutional affiliations.

Copyright: © 2022 by the authors. Licensee MDPI, Basel, Switzerland. This article is an open access article distributed under the terms and conditions of the Creative Commons Attribution (CC BY) license (https://creativecommons.org/licenses/by/4.0/).

1. Introduction

Nanotechnology is an important paradigm of modern science dealing with fabrication, synthesis, characterization, strategies, and manipulation of particle shapes and sizes ranging from 1 to 100 nm and exploring their multifaceted properties [1]. Nanoparticles (NPs) possess distinctive physical, optical, electric, thermal, chemical, and biological properties, unlike their bulk materials, due to the high surface area to volume ratio (small size) and quantum effects. The properties mentioned earlier are all massively correlated with their shape, size, charge, temperature, and surface coatings, among other materials [2]. The potentiality of NPs has been explored globally in various fields, and it is easily noticeable for several applications such as cosmetics, textile, and food packaging industries, drug and gene delivery, cancer therapy, nano-medicine, bio-imaging, bio-detection of pathogens, biomedical devices, and environmental protection and energy sectors [3,4].

For the synthesis of nanoparticles, diffusion, evaporation–condensation, electrolysis, sputter deposition, pyrolysis, and plasma arcing are among the most common physical techniques. However, low production rate, costly operations, and high energy consumption are the major drawbacks of these processes [5]. In chemical synthesis techniques, the uses of toxic chemicals pose risks such as carcinogenicity and environmental toxicity. Due to the use of hazardous substances such as reductants, chemical agents, and stabilizers, the toxicity issues are quite prominent [6]. The use of toxic solvents and chemical contaminations restrict the clinical and biomedical applications of nanoparticles [7].

In recent decades, physicochemical synthesis has been supplanted by a new era of green synthesis for the development of nonhazardous nanoparticles [5] to counteract the disadvantages of conventional methods. Several components, including the reduction in derivatives or pollution, the prevention or minimization of waste, and the use of a safer (or non-toxic) solvent in addition to renewable feedstock [8,9], can explain the fundamental principles of green synthesis.

Although NPs have an optimistic future in nano biomedicine, remarkable attempts are required to comprehend the intricate mechanisms underlying their biological functions and horrific or adverse effects [10,11]. Among nanoparticles, various metal and metal oxide NPs were fabricated from copper oxide, gold, palladium, selenium, silver, zinc oxide, etc., by implementing different protocols. In addition to these metal nanoparticles, silver nanoparticles (AgNPs) have played a prominent role in wound healing due to their innate therapeutical properties [12,13].

Phyto-synthesis of AgNPs can be achieved from roots, stems, leaves, fruits, flowers and flower buds, seeds, and bark. They are inexpensive compared to microbial fabrication using bacteria, fungi, algae, etc. Microbe-mediated synthesis is relatively expensive and tedious because of the heresy of cultures [14,15]. Plant-mediated synthesis is a fast, reliable-safe, low-cost, environmentally friendly, and single-step approach. Numerous bioactive secondary metabolites, such as alkaloids, tannins, saponins, terpenoids, anthraquinones, etc., are found in plant extracts, which act as reducing, stabilizing, and capping materials in the synthesis of nanoparticles [16]. The common properties of these biosynthesized AgNPs have been investigated for use in biomedical diagnostics, antimicrobials, anticancer agents, molecular sensing, and labeling of biological systems [9].

Pancreatic cancer is the fourth leading cause of cancer deaths globally, accounting for approximately 3% of all human malignant tumors [17]. According to data, the number of people diagnosed with pancreatic cancer rose from 277,000 in 2008 to 338,000 in 2012. Approximately 266,000 people died from pancreatic cancer in 2008, and this number rose to 331,000 in 2012. It is predicted that this could increase by five times in the future [18]. Pancreatic ductal adenocarcinoma (PANC-1) is the most severe and common form of pancreatic cancer, with the current objective being to alleviate disease-related symptoms and prolong survival. Currently available therapeutic options include chemotherapy, immunotherapy, surgery, radiation, and the use of targeted drugs, all of which result in significant adverse effects in cancer patients [19].

Hence, researchers have been developing alternative drugs to expand the cancer case burden worldwide; many research communities have synthesized numerous varieties of NPs through biological approach. AgNPs play a prominent role in therapeutic applications, such as through anticancer, antidiabetic, antioxidant, antimicrobial, and antiviral activities. The existence of several phytoconstituents on the surface of AgNPs derived from the green approach is allocated to their excellent antibacterial and anticancer activities [11,20]. It has been presented that plant-mediated AgNPs have been shown to have a potentially broad spectrum of cytotoxic potential and remarkable selectivity towards the tumor cells in a concentration-dependent manner against various types of carcinoma cells. AgNPs derived from the *Tussilago farfara* flower bud extract exhibited cytotoxic activity against AGS, HT-29, and PANC-1 cell lines [21], whereas AgNPs synthesized from the *Salvia miltiorrhiza* leaf aqueous extract exhibited anticancer activity against LNCaP cell lines [22]. *Zingiber officinale* extract bio-fabricated AgNPs showed cytotoxic potential against PANC-1, AsPC-1, and MIA PaCa-2 cell lines [23]. The flower extract of *Abelmoschus esculentus*-mediated AgNPs revealed cytotoxic activity against A-549 and TERT-4 cell lines [24]. AgNPs derived from the *Datura inoxia* flower extract had a potent anti-proliferative effect on the MCF-7 cells [25].

The genus *Leonotis* has 14 species widely distributed in tropical regions and is represented by 1 species in India, *Leonotis nepetifolia* (L.) R.Br. It is an herbaceous plant or subshrub that belongs to the family Lamiaceae. Its leaves and roots are used to treat fever, cough, skin infection, stomachache, rheumatism, bronchial asthma, and kidney dysfunction in various Indian traditional systems of medicine such as Ayurveda, unani, and siddha, and are used for the same purposes in various countries such as Brazil, Canada, Madagascar, and many African nations [26]. Some compounds isolated from this herb exhibited biological activities, including antibacterial, antifungal, antidiabetic, and anti-proliferative properties [27,28]. The use of the plant's leaves and roots is well-documented, whereas the flower and flower buds have not been extensively studied.

In the current research work, we conceptualized to synthesize silver nanoparticles from the *L. nepetifolia* flower bud extract and evaluate the characterizations, antioxidant efficacy, antimicrobial potential, and *in vitro* cytotoxicity on a pancreatic ductal adenocarcinoma (PANC-1) cell line of these nanoparticles, as well as apoptotic studies through flow cytometry.

2. Materials and Methods

2.1. Collection of Plant Material and Chemicals

Leonotis nepetifolia (L.) R.Br. flower buds were collected from September to December from the surroundings of Karnatak University campus, Dharwad (15°43′59.6″ N 74°98′26.8″ E), and comparing the specimens to the herbarium collection of animals and plants at Karnatak University's departmental herbarium museum in Dharwad, Karnataka, India, confirmed the specimens' authenticity. The chemicals and reagents, silver nitrate ($AgNO_3$), dimethyl sulfoxide (DMSO), 2,2-diphenyl-1-picrylhydrazyl (DPPH), α-naphthol, methanol, ascorbic acid (Vitamin C), nutrient broth, agar, standard antibiotics (streptomycin and nystatin), 3-(4,5-dimethylthiazol-2-yl)-2,5-diphenyltetrazolium bromide (MTT), Dulbecco's modified eagle's medium (DMEM), fetal bovine serum (FBS), D-PBS buffer, and FITC Annexin V/Propidium Iodide (PI), were of AR grade and procured from SD-fine chemicals and Hi-media Pvt. Ltd. (Mumbai, India). Pathogenic strains *S. aureus* (MTCC 6908), *B. subtilis* (MTCC 6633), *E. coli* (MTCC 40), *P. aeruginosa* (MTCC 9027), *C. glabrata* (MTCC 3019), and *C. albicans* (MTCC 227) were procured from IMTECH, Chandigarh, India. The pancreatic ductal adenocarcinoma (PANC-1) cell line was obtained from NCCS, Pune, India.

2.2. Preparation of Flower Bud Extract

The healthy unopened flower buds (Figure 1a,b) were collected and washed thoroughly with tap water, later with demineralized water, and dried in the shade. Then, they were powdered using an electric blender and stored in sterile polythene bags. Finely powdered, approximately weighed (10 g), and minced with 100 mL of Milli-Q water, the suspension was boiled for 1 h at 60 °C, cooled down, and filtered using Whatman No. 1 filter paper. The obtained filtrate was kept at 4 °C for further investigations [29].

2.3. Biosynthesis of AgNPs

For biosynthesis of AgNPs, 1 mM silver nitrate ($AgNO_3$) was prepared by dissolving 0.169 g in 1000 mL of deionized water. The obtained flower bud extract and $AgNO_3$ solution was mixed at a ratio of 1:4 (v/v) in a 1000 mL Erlenmeyer flask for synthesis optimization (the pH of which was set to 9). The final solution was incubated for 24 h at room temperature in a dark chamber. The reduction process gradually transformed the orange solution into a dark brownish color. Thus, the change in color affirmed the accomplishment of AgNPs. Further, LnFb-AgNPs were obtained by centrifuging the biosynthesized AgNPs at 10,000 rpm for about 20 min. The collected LnFb-AgNPs were then oven-dried and stored for later experimental analysis [30]. The biosynthesis process is schematically represented in Figure 2.

Figure 1. (a) Plant of *L. nepetifolia* with flowering twig, and (b) shade-dried flower buds.

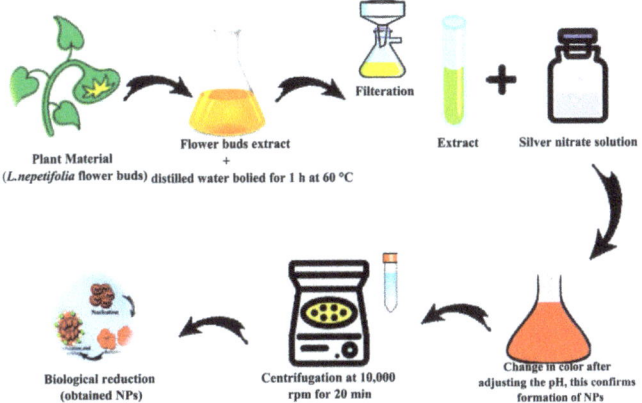

Figure 2. Schematic representation of biosynthesis of LnFb-AgNPs.

2.4. Characterizations of Biosynthesized LnFb-AgNPs

Following the synthesis and refinement of LnFb-AgNPs, their size, shape, and morphological characteristics were confirmed using several experimental techniques.

2.4.1. UV-Visible Spectrophotometric Analysis

The UV-visible spectra of biosynthesized nanoparticles were obtained by configuring a double-beam UV-visible spectrophotometer (METASH UV-9600A, Shanghai, China) with a 1 nm resolution, placing the test solution in a quartz cuvette, and analyzing its optical density at wavelengths between 300 and 600 nm. The graph was constructed by plotting wavelength (X-axis) versus absorbance (Y-axis) [31].

2.4.2. Fourier Transform Infrared (FT-IR) Spectroscopy Analysis

To investigate the bioactive compounds on the synthesized LnFb-AgNPs and the *Leonotis nepetifolia* flower bud extract, Fourier transform infrared (FTIR) spectroscopy analysis was performed. Using an FTIR spectrophotometer (NICOLET 6700, Thermo Fisher Scientific, Waltham, MA, USA), the FTIR spectra of *L. nepetifolia* flower bud extract and LnFb-AgNPs in KBr pellets were measured between 400 and 4000 cm^{-1}. Briefly, the *L. nepetifolia* flower bud extract and dried powder form of biosynthesized LnFb-AgNPs from the flower bud extract were combined with potassium bromide (Kbr) to produce a pellet, which was then examined for the presence of IR spectral bands. To decipher the functional groups present in the sample, spectral data between 400 and 4000 cm^{-1} in resolution were collected [31].

2.4.3. X-ray Diffraction (XRD) Analysis

The dried powder form of biosynthesized LnFb-AgNPs was analyzed for XRD to determine the grain size and crystallinity by placing the AgNPs in a sample holder and then placing that in an X-ray diffractometer (Rigaku Miniflex 600, Smart-Lab SE, Tokyo, Japan) to record the spectral patterns by employing a current of 30 mA with Cu Kα radiation with an angle of 2θ ranging from 30° to 90° operating at 40 kV [32].

2.4.4. Scanning Electron Microscopy (SEM) Analysis and Energy Dispersive X-ray Spectroscopy (EDS) Analysis

Using scanning electron microscopy (SEM) coupled with an energy dispersive X-ray (EDS) instrument, the topology and elemental compositions of LnFb-AgNPs were determined using (JEOL, JSM IT 500LA, Peabody, MA, USA). Briefly, LnFb-AgNPs were placed on the stub using carbon tape, then fixed and covered with gold using sputtering, and the loaded stub was placed in the instrument chamber for analysis [30].

2.4.5. Transmission Electron Microscopy (TEM) Analysis

To determine the size and structure of LnFb-AgNPs, images from a 300 kV transmission electron microscope (TEM) (FEI, TECNAI G2, F30, Beijing, China) were acquired. Before analysis, approximately 5 µL of LnFb-AgNPs were deposited on the TEM copper grid, followed by the application of a carbon tape coating and drying in desiccation for 48 h [31].

2.4.6. Zeta Potential Analysis and Dynamic Light Scattering (DLS) Analysis

In order to determine the surface charge and stability in a solution, LnFb-AgNPs were centrifuged for 20 min at 8000 rpm to collect the supernatant, then diluted with MilliQ water and ultrasonified for 15 min. Subsequently, the solution was analyzed with a zeta analyzer. Using nano-analyzer equipment (Horiba scientific nanoparticle analyzer SZ-100, Kyoto, Japan), a DLS analysis was performed on the supernatant solution to determine the dispersal pattern, size, and surface charge of silver nanoparticles [30,31].

2.4.7. Thermo Gravimetric (TGA) Analysis

To investigate the thermal behavior and nature of biosynthesized LnFb-AgNPs, thermo gravimetric analysis (TGA) was performed; a known quantity of weighed sample was placed in a furnace, and the heating temperature was increased gradually by passing inert gas over the sample. TGA analysis was conducted with an increasing heat rate of 10 °C per min from room temperature (RT-27 °C) to 600 °C using a TA instrument (SDT Q 600, New Castle, DE, USA) [32,33].

2.5. Antioxidant Activity of Biosynthesized LnFb-AgNPs

With some minor modifications described by Singh et al. (2021), we used the DPPH free radical assay to evaluate the antioxidant activity of the *L. nepetifolia* flower bud extract and biosynthesized LnFb-AgNPs [34]. In brief, samples of the same volume measured (0.2 mL) at different concentrations (100 to 1000 µg/mL) were added to 2 mL of DPPH solution (0.003% DPPH prepared in methanol), with ascorbic acid serving as a standard. After that, the tubes were stored in the dark for 30 min at room temperature. After incubation, a UV-Visible spectrophotometer was used to measure the absorbance of the solutions at 517 nm (METASH UV-9600A, Shanghai, China). A decrease in absorbance indicated a reduction in DPPH free radicals in the solution, calculated by the following formulae.

$$\text{Scavenging activity } (\%) = \frac{A_0 - A_1}{A_0} \times 100$$

where A_0 = absorbance of DPPH and A_1 = absorbance of the experimental sample.

2.6. Antimicrobial Activity of Biosynthesized LnFb-AgNPs

Using the agar well diffusion method, the antimicrobial activity of biosynthesized LnFb-AgNPs was evaluated against selected pathogenic bacteria and fungi. Four bacterial strains were selected: two Gram-negative, *E. coli* (MTCC 40) and *P. aeruginosa* (MTCC 9027), two Gram-positive, *S. aureus* (MTCC 6908) and *B. subtilis* (MTCC 6633), and two fungal strains, *C. albicans* (MTCC 227) and *C. glabrata* (MTCC 3019). Pure cultures of pathogens were sub-cultured on their respective agar medium. The agar plates were swabbed with a bacteria and fungi suspension using cotton swabs. A gel-hole punch was used to bore 6 mm wells on 4 mm thick agar plates (pH was set to 7.4). Subsequently, four different concentrations (25 to 100 µg/µL) of LnFb-AgNPs were laden inside the wells. As positive controls, streptomycin and nystatin were used, and then culture plates were incubated at 37 °C overnight. In accordance with the standard antibiotic zone of inhibition chart, the diameter of a zone of inhibition (around the wells, mm) was measured [35].

2.7. In Vitro Anticancer Activity of Biosynthesized LnFb-AgNPs

Pancreatic ductal adenocarcinoma (PANC-1, cell line obtained from NCCS, Pune, India) cells were cultured for 24 h at 37 °C, 95% humidity, and 5% CO_2 atmospheric condition to promote cell proliferation on DMEM supplemented with 10% FBS. After incubation, the cells were seeded at a density of 20,000 cells/well in 200 µL of medium in a 96-well plate. In contrast, biosynthesized LnFb-AgNPs (12.5–200 µg/mL) were added to wells with PANC-1 cells and incubated at 37 °C for 24 h. The assay included a positive control with doxorubicin (4 µM/mL), and cells lacking LnFb-AgNPs were used as a negative control. Then, MTT solution was prepared in a growth medium in order to determine the viability of the cells. A freshly prepared MTT solution (200 µL, 0.5 mg/mL) was added to each well containing cell culture and incubated at 37 °C (4 to 5 h). Following post-incubation, formazan crystals were minced in 100 µL of DMSO, and viable cells were measured at 570 nm using a microplate reader (ELX-800, BioTek, Winooski, VT, USA) [36,37]. Final results were expressed as an IC_{50} value.

2.8. Annexin V/PI FITC Assay for Apoptotic Analysis

Analyses of apoptotic/necrosis cells were evaluated by staining with Annexin V-FITC/PI assay (Annexin V-FITC apoptosis kit, BD Biosciences, Franklin Lakes, NJ, USA) and the results were determined according to the manufacturer's protocol. PANC-1 (pancreatic ductal adenocarcinoma) cells were laden in a 6-well plate (at a density of 0.5×10^6 cells/2 mL) and incubated overnight at 37 °C in a CO_2 incubator. The cells were then treated with the IC_{50} concentration of LnFb-AgNPs (35.84 µg/mL) following incubation. The cells were accumulated by EDTA-trypsinization, washed twice with PBS solution, and then 5 µL of Annexin V binding buffer was added. The cells were then vortexed gently and incubated for 15 min at room temperature (27 °C) in the dark. Later, 5 µL of Propidium Iodide and 400 µL of 1X binding buffer were added to each tube and delicately vortexed. In addition, a flow cytometer (BD FACS Calibur) was used to analyze the test samples in accordance with the method developed by O'Brien and Bolton (1995). The software BD Cell Quest Pro Ver.6.0 was used to calculate the results of the experiment [38].

2.9. Statistical Analysis

All analyses were carried out in three sets, and the data were presented as mean standard deviation (SD). SPSS v17 and Origin 2022b software were used to perform the statistical analysis.

3. Results and Discussion

3.1. Phyto-Synthesis of LnFb-AgNPs and Their Characterizations

In the current study, we synthesized the AgNPs from an aqueous extract of *L. nepetifolia* flower buds. A change in extract color confirmed the synthesis following the addition of $AgNO_3$ (1 mM, 1:4); after 24 h of incubation at RT (27 °C), at pH 9.0, the suspension turned from orange to dark brown (Figure 3a–c). In the range of 300 to 600 nm, the characteristic SPR absorption spectrum of the synthesized LnFb-AgNPs was observed at 418 nm (Figure 3d), confirming the synthesis of LnFb-AgNPs. The synthesis was affirmed by the transformation of the solution's color from orange to a dark brownish color, a phenomenon attributed to SPR. The absorption spectrum peaks of the biosynthesized LnFb-AgNPs occurred at 418 nm with a high absorbance value, which is characteristic of silver nanoparticles. In general, distinctive AgNPs show characteristic surface plasmon resonance at wavelengths ranging from 400 to 450 nm; the same was also observed in the current investigation. The spectral absorption peaks rise as extract concentration and time increase. Therefore, flower bud extract concentrations provide the time-dependent optimal amount of phytoconstituents required to reduce silver ions (Ag^+) into AgNPs [39,40].

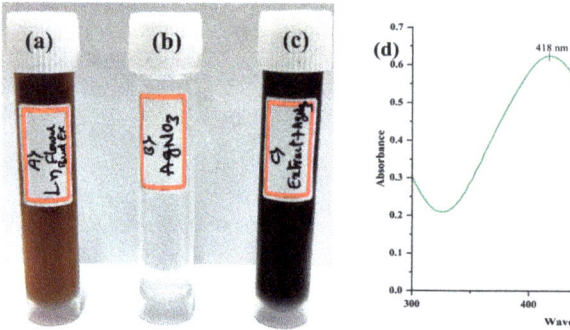

Figure 3. Change in color of the solution: (**a**) flower bud extract; (**b**) $AgNO_3$ solution; (**c**) flower bud extract + $AgNO_3$; (**d**) UV-Vis absorption spectrum of biosynthesized LnFb-AgNPs from *L. nepetifolia* flower bud extract.

The pH value always plays a crucial role in a reaction mixture. The color of the reaction mixture, the intensity of the SPR peak, and the size and shape of the nanoparticles were found to be pH-dependent. The absorbance peak of the LnFb-AgNPs increases with the pH of the suspension, with maximum production occurring at pH 9 [41]. Similar outcomes were reported by Rajesh et al. (2016), where *Couroupita guianensis* Aubl. flower bud extract mediated AgNPs showed a change in color from yellowish to brown at 420 nm [42]. Pereira et al. (2020) reported that silver nanoparticles synthesized using different parts of *Handroanthus heptaphyllus* displayed comparable SPR peaks at 450 nm at pH 8–10 [43].

3.2. FTIR Spectroscopy Analysis

Diversified secondary metabolites stabilize silver nanoparticles synthesized through biomimetic methods via a molecular level of interaction with metallic surfaces. The nature of the functional groups' interplay between synthesized materials was studied using FTIR analysis. The FTIR spectra of the *L. nepetifolia* flower bud extract exhibited peaks at 533, 574, 616, 790, 840, 1110, 1245, 1330, 1384, 1515, 1608, 2930, and 3367 cm^{-1}, whereas the LnFb-AgNPs showed major peaks at 536, 617, 780, 826, 110, 1384, 1615, 2926, and 3307 cm^{-1}. As a result of bio-reduction, the FTIR spectra of the extract and the LnFb-AgNPs exhibited minute shifts in peak positions. As shown, the spectrum of the *L. nepetifolia* flower bud extract differed little from the spectrum of the LnFb-AgNPs (Figure 4). The results obtained in the flower bud extract exhibited a peak at 574 cm^{-1}, a weak peak for a (C-Br) stretching halo compound, and 840 cm^{-1} for (C-C) bending alkene. In the FTIR spectrum of the LnFb-AgNPs, a medium peak for (C-O) aliphatic ether was shifted to lower wave numbers at 536 cm^{-1} (C-Br), 826 cm^{-1} (C-Cl), and 1100 cm^{-1} (C-O), respectively. In contrast to the normal weak peak at 1515 cm^{-1} for the (N-O) stretching nitro compound and the broad peak at 2930 cm^{-1} for (C-H) stretching alkane, these peaks have shifted to higher wave numbers, i.e., 1615 cm^{-1} (C=C) and 2976 cm^{-1} (O-H). In the FTIR spectrum of the LnFb-AgNPs, the peak at 1245 cm^{-1} for (C-N) stretching of amine disappeared. The analysis of the FTIR spectra of the flower bud extract and LnFb-AgNPs revealed the presence of various functional groups, including phenolic/hydroxyl, alkenes, alkanes, and amines, which are involved in the interactions between bio-molecules and metal particles. Due to its electron-donating property, the flower bud extract contains several phytoconstituents that may facilitate the reduction and stabilization of silver ions (Ag$^+$) to (Ag^0) and the formation of AgNPs [44,45]. In a previous study, the mechanism of adsorption and capping of silver nanoparticles by plant extracts can be explained by the interactions and coordination of different carbonyl bonds and adjacent electron transfer to AgNPs [46]. Similarly, Algebaly et al. (2020) reported that the phenolic compounds present in the aqueous extracts act as a reducing, capping, and stabilizing agent to convert silver nitrate into AgNPs [47].

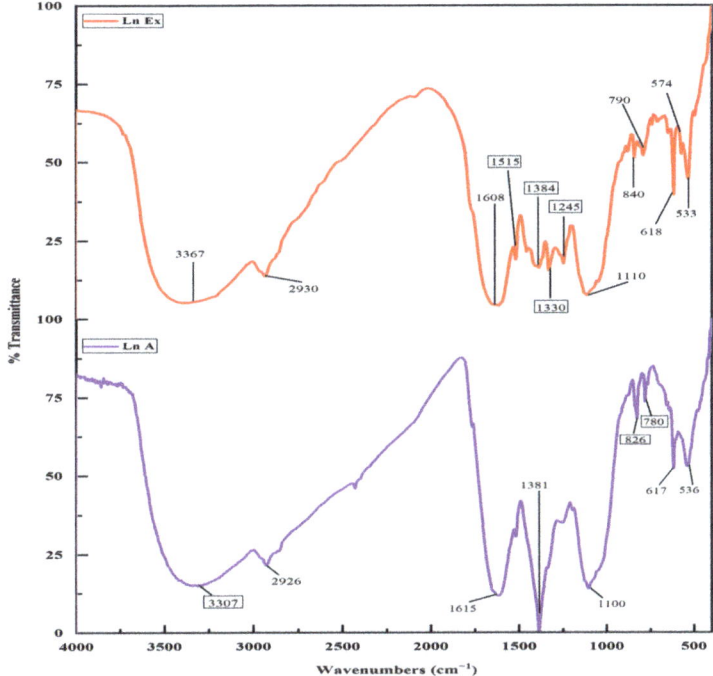

Figure 4. FTIR spectrum of flower bud extract and biosynthesized LnFb-AgNPs.

3.3. XRD Analysis

The XRD patterns of the biosynthesized LnFb-AgNPs are displayed in Figure 5. Bragg reflection peaks were found in all patterns of synthesized silver nanoparticles, located at 2θ values of 38.10°, 44.22°, 64.44°, and 77.37°, which correspond to the (111), (200), (220), and (311) crystallographic planes of Ag's face-centered cubic (fcc) structure of AgNPs. The obtained results were confirmed after matching with the standard silver card JCPDS reference code 04-0783. The crystalline nature of the biosynthesized LnFb-AgNPs was confirmed by the sharp diffraction peaks. The observed peaks in the XRD analysis clearly indicated that the silver ions (Ag$^+$) had completely reduced to (Ag^0) by stabilizing and reducing compounds in the aqueous extracts. The XRD analysis also exhibited some extra peaks that were unassigned. The presence of bioorganic phases on the particle surfaces accounted for the additional unassigned peaks [48]. This result can be compared to the findings of Ajitha et al. (2019), in which AgNPs synthesized using *Syzygium aromaticum* (clove) extract displayed the same patterns as AgNPs produced under optimal conditions. The AgNPs produced four dominant diffraction peaks positioned at 2θ, 38.0°, 44.1°, 64.4°, and 77.4°, which were correlated with the lattice planes of the face-centered cubic (fcc) structure and crystalline nature of metallic silver [49].

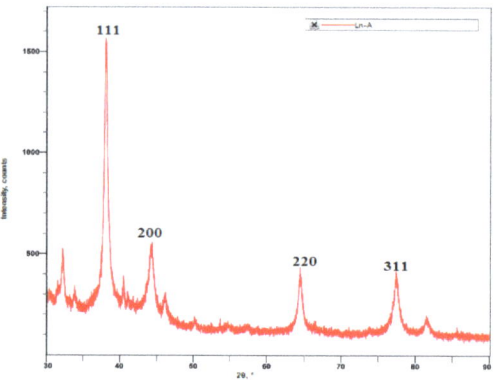

Figure 5. XRD spectra of biosynthesized LnFb-AgNPs from *L. nepetifolia* flower bud extract.

3.4. SEM and EDS Analysis

The LnFb-AgNPs exhibited a granular morphology, as shown in Figure 6a, which was captured by SEM. The image revealed that the NPs were polydispersed and spherical, with minor agglomeration. An EDS analysis at 5 keV was employed to assess the purity and different elemental components. In Figure 6b, the EDS spectrum exhibited an elemental peak signal from silver metal at 3.09 keV and the presence of 35.34% of a sliver; typically, metallic AgNPs display an optical absorption peak at approximately 3 keV due to their SPR. Other than silver metal, a few peaks are attributed to the presence of conjugated bio-molecules over the surface AgNPs or chlorine on the glass slide used while preparing the samples for analysis. The biosynthesis of AgNPs using the extract from the aerial parts of plant formed a product with many dimensions; due to the variable values of NPs, they are distinct, with considerable alteration due to the optical and electronic properties of metallic NPs [25,50]. Similar results were obtained by Allafchian et al. (2018), who explained that the synthesis of NPs using a *Glaucium corniculatum* (L.) curtis extract yielded spherical, polydispersed, and agglomeration-free nanoparticles with a single peak at 3 keV, confirming the presence of silver metal [51].

3.5. TEM Analysis

The TEM analysis revealed the NPs' size, shape, texture, surface morphology, and distribution. Figure 7a,b are TEM micrographs of the LnFb-AgNPs, demonstrating the formation of polydisperse, spherical AgNPs with minor aggregation. The size of the LnFb-AgNPs, as depicted by the TEM image, ranged from 10–45 nm; the TEM micrographs disclosed the average particle size to be 24.50 nm. In the focusing zone, the SAED pattern (Figure 7c) revealed the crystalline nature and distribution of LnFb-AgNPs. A crystalline analysis of LnFb-AgNPs at 5 1/nm resolution revealed crystal lattice fringes with a d-spacing value of 0.207 nm. However, LnFb-AgNPs were evenly spread over the surface with minor agglomeration. This is in accordance with the result of Rajesh et al. (2016), who determined that *Couroupita guianensis* Aubl. flower bud extract mediated AgNPs had sizes of synthesized NPs that ranged from 5 to 40 nm, with an average of approximately

17 nm [42]. Similarly, the NPs synthesized using a clove bud extract had a polydispersed nature and nanoparticle sizes ranging from 10 to 50 nm without agglomeration [52].

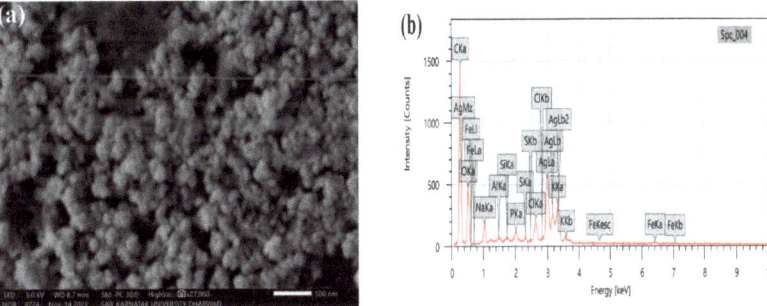

Figure 6. (a) SEM micrograph and (b) EDS spectrum of biosynthesized LnFb-AgNPs from *L. nepetifolia* flower bud extract.

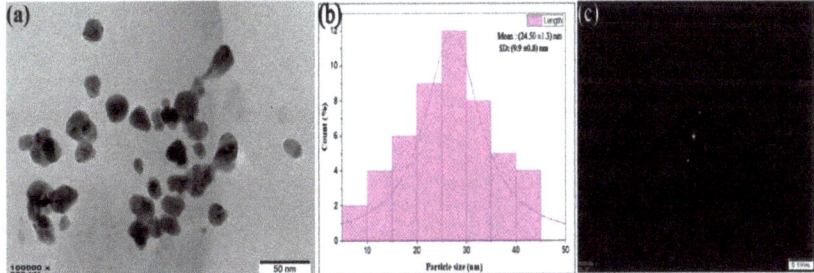

Figure 7. (a) TEM image, (b) histogram showing particle size distribution, and (c) SAED pattern image of biosynthesized LnFb-AgNPs from *L. nepetifolia* flower bud extract.

3.6. Zeta Potential and DLS Analysis

The zeta potential of the prepared LnFb-AgNPs was dispersed in an aqueous colloidal solution at ambient temperature and exhibited a negative value of −31.4 mV (Figure 8a), indicating a high negative surface charge. The stability of the LnFb-AgNPs is a result of their relative surface charges, which prevents agglomeration. Therefore, the LnFb-AgNPs in the synthesized colloidal medium are exceptionally stable and evenly dispersed. The negative zeta potential value may be attributable to the capping of phyto-organic components present in the *L. nepetifolia* flower bud extract, as well as the electrostatic stabilization of LnFb-AgNPs in the colloidal medium. The negative values indicate that the NPs are stable. In a DLS analysis, nanoparticle hydrodynamic size is measured by scattered light from the nanocore and phytomolecule cloud as a function of time. While TEM measures the diameter of individual particles, consequently, the nanoparticle size estimated by DLS analysis is slightly larger than the actual size determined by TEM. The average size of LnFb-AgNPs is 98.5 nm and depicted in Figure 8b, which is significantly bigger than the TEM findings. Similarly, El-Aswar et al. (2019) reported that the zeta potential of the NPs synthesized from a *Haplophyllum tuberculatum* extract was −42 mV, and the nanoparticle hydrodynamic size measured by scattered light was 86.3 nm. Zeta potential values greater than −30 mV or greater than +30 mV are typically considered stable; this is due to the negatively charged electrostatic repulsive forces, which likely create a greater energy barrier to preserve the silver nanoparticles in the colloidal solution without coagulation [53]. The greater negative surface charge value, according to Ardestani et al. (2016), is due to the constructive functional bioactive phytoconstituents as a capping agent in the plant extract [54].

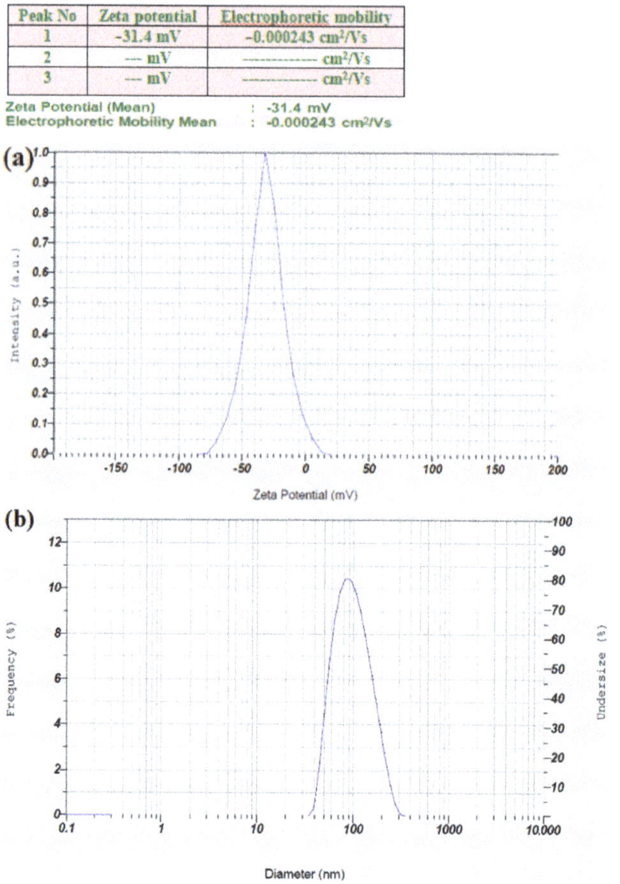

Peak No	Zeta potential	Electrophoretic mobility
1	-31.4 mV	-0.000243 cm²/Vs
2	--- mV	------------ cm²/Vs
3	--- mV	------------ cm²/Vs

Zeta Potential (Mean) : -31.4 mV
Electrophoretic Mobility Mean : -0.000243 cm²/Vs

Figure 8. (a) Zeta potential analysis graph and (b) dynamic light scattering analysis of biosynthesized LnFb-AgNPs from *L. nepetifolia* flower bud extract.

3.7. TGA Analysis

Using thermo gravimetric analysis, the thermal behavior and stability of biosynthesized LnFb-AgNPs were evaluated. The TGA curve (Figure 9) reveals that the biosynthesized LnFb-AgNPs were extremely steady and stable at temperatures ranging from 27 °C to 600 °C, with minimal weight loss. Between 43 °C and 208 °C, 209 °C and 307 °C, and 308 °C and 448 °C, the LnFb-AgNPs exhibited three significant weight losses of 4.72%, 22.61%, and 7.59%, for a total weight loss of approximately 35%. The initial weight loss was caused by the evaporation of the AgNPs' moisture content. Similarly, the desorption of organic bioactive phytochemicals, which act as conjugated molecules on the surface of silver nanoparticles, was primarily responsible for the two subsequent degradations. It indicates that bioactive phytoconstituents present in nanoparticles are accountable for their reduction and stabilization [55]. This result can be compared to the findings of Moteriya and Chanda (2017), who discovered that the *Caesalpinia pulcherrima*-flower-extract-mediated NPs exhibited a constant weight loss in temperatures from 0–800 °C, and the total weight loss up to 800 °C for the synthesized AgNPs is approximately 71.68% [56].

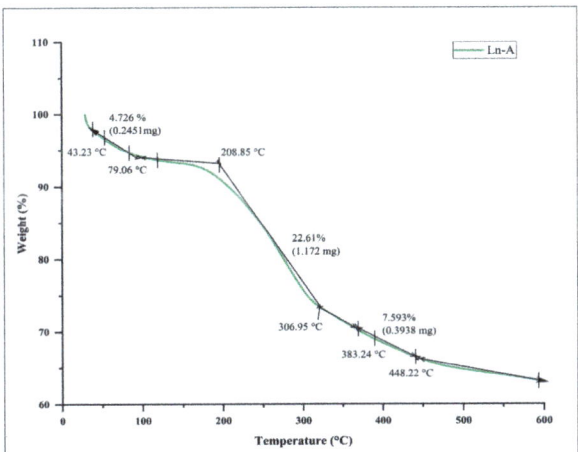

Figure 9. Thermo gravimetric analysis curve of biosynthesized LnFb-AgNPs from *L. nepetifolia* flower bud extract.

3.8. Antioxidant Activity

The antioxidant activity of the flower bud aqueous extract and biosynthesized LnFb-AgNPs was evaluated using the DPPH free radical scavenging assay. DPPH is a stable compound and conventional free radical that can be reduced by accepting hydrogen or an electron from ions that donate them. The antioxidant-reducing potentiality of the biosynthesized AgNPs was determined by visualizing the change in color formation. The DPPH assay demonstrated that the LnFb-AgNPs inhibited oxidative stress more effectively than flower buds; the aqueous extract is depicted in (Figure 10) and ascorbic acid served as the standard. The scavenging activity of the LnFb-AgNPs was found to increase with increasing concentration. The LnFb-AgNPs were shown in serial concentrations of 100, 200, 400, 600, 800, and 1000 µg/mL, with inhibition percentages of 42.3 ± 2.02%, 51.3 ± 1.04%, 59.19 ± 1.95%, 67.9 ± 1.01%, 75 ± 1%, and 81 ± 1.12%, respectively. The antioxidant activity of biosynthesized silver nanoparticles could be associated with the plant-derived functional groups attached to them. The DPPH free radical scavenging activity of the LnFb-AgNPs was found to be dose-dependent. To investigate the scavenging activity of the flower bud extract, the ability of DPPH to readily accept a H^+ or e^- from an antioxidant moiety under stable conditions was analyzed. Compared to the flower bud extract, the LnFb-AgNPs exhibited moderate reducing power. This activity is due to the presence of phenolic compounds from the extract as capping agents and stabilizers on the surface of the LnFb-AgNPs. The antioxidant capacity demonstrates the ability of silver nanoparticles to transfer electrons and neutralize reactive DPPH radicals in the reaction solution [57,58]. The results of our AgNPs derived from the *L. nepetifolia* flower bud extract demonstrate their significant antioxidant potential when compared to previously published data where AgNPs where synthesized from aqueous corn leaves. When the concentration of the AgNPs was increased, the extract showed a dose-dependent reduction in free radicals that was proportional to the dose. At 100 µg/mL, it displayed a moderate DPPH radical scavenging activity of 36.3%, while at 1000 µg/mL, it displayed 89.01% [59].

3.9. Antimicrobial Activity

The antimicrobial efficacy of the biosynthesized LnFb-AgNPs was evaluated using the agar well diffusion method; the zone of inhibition was observed as a distinct circular zone on microbial culture plates. The LnFb-AgNPs were treated against pathogenic organisms such as Gram-positive bacteria, Gram-negative bacteria, and fungi. Our results displayed that the AgNPs capped with the *L. nepetifolia* flower bud extract exhibited maximum restriction in the growth of *P. aeruginosa*, *S. aureus*, and *C. glabrata*, with zones of inhibition measuring 23 ± 0.8 mm, 25 ± 0.2 mm, and 23 ± 1 mm, respectively, while *E. coli*, *B. subtilis*, and *C. albicans* exhibited minimal sensitivity with zones of inhibition measuring 20 ± 0.7 mm, 20.5 ± 0.5 mm, and 21.5 ± 0.7 mm when treated with a 100 μL concentration of LnFb-AgNPs. The LnFb-AgNPs treated against different pathogens are depicted in Figure 11 a–f and graphical representation for the same is displayed in Figure 11g. Global biomedical systems have recently been impacted by the emergence of pathogens with multidrug resistance. Therefore, the promising effects of silver nanoparticles on numerous pathogenic microorganisms could aid in the development of new antibacterial agents to combat pathogenic microorganisms [60]. The precise cause of AgNPs' antibacterial sensitivity mechanism on pathogenic microorganisms is relatively well understood. Few studies have explained how AgNPs become attached to the surface of bacterial cell membranes by forming bonds with sulfur–phosphorous-containing compounds, thereby altering and interfering with the cell's vital functions, including permeability and respiration. Thus, this leads to the degradation of enzymes, inactivation of cellular proteins, and damage of DNA. Finally, cell death occurs [35,47,61]. The smaller-sized NPs exhibited greater sensitivity against *E. coli*, *B. subtilis*, *K. pneumonia*, and *P. aeruginosa*, according to Singh (2014), who reported that dose-size-dependent AgNPs could be more effective against multidrug-resistant bacteria [62]. In a previous study, Lee et al. (2019) observed that AgNPs fabricated from a *Tussilago farfara* flower bud extract inhibited the growth of tested microbial strains in a significant manner [21].

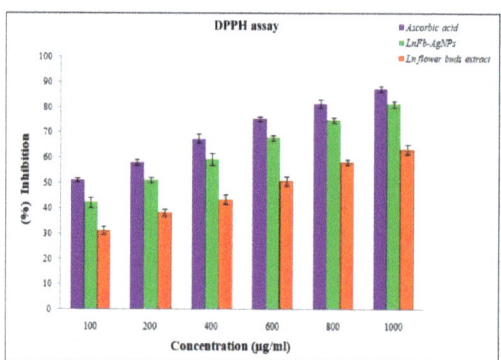

Figure 10. Graph showing antioxidant activity of biosynthesized LnFb-AgNPs from *L. nepetifolia* flower bud extract.

3.10. In Vitro Anticancer Activity

Biosynthesized LnFb-AgNPs were tested for their in vitro cytotoxicity against PANC-1 (pancreatic ductal adenocarcinoma) cells using the MTT assay. In this technique, the yellow-colored dye MTT (3-(4,5 dimethyithiazol2-yl)-2,5-diphenyltetrazolium bromide) is diminished by the mitochondrial enzyme succinate dehydrogenase, resulting in the formation of formazan (bluish-purple-colored) crystals. Since the assay is a colorimetric assessment, the results were recorded in absorbance at 570 nm. After 24 h of incubation with LnFb-AgNPs, there was a dose-dependent decrease in the relative cell viability (%) of PANC-1 cells. As the concentration of LnFb-AgNPs increased from 12.5 μg/mL to 200 μg/mL, cell viability percentages decreased proportionally. Figure 12a–g illustrate the morphological changes of cells treated with LnFb-AgNPs and positive and negative controls. Post incubation, the viability of cancer cells was 72.01%, 64.41%, 46.23%, and 21.40% at 12.5, 25, 50, and 100 μg/mL of LnFb-AgNPs, respectively, and it was reduced to a cell viability of 1.95% when treated with 200 μg/mL of AgNPs. Using the dose-dependent curve (Figure 12h), the IC_{50} value was calculated as 35.84 μg/mL. Kanniah et al. (2021) reported a similar finding, stating that green-synthesized AgNPs inhibit the viability of the PANC-1 cell line at concentrations ranging from 10 to 200 μg/mL [63]. Comparing our findings to those of Wang et al. (2021), the AgNPs synthesized from a *Zingiber officinale* leaf aqueous extract exhibited an IC$_{50}$ of 295 μg/mL [23]. There are numerous reports on the anti-proliferative potential of AgNPs against various types of cancer cells. The potential mechanism underlying the cytotoxic potential of AgNPs against cancer cell lines is depicted in Figure 13. It is suggested that the cytotoxic potential of AgNPs is mostly due to oxidative stress and apoptosis via a caspase-dependent pathway, resulting in DNA damage and mitochondrial dysfunction and ultimately cell death [54,56,64]. In another finding reported by Shameli et al. (2021), it was revealed that the anticancer activity of a *D. regia* extract and AgNPs intensified with increasing extract doses and time. In comparison to cancerous cells, the *D. regia* extract and AgNPs had minimal effects on normal cells and did not inhibit normal cells. The differential sensitivity of MCF-7 and Panc-1 cancer cells and normal cells to the combination of the *D. regia* extract and AgNPs suggests that this combination is a promising candidate for cancer treatments [65]. Balkrishna et al. (2020) reported a similar finding, stating that even at the lowest concentration tested, AgNO$_3$ is toxic to normal cells, whereas *Putranjiva roxburghii*-seed-extract-mediated silver nanoparticles (PJSNPs) exhibited no toxicity at the same concentration. This highlights the significance of the nanonization of AgNO$_3$ to PJSNPs [66]. According to Barcinska et al. (2018) in a previously reported study, the effect of AgNPs on PANC-1 was significantly greater than on nontransformed pancreatic cells. They evaluated the contribution of oxidative and nitro-oxidative stress to AgNP-induced cytotoxicity against human pancreatic adenocarcinoma cells due to their crucial role in cancer cell death. The addition of AgNPs to PANC-1 cells increased the production of reactive oxygen species (ROS). Furthermore, this increase was more pronounced in cancer cells than in normal cells of the same tissue [67].

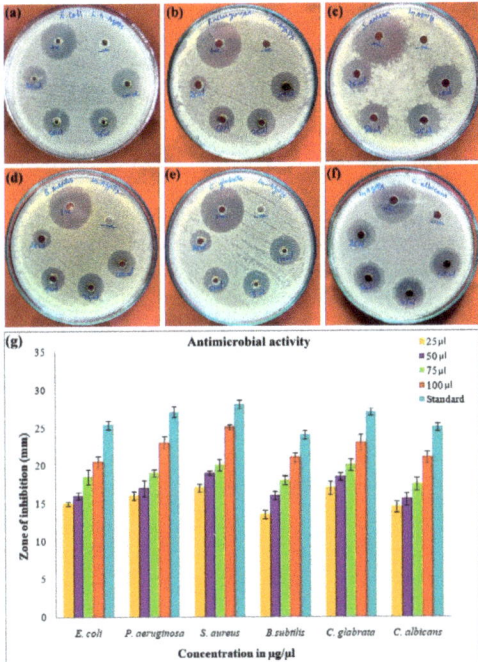

Figure 11. Antimicrobial activity of different concentrations of biosynthesized LnFb-AgNPs from *L. nepetifolia* flower bud extract: (**a**) *E. coli*, (**b**) *P. aeruginosa*, (**c**) *S. aureus*, (**d**) *B. subtilis*, (**e**) *C. glabrata*, and (**f**) *C. albicans*. (**g**) Graphical representation of antimicrobial activity of inhibition zones of LnFb-AgNPs against tested pathogens.

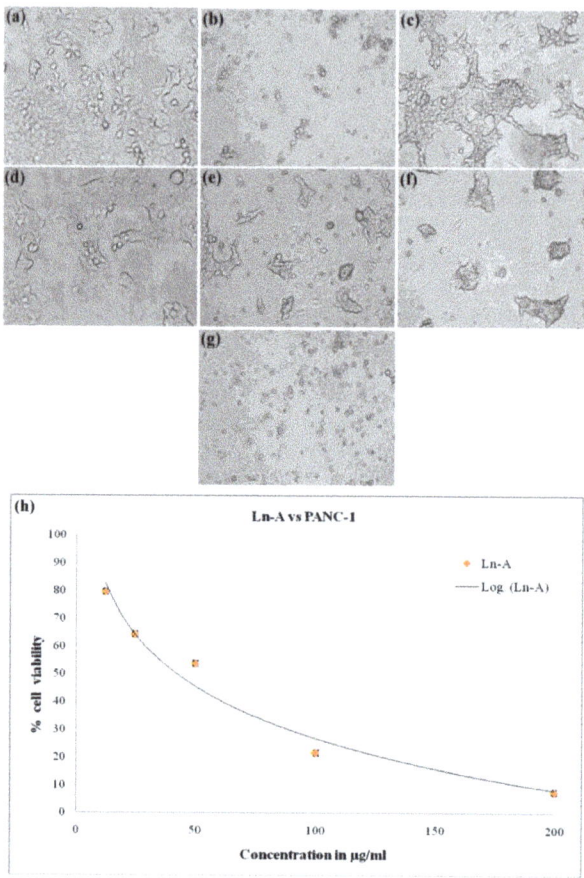

Figure 12. Anticancer assay of different concentration of biosynthesized LnFb-AgNPs from *L. nepetifolia* flower bud extract: (a) negative control, (b) positive control, (c) 12.5 µg/mL, (d) 25 µg/mL, (e) 50 µg/mL, (f) 100 µg/mL, and (g) 200 µg/mL. (h) Graphical representation showing comparative % of cell viability.

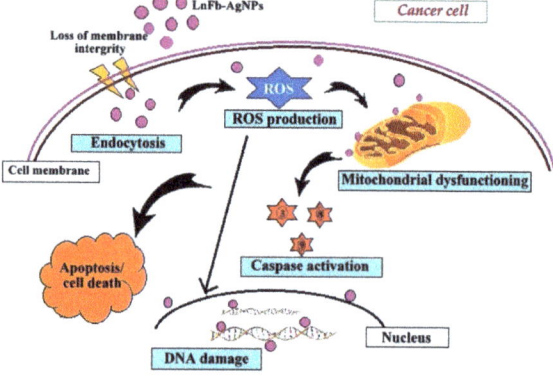

Figure 13. Diagrammatic representation of possible model mechanism of anticancer activity for biosynthesized LnFb-AgNPs from *L. nepetifolia* flower bud extract.

3.11. Apoptosis/Necrosis Studies

To determine the cytotoxicity caused by silver nanoparticles capped with *L. nepetifolia* flower bud extract induced through an apoptotic pathway, Annexin V-mediated apoptosis of PANC-1 cells was studied by staining the cells with FITC Annexin V and PI, followed by fluorescence-activated cell sorting (FACS) detection flow cytometry. Flow cytometry results are displayed in Figure 14a,b. In the flow cytometry plots, the upper left quadrant (Q1) represents the percentage of dead cells (2.59%), (Q2) the upper right quadrant represents the percentage of late apoptosis (38.3%), the lower right quadrant (Q3) displays the percentage of early apoptosis (10.71%), and the last lower left quadrant (Q4) indicates the presence of viable cells percentage (48.4%). These results were obtained after 24 h of treatment with an IC$_{50}$ concentration of 35.84 µg/mL of biosynthesized LnFb-AgNPs. The cell cycle analysis with the markers M1 and M2 is displayed in Figure 13c,d. The untreated cells expressed 99.72% of viable cells to M1 and 0.28% to M2 (Figure 14c). A total of 24.19% of viable PANC-1 cancer cells corresponded to M1, while 75.81% of damaged cells corresponded to M2 (Figure 14d). The treated cells displayed significant early and late apoptosis cell populations against PANC-1 cells, whereas untreated cells did not display any significant apoptosis. In addition, to investigate the other inhibition mechanisms, the rate of apoptosis/necrosis in the treated PANC-1 cell line was determined by an Annexin V-FITC/PI apoptosis detection assay via the flow cytometry method. Indicating the percentage of early and late apoptosis in the treated cell line, the results demonstrated that the biosynthesized LnFb-AgNPs could induce apoptosis. Similarly, Ardestani et al. (2016) reported that when AGS cancer cells were treated with 21.05 µg/mL AgNPs, the biosynthesized AgNPs exhibited 11.79% early apoptosis and 32.70% late apoptosis [54]. Ayromlou et al. (2019) also reported that the *Scorzonera calyculata* aerial part extract mediated synthesis of AgNPs showed 60% induced apoptosis against the A549 lung cancer cell line [68].

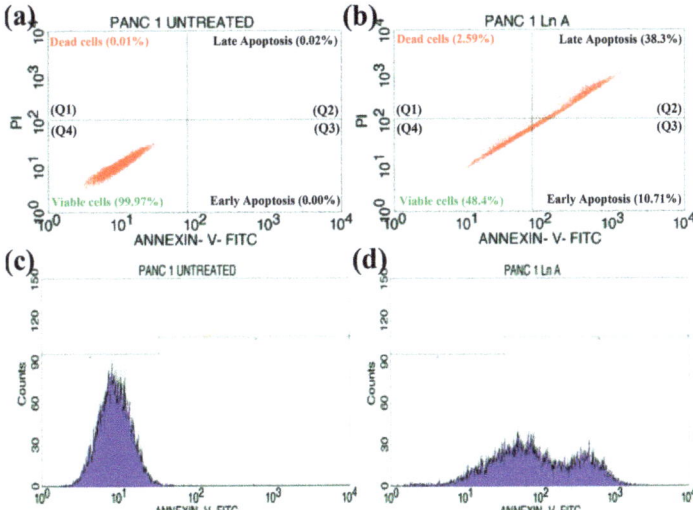

Figure 14. Quadrangular plot representing the Annexin V/PI expression in PANC-1 cancer cells: (a) untreated cells, (b) cells treated with LnFb-AgNPs, (c) cell cycle analysis of untreated cells, and (d) cells treated with LnFb-AgNPs analyzed by using flow cytometry.

4. Conclusions

In the current study, an environmentally friendly, reliable, safe, and low-cost synthesis of AgNPs was accomplished using an *L. nepetifolia* flower bud aqueous extract. UV-Vis, FTIR, XRD, SEM, EDS, TEM, TGA, and Zeta potential with DLS analysis were used to characterize the phyto-synthesized LnFb-AgNPs. In UV-Vis absorption spectra, the SPR peak was detected at 418 nm. Bio-molecules were accountable for reducing and capping of silver nanoparticles, which were revealed by FTIR analysis. The XRD pattern demonstrated that the LnFb-AgNPs had a face-centered cubic crystalline structure. EDS, zeta potential, and thermo gravimetric techniques were utilized to study elemental analysis, particle stabilization, and thermal behavior. The biosynthesized LnFb-AgNPs were spherical in shape and the average particle size was 24.50 nm, and this was confirmed by TEM analysis. The DPPH free radical scavenging assay showed that the LnFb-AgNPs exhibited significantly higher antioxidant activity than the raw flower bud extract. The biosynthesized LnFb-AgNPs were found to have a pronounced antimicrobial activity towards pathogenic bacteria and fungi strains. In addition, the LnFb-AgNPs exhibited a potent cytotoxic effect against the PANC-1 cancer cell line with an IC_{50} value of 35.84 µg/mL. In addition, the apoptosis/necrosis of cancer cells was assessed using the flow cytometry method. The PANC-1 cell line exhibited significant early (10.71% of the cell population) and late (38.3% of the cell population) apoptosis. As an outcome, this research will have a significant impact on the development of improved AgNP products for the pharmaceutical, biotechnological, biomedical, and nanotechnology industries, as well as the identification of advanced drugs to treat the problem of tumor-causing cancer cells using green nanotechnology. Hence, it was suggested that the *L. nepetifolia* flower bud extract mediated synthesis of AgNPs could be used in treating pancreatic cancer after further in vivo studies.

Author Contributions: Conceptualization, S.K.N. (Shashiraj Kariyellappa Nagaraja) and S.N.; Investigation, S.K.N. (Shashiraj Kariyellappa Nagaraja); Methodology, S.K.N. (Shaik Kalimulla Niazi) and S.N.; Supervision, S.N.; Validation and Visualization, S.K.N. (Shashiraj Kariyellappa Nagaraja), R.A.A. and A.B.; Data Curation, A.B. and R.A.A.; Writing—Original Draft, S.K.N. (Shashiraj Kariyellappa Nagaraja); Writing—Review and Editing, S.K.N. (Shashiraj Kariyellappa Nagaraja), S.N., and S.K.N. (Shaik Kalimulla Niazi). All authors have read and agreed to the published version of the manuscript.

Funding: This research received no external funding.

Institutional Review Board Statement: Not applicable.

Informed Consent Statement: Not applicable.

Data Availability Statement: All data generated or analyzed during this study are included in this published article.

Acknowledgments: The authors are thankful to the Department of P.G. Studies in Botany for providing the laboratory facility and to the Sophisticated Analytical Instrumentation Facility (SAIF), University Scientific Instrumentation Centre (USIC), Karnatak University, Dharwad, for extending necessary instrumentation facilities.

Conflicts of Interest: The authors declare no conflict of interest.

References

1. Chandraker, S.K.; Ghosh, M.K.; Lal, M.; Shukla, R. A Review on Plant-Mediated Synthesis of Silver Nanoparticles, Their Characterization and Applications. *Nano. Express* **2021**, *2*, 022008. [CrossRef]
2. Khan, I.; Saeed, K.; Khan, I. Nanoparticles: Properties, Applications and Toxicities. *Arab. J. Chem.* **2019**, *12*, 908–931. [CrossRef]
3. Patra, J.K.; Das, G.; Fraceto, L.F.; Campos, E.V.R.; Rodriguez-Torres, M.d.P.; Acosta-Torres, L.S.; Diaz-Torres, L.A.; Grillo, R.; Swamy, M.K.; Sharma, S.; et al. Nano Based Drug Delivery Systems: Recent Developments and Future Prospects. *J. Nanobiotechnol.* **2018**, *16*, 71. [CrossRef] [PubMed]
4. Jebril, S.; Fdhila, A.; Dridi, C. Nanoengineering of Eco-Friendly Silver Nanoparticles Using Five Different Plant Extracts and Development of Cost-Effective Phenol Nanosensor. *Sci. Rep.* **2021**, *11*, 22060. [CrossRef]
5. Iravani, S.; Korbekandi, H.; Mirmohammadi, S.V.; Zolfaghari, B. Synthesis of Silver Nanoparticles: Chemical, Physical and Biological Methods. *Res. Pharm. Sci.* **2014**, *9*, 385–406. [PubMed]
6. Gupta, R.; Xie, H. Nanoparticles in Daily Life: Applications, Toxicity and Regulations. *J. Environ. Pathol. Toxicol. Oncol.* **2018**, *37*, 209–230. [CrossRef] [PubMed]
7. Hua, S.; de Matos, M.B.C.; Metselaar, J.M.; Storm, G. Current Trends and Challenges in the Clinical Translation of Nanoparticulate Nanomedicines: Pathways for Translational Development and Commercialization. *Front. Pharmacol.* **2018**, *9*, 790. [CrossRef]

8. Basavarajappa, D.S.; Kumar, R.S.; Almansour, A.I.; Chakraborty, B.; Bhat, M.P.; Nagaraja, S.K.; Hiremath, H.; Perumal, K.; Nayaka, S. Biofunctionalized Silver Nanoparticles Synthesized from *Passiflora vitifolia* Leaf Extract and Evaluation of Its Antimicrobial, Antioxidant and Anticancer Activities. *Biochem. Eng. J.* **2022**, *187*, 108517. [CrossRef]
9. Burdusel, A.-C.; Gherasim, O.; Grumezescu, A.M.; Mogoanta, L.; Ficai, A.; Andronescu, E. Biomedical Applications of Silver Nanoparticles: An Up-to-Date Overview. *Nanomaterials* **2018**, *8*, 681. [CrossRef]
10. Ranjani, S.; Matheen, A.; Antony Jenish, A.; Hemalatha, S. Nanotechnology Derived Natural Poly Bio-Silver Nanoparticles as a Potential Alternate Biomaterial to Protect against Human Pathogens. *Mater. Lett.* **2021**, *304*, 130555. [CrossRef]
11. Nagaraja, S.K.; Kumar, R.S.; Chakraborty, B.; Hiremath, H.; Almansour, A.I.; Perumal, K.; Gunagambhire, P.V.; Nayaka, S. Biomimetic Synthesis of Silver Nanoparticles Using *Cucumis sativus* Var. *Hardwickii* Fruit Extract and Their Characterizations, Anticancer Potential and Apoptosis Studies against Pa-1 (Human Ovarian Teratocarcinoma) Cell Line via Flow Cytometry. *Appl. Nanosci.* **2022**. [CrossRef]
12. Nayaka, S.; Bhat, M.P.; Chakraborty, B.; Pallavi, S.S.; Airodagi, D.; Muthuraj, R.; Halaswamy, H.M.; Dhanyakumara, S.B.; Shashiraj, K.N.; Kupaneshi, C. Seed Extract-Mediated Synthesis of Silver Nanoparticles from *Putranjiva roxburghii* Wall., Phytochemical Characterization, Antibacterial Activity and Anticancer Activity Against MCF-7 Cell Line. *Int. J. Pharm. Sci.* **2020**, *82*, 260–269. [CrossRef]
13. Veeraraghavan, V.P.; Periadurai, N.D.; Karunakaran, T.; Hussain, S.; Surapaneni, K.M.; Jiao, X. Green Synthesis of Silver Nanoparticles from Aqueous Extract of *Scutellaria barbata* and Coating on the Cotton Fabric for Antimicrobial Applications and Wound Healing Activity in Fibroblast Cells (L929). *Saud. J. Biol. Sci.* **2021**, *28*, 3633–3640. [CrossRef] [PubMed]
14. Saif, S.; Tahir, A.; Chen, Y. Green Synthesis of Iron Nanoparticles and Their Environmental Applications and Implications. *Nanomaterials* **2016**, *6*, 209. [CrossRef] [PubMed]
15. Dutta, R.; Ahmad, N.; Bhatnagar, S.; Ali, S.S. Phytofabrication of Bioinduced Silver Nanoparticles for Biomedical Applications. *IJN* **2015**, *10*, 7019. [CrossRef]
16. Bhagat, M.; Anand, R.; Datt, R.; Gupta, V.; Arya, S. Green Synthesis of Silver Nanoparticles Using Aqueous Extract of *Rosa brunonii* Lindl and Their Morphological, Biological and Photocatalytic Characterizations. *J. Inorg. Organomet. Polym.* **2019**, *29*, 1039–1047. [CrossRef]
17. Jemal, A.; Siegel, R.; Xu, J.; Ward, E. Cancer Statistics, 2010. *CA A Cancer J. Clin.* **2010**, *60*, 277–300. [CrossRef]
18. Ferlay, J.; Soerjomataram, I.; Dikshit, R.; Eser, S.; Mathers, C.; Rebelo, M.; Parkin, D.M.; Forman, D.; Bray, F. Cancer Incidence and Mortality Worldwide: Sources, Methods and Major Patterns in GLOBOCAN 2012: Globocan 2012. *Int. J. Cancer.* **2015**, *136*, E359–E386. [CrossRef]
19. Yang, F.; Jin, C.; Jiang, Y.; Li, J.; Di, Y.; Ni, Q.; Fu, D. Liposome Based Delivery Systems in Pancreatic Cancer Treatment: From Bench to Bedside. *Cancer. Treat. Rev.* **2011**, *37*, 633–642. [CrossRef]
20. Tadele, K.T.; Abire, T.O.; Feyisa, T.Y. Green synthesized silver nanoparticles using plant extracts as promising prospect for cancer therapy: A review of recent findings. *J. Nanomed.* **2021**, *4*, 1040.
21. Lee, Y.J.; Song, K.; Cha, S.-H.; Cho, S.; Kim, Y.S.; Park, Y. Sesquiterpenoids from *Tussilago farfara* Flower Bud Extract for the Eco-Friendly Synthesis of Silver and Gold Nanoparticles Possessing Antibacterial and Anticancer Activities. *Nanomaterials* **2019**, *9*, 819. [CrossRef]
22. Zhang, K.; Liu, X.; Samuel Ravi, S.O.A.; Ramachandran, A.; Aziz Ibrahim, I.A.; Nassir, A.M.; Yao, J. Synthesis of Silver Nanoparticles (AgNPs) from Leaf Extract of *Salvia miltiorrhiza* and Its Anticancer Potential in Human Prostate Cancer LNCaP Cell Line. *Arti. Cells. Nanomed. Biotechnol.* **2019**, *47*, 2846–2854. [CrossRef]
23. Wang, Y.; Chinnathambi, A.; Nasif, O.; Alharbi, S.A. Green Synthesis and Chemical Characterization of a Novel Anti-Human Pancreatic Cancer Supplement by Silver Nanoparticles Containing *Zingiber officinale* Leaf Aqueous Extract. *Arab. J. Chem.* **2021**, *14*, 103081. [CrossRef]
24. Devanesan, S.; AlSalhi, M.S. Green Synthesis of Silver Nanoparticles Using the Flower Extract of *Abelmoschus esculentus* for Cytotoxicity and Antimicrobial Studies. *IJN* **2021**, *16*, 3343–3356. [CrossRef]
25. Gajendran, B.; Durai, P.; Varier, K.M.; Liu, W.; Li, Y.; Rajendran, S.; Nagarathnam, R.; Chinnasamy, A. Green Synthesis of Silver Nanoparticle from *Datura inoxia* Flower Extract and Its Cytotoxic Activity. *BioNanoScience* **2019**, *9*, 564–572. [CrossRef]
26. Veerabadran, U.; Venkatraman, A.; Souprayane, A.; Narayanasamy, M.; Perumal, D.; Elumalai, S.; Sivalingam, S.; Devaraj, V.; Perumal, A. Evaluation of Antioxidant Potential of Leaves of *Leonotis nepetifolia* and Its Inhibitory Effect on MCF7 and Hep2 Cancer Cell Lines. *Asian. Pac. J. Trop. Dis.* **2013**, *3*, 103–110. [CrossRef]
27. Casiglia, S.; Bruno, M.; Senatore, F. Activity against Microorganisms Affecting Cellulosic Objects of the Volatile Constituents of *Leonotis nepetaefolia* from Nicaragua. *Nat. Prod. Commun.* **2014**, *9*, 1934578X1400901. [CrossRef]
28. Nagaraja, S.K.; Nayaka, S.; Kumar, R.S. Phytochemical analysis, GC-MS Profiling, and In Vitro Evaluation of Biological Applications of Different Solvent Extracts of *Leonotis nepetifolia* (L.) R.Br. Flower Buds. *Appl. Biochem. Biotechnol.* **2022**, *194*. [CrossRef]
29. Chakraborty, B.; Kumar, R.S.; Almansour, A.I.; Kotresha, D.; Rudrappa, M.; Pallavi, S.S.; Hiremath, H.; Perumal, K.; Nayaka, S. Evaluation of Antioxidant, Antimicrobial and Antiproliferative Activity of Silver Nanoparticles Derived from *Galphimia glauca* Leaf Extract. *J. King. Saud. Univ.-Sci.* **2021**, *33*, 101660. [CrossRef]

30. Chand, K.; Cao, D.; Eldin Fouad, D.; Hussain Shah, A.; Qadeer Dayo, A.; Zhu, K.; Nazim Lakhan, M.; Mehdi, G.; Dong, S. Green Synthesis, Characterization and Photocatalytic Application of Silver Nanoparticles Synthesized by Various Plant Extracts. *Arab. J. Chem.* **2020**, *13*, 8248–8261. [CrossRef]
31. Bhat, M.; Chakraborty, B.; Kumar, R.S.; Almansour, A.I.; Arumugam, N.; Kotresha, D.; Pallavi, S.S.; Dhanyakumara, S.B.; Shashiraj, K.N.; Nayaka, S. Biogenic Synthesis, Characterization and Antimicrobial Activity of *Ixora brachypoda* (DC) Leaf Extract Mediated Silver Nanoparticles. *J. King. Saud. Uni.-Sci.* **2021**, *33*, 101296. [CrossRef]
32. Kambale, E.K.; Nkanga, C.I.; Mutonkole, B.-P.I.; Bapolisi, A.M.; Tassa, D.O.; Liesse, J.-M.I.; Krause, R.W.M.; Memvanga, P.B. Green Synthesis of Antimicrobial Silver Nanoparticles Using Aqueous Leaf Extracts from Three Congolese Plant Species (*Brillantaisia patula, Crossopteryx febrifuga* and *Senna siamea*). *Heliyon* **2020**, *6*, e04493. [CrossRef]
33. Ndikau, M.; Noah, N.M.; Andala, D.M.; Masika, E. Green Synthesis and Characterization of Silver Nanoparticles Using *Citrullus lanatus* Fruit Rind Extract. *Int. J. Analy. Chem.* **2017**, *2017*, 8108504. [CrossRef]
34. Singh, R.; Hano, C.; Nath, G.; Sharma, B. Green Biosynthesis of Silver Nanoparticles Using Leaf Extract of *Carissa carandas* L. and Their Antioxidant and Antimicrobial Activity against Human Pathogenic Bacteria. *Biomolecules* **2021**, *11*, 299. [CrossRef]
35. Aritonang, H.F.; Koleangan, H.; Wuntu, A.D. Synthesis of Silver Nanoparticles Using Aqueous Extract of Medicinal Plants' (*Impatiens balsamina* and *Lantana camara*) Fresh Leaves and Analysis of Antimicrobial Activity. *Int. J. Microbiol.* **2019**, *2019*, 8642303. [CrossRef]
36. Mosmann, T. Rapid colorimetric assay for cellular growth and survival: Application to proliferation and cytotoxicity assays. *J. Immunol. Methods.* **1983**, *65*, 55–63. [CrossRef]
37. Rudrappa, M.; Rudayni, H.A.; Assiri, R.A.; Bepari, A.; Basavarajappa, D.S.; Nagaraja, S.K.; Chakraborty, B.; Swamy, P.S.; Agadi, S.N.; Niazi, S.K.; et al. *Plumeria alba*-Mediated Green Synthesis of Silver Nanoparticles Exhibits Antimicrobial Effect and Anti-Oncogenic Activity against Glioblastoma U118 MG Cancer Cell Line. *Nanomaterials* **2022**, *12*, 493. [CrossRef]
38. O'Brien, M.C.; Bolton, W.E. Comparison of Cell Viability Probes Compatible with Fixation and Permeabilization for Combined Surface and Intracellular Staining in Flow Cytometry. *Cytometry* **1995**, *19*, 243–255. [CrossRef]
39. Kim, Y.J.; Singh, P.; Yang, D.-C.; Singh, H.; Wang, C.; Farh, M.E.-A.; Hwang, K.H. Biosynthesis, Characterization, and Antimicrobial Applications of Silver Nanoparticles. *IJN* **2015**, *10*, 2567. [CrossRef]
40. Gnanakani, P.E.; Santhanam, P.; Premkumar, K.; Eswar Kumar, K.; Dhanaraju, M.D. Nannochloropsis Extract–Mediated Synthesis of Biogenic Silver Nanoparticles, Characterization and In Vitro Assessment of Antimicrobial, Antioxidant and Cytotoxic Activities. *Asian. Pac. J. Cancer. Prev.* **2019**, *20*, 2353–2364. [CrossRef]
41. Handayani, W.; Ningrum, A.S.; Imawan, C. The Role of pH in Synthesis Silver Nanoparticles Using *Pometia pinnata* (Matoa) Leaves Extract as Bioreductor. *J. Phys. Conf. Ser.* **2020**, *1428*, 012021. [CrossRef]
42. Rajesh Kumar, T.V.; Murthy, J.S.R.; Narayana Rao, M.; Bhargava, Y. Evaluation of Silver Nanoparticles Synthetic Potential of *Couroupita guianensis* Aubl., Flower Buds Extract and Their Synergistic Antibacterial Activity. *3 Biotech* **2016**, *6*, 92. [CrossRef] [PubMed]
43. Pereira, T.M.; Polez, V.L.P.; Sousa, M.H.; Silva, L.P. Modulating Physical, Chemical, and Biological Properties of Silver Nanoparticles Obtained by Green Synthesis Using Different Parts of the Tree *Handroanthus heptaphyllus* (Vell.) Mattos. *Colloid Interface Sci. Commun.* **2020**, *34*, 100224. [CrossRef]
44. Trivedi, A.; Sethiya, N.K.; Mishra, S.H. Preliminary pharmacognostic and phytochemical analysis of Grantbika (*Leonotis nepetifolia*): An ayurvedic herb. *Indian J. Tradit. Knowl.* **2011**, *10*, 682–688.
45. Salari, S.; Esmaeilzadeh Bahabadi, S.; Samzadeh-Kermani, A.; Yosefzaei, F. In-Vitro Evaluation of Antioxidant and Antibacterial Potential of GreenSynthesized Silver Nanoparticles Using *Prosopis farcta* Fruit Extract. *Iran. J. Pharm. Res.* **2019**, *18*, 430–455.
46. Muthukrishnan, S.; Bhakya, S.; Senthil Kumar, T.; Rao, M.V. Biosynthesis, Characterization and Antibacterial Effect of Plant-Mediated Silver Nanoparticles Using *Ceropegia thwaitesii*—An Endemic Species. *Ind. Crops Prod.* **2015**, *63*, 119–124. [CrossRef]
47. Algebaly, A.S.; Mohammed, A.E.; Abutaha, N.; Elobeid, M.M. Biogenic Synthesis of Silver Nanoparticles: Antibacterial and Cytotoxic Potential. *Saud. J. Biol. Sci.* **2020**, *27*, 1340–1351. [CrossRef]
48. Awwad, A.M.; Salem, N.M.; Abdeen, A.O. Green Synthesis of Silver Nanoparticles Using Carob Leaf Extract and Its Antibacterial Activity. *Int. J. Indus. Chem.* **2013**, *4*, 29. [CrossRef]
49. Ajitha, B.; Reddy, Y.A.K.; Lee, Y.; Kim, M.J.; Ahn, C.W. Biomimetic Synthesis of Silver Nanoparticles Using *Syzygium aromaticum* (Clove) Extract: Catalytic and Antimicrobial Effects. *Appl. Organometal. Chem.* **2019**, *33*, e4867. [CrossRef]
50. Chen, S.; Webster, S.; Czerw, R.; Xu, J.; Carroll, D.L. Morphology Effects on the Optical Properties of Silver Nanoparticles. *J. Nanosci. Nanotech.* **2004**, *4*, 254–259. [CrossRef]
51. Allafchian, A.R.; Jalali, S.A.H.; Aghaei, F.; Farhang, H.R. Green Synthesis of Silver Nanoparticles Using *Glaucium corniculatum* (L.) Curtis Extract and Evaluation of Its Antibacterial Activity. *IET Nanobiotechnol.* **2018**, *12*, 574–578. [CrossRef] [PubMed]
52. Lakhan, M.N.; Chen, R.; Shar, A.H.; Chand, K.; Shah, A.H.; Ahmed, M.; Ali, I.; Ahmed, R.; Liu, J.; Takahashi, K.; et al. Eco-Friendly Green Synthesis of Clove Buds Extract Functionalized Silver Nanoparticles and Evaluation of Antibacterial and Antidiatom Activity. *J. Microbiol. Methods* **2020**, *173*, 105934. [CrossRef] [PubMed]
53. El-Aswar, E.I.; Zahran, M.M.; El-Kemary, M. Optical and Electrochemical Studies of Silver Nanoparticles Biosynthesized by *Haplophyllum tuberculatum* Extract and Their Antibacterial Activity in Wastewater Treatment. *Mater. Res. Express* **2019**, *6*, 105016. [CrossRef]

54. Ardestani, M.S.; Sadat Shandiz, S.A.; Salehi, S.; Ghanbar, F.; Darvish, M.R.; Mirzaie, A.; Jafari, M. Phytosynthesis of Silver Nanoparticles Using *Artemisia marschalliana* Sprengel Aerial Part Extract and Assessment of Their Antioxidant, Anticancer, and Antibacterial Properties. *IJN* **2016**, *11*, 1835. [CrossRef]
55. Mittal, A.K.; Bhaumik, J.; Kumar, S.; Banerjee, U.C. Biosynthesis of Silver Nanoparticles: Elucidation of Prospective Mechanism and Therapeutic Potential. *J. Colloid Interface Sci.* **2014**, *415*, 39–47. [CrossRef]
56. Moteriya, P.; Chanda, S. Synthesis and Characterization of Silver Nanoparticles Using *Caesalpinia pulcherrima* Flower Extract and Assessment of Their in Vitro Antimicrobial, Antioxidant, Cytotoxic, and Genotoxic Activities. *Artif. Cells Nanomed. Biotechnol.* **2017**, *45*, 1556–1567. [CrossRef]
57. Rasheed, T.; Bilal, M.; Iqbal, H.M.N.; Li, C. Green Biosynthesis of Silver Nanoparticles Using Leaves Extract of *Artemisia vulgaris* and Their Potential Biomedical Applications. *Colloids Surf. B Biointerfaces* **2017**, *158*, 408–415. [CrossRef]
58. Yousaf, H.; Mehmood, A.; Ahmad, K.S.; Raffi, M. Green Synthesis of Silver Nanoparticles and Their Applications as an Alternative Antibacterial and Antioxidant Agents. *Mater. Sci. Eng. C* **2020**, *112*, 110901. [CrossRef]
59. Patra, J.K.; Baek, K.-H. Antibacterial Activity and Synergistic Antibacterial Potential of Biosynthesized Silver Nanoparticles against Foodborne Pathogenic Bacteria along with Its Anticandidal and Antioxidant Effects. *Front. Microbiol.* **2017**, *8*, 167. [CrossRef]
60. Rajeshkumar, S.; Malarkodi, C. In Vitro Antibacterial Activity and Mechanism of Silver Nanoparticles against Foodborne Pathogens. *Bioinorg. Chem. Appl.* **2014**, *2014*, 581890. [CrossRef]
61. Ruden, S.; Hilpert, K.; Berditsch, M.; Wadhwani, P.; Ulrich, A.S. Synergistic Interaction between Silver Nanoparticles and Membrane-Permeabilizing Antimicrobial Peptides. *Antimicrob. Agents Chemother.* **2009**, *53*, 3538–3540. [CrossRef] [PubMed]
62. Singh, K. Antibacterial Activity of Synthesized Silver Nanoparticles from *Tinospora cordifolia* against Multi Drug Resistant Strains of Pseudomonas Aeruginosa Isolated from Burn Patients. *J. Nanomed. Nanotechnol.* **2014**, *5*, 1000192. [CrossRef]
63. Kanniah, P.; Chelliah, P.; Thangapandi, J.R.; Gnanadhas, G.; Mahendran, V.; Robert, M. Green Synthesis of Antibacterial and Cytotoxic Silver Nanoparticles by *Piper nigrum* Seed Extract and Development of Antibacterial Silver Based Chitosan Nanocomposite. *Int. J. Biol. Macromol.* **2021**, *189*, 18–33. [CrossRef] [PubMed]
64. Donga, S.; Chanda, S. Facile Green Synthesis of Silver Nanoparticles Using *Mangifera indica* Seed Aqueous Extract and Its Antimicrobial, Antioxidant and Cytotoxic Potential (3-in-1 System). *Arti. Cells. Nanomed. Biotechnol.* **2021**, *49*, 292–302. [CrossRef]
65. Shameli Rajiri, M.; Aminsalehi, M.; Shahbandeh, M.; Maleki, A.; Jonoubi, P.; Rad, A.C. Anticancer and Therapeutic Potential of *Delonix regia* Extract and Silver Nanoparticles (AgNPs) against Pancreatic (Panc-1) and Breast (MCF-7) Cancer Cell. *Toxicol. Environ. Health Sci.* **2021**, *13*, 45–56. [CrossRef]
66. Balkrishna, A.; Sharma, V.K.; Das, S.K.; Mishra, N.; Bisht, L.; Joshi, A.; Sharma, N. Characterization and Anti-Cancerous Effect of *Putranjiva roxburghii* Seed Extract Mediated Silver Nanoparticles on Human Colon (HCT-116), Pancreatic (PANC-1) and Breast (MDA-MB 231) Cancer Cell Lines: A Comparative Study. *IJN* **2020**, *15*, 573–585. [CrossRef]
67. Barcinska, E.; Wierzbicka, J.; Zauszkiewicz-Pawlak, A.; Jacewicz, D.; Dabrowska, A.; Inkielewicz-Stepniak, I. Role of Oxidative and Nitro-Oxidative Damage in Silver Nanoparticles Cytotoxic Effect against Human Pancreatic Ductal Adenocarcinoma Cells. *Oxidative Med. Cell. Longev.* **2018**, *2018*, 8251961. [CrossRef]
68. Ayromlou, A.; Masoudi, S.; Mirzaie, A. *Scorzonera calyculata* Aerial Part Extract Mediated Synthesis of Silver Nanoparticles: Evaluation of Their Antibacterial, Antioxidant and Anticancer Activities. *J. Clust. Sci.* **2019**, *30*, 1037–1050. [CrossRef]

Article

Novel Carboxymethyl Cellulose-Based Hydrogel with Core–Shell Fe$_3$O$_4$@SiO$_2$ Nanoparticles for Quercetin Delivery

Mohammad Mahdi Eshaghi [1], Mehrab Pourmadadi [1], Abbas Rahdar [2,*] and Ana M. Díez-Pascual [3,*]

[1] Department of Biotechnology, School of Chemical Engineering, College of Engineering, University of Tehran, Tehran 1417935840, Iran
[2] Department of Physics, Faculty of Sciences, University of Zabol, Zabol 538-98615, Iran
[3] Universidad de Alcalá, Facultad de Ciencias, Departamento de Química Analítica, Química Física e Ingeniería Química, Ctra. Madrid-Barcelona, Km. 33.6, 28805 Alcalá de Henares, Madrid, Spain
* Correspondence: a.rahdar@uoz.ac.ir (A.R.); am.diez@uah.es (A.M.D.-P.)

Abstract: A nanocomposite composed of carboxymethyl cellulose (CMC) and core–shell nanoparticles of Fe$_3$O$_4$@SiO$_2$ was prepared as a pH-responsive nanocarrier for quercetin (QC) delivery. The nanoparticles were further entrapped in a water-in-oil-in-water emulsion system for a sustained release profile. The CMC/Fe$_3$O$_4$@SiO$_2$/QC nanoparticles were characterized using dynamic light scattering (DLS), Fourier transform infrared spectroscopy (FTIR), X-ray diffraction (XRD), a field emission scanning electron microscope (FE-SEM), and a vibrating sample magnetometer (VSM) to obtain insights into their size, stability, functional groups/chemical bonds, crystalline structure, morphology, and magnetic properties, respectively. The entrapment and loading efficiency were slightly improved after the incorporation of Fe$_3$O$_4$@SiO$_2$ NPs within the hydrogel network. The dialysis method was applied for drug release studies. It was found that the amount of QC released increased with the decrease in pH from 7.4 to 5.4, while the sustained-release pattern was preserved. The A549 cell line was chosen to assess the anticancer activity of the CMC/Fe$_3$O$_4$@SiO$_2$/QC nanoemulsion and its components for lung cancer treatment via an MTT assay. The L929 cell line was used in the MTT assay to determine the possible side effects of the nanoemulsion. Moreover, a flow cytometry test was performed to measure the level of apoptosis and necrosis. Based on the obtained results, CMC/Fe$_3$O$_4$@SiO$_2$ can be regarded as a novel promising system for cancer therapy.

Keywords: carboxymethyl cellulose; quercetin; Fe$_3$O$_4$ nanoparticles; core–shell nanoparticles; double-emulsion system

1. Introduction

Cancer is among the most commonly occurring diseases worldwide. In the United States, cancer is identified as the second most frequent cause of death. Statistics indicate that lung cancer accounts for largest share of cancer-related deaths in both men and women in the United States [1]. Even though chemotherapy is the prominent approach for curing various types of cancer, including lung cancer, its efficacy is limited due to factors such as drug resistance, non-specific targeting, severe side effects on healthy cells, and low bioavailability [2]. The mentioned drawbacks of chemotherapy indicate the need to use of less toxic drugs in conjunction with smart drug delivery systems, which can lead to less side effects and enhanced therapeutic effects [3].

Plant-based polyphenols have been extensively studied for their potential usage in cancer treatment. Among these natural compounds, quercetin (QC) has been recognized as a promising member in terms of antineoplastic activity. QC belongs to the flavonol family, which is a class of flavonoids. Studies have proved that quercetin has anti-oxidative, anti-inflammatory, and anticancer activity. Its anticancer activity is shown via different mechanisms, including metastasis inhibition, apoptosis induction, the disturbing of cell proliferation, angiogenesis disruption, oxidative stress suppression, and the influencing of autophagy [4–9]. Regarding the anti-tumor activity of QC against lung cancer tumors, studies have indicated that quercetin shows anti-proliferation and anti-metastasis behavior towards the A549 cell line by affecting the cytoskeleton of cancerous cells. The anti-proliferative behavior of quercetin towards A549 cells has been associated with a disturbance in cytokinesis during mitosis as a result of cytoskeleton components getting eliminated from the cytoplasm by quercetin [4,10]. It has been shown that QC can also improve radio-sensitivity of non-small cell lung cancer [11]. Further studies have proven that despite the complicated correlation between autophagy and apoptosis in different cancerous cell lines, autophagy inhibition can limit quercetin-induced apoptosis on the A549 cell line. Hence, in order for quercetin to be an effective therapeutic agent for lung cancer, it needs to be used in parallel with an initiator of autophagy [12].

The therapeutic effects of QC and similar flavonoid compounds have been limited by drawbacks such as low bioavailability [13], instability [14], poor solubility [15], and inefficient biodistribution [16]. To overcome these limitations, novel drug delivery platforms for enhanced therapeutic efficiency need to be developed. QC-loaded nanoparticles offer several advantages, such as long circulation time, controlled release, improved entrapment efficiency, and stability. Among different NPs, polymeric ones possess advantageous characteristics for drug delivery. They are mostly non-toxic, biodegradable, biocompatible, and stable [17–19]. About a hundred years ago, carboxymethyl cellulose (CMC) was manufactured in Germany for the first time. It is a highly water-soluble biological macromolecule. Apart from the abovementioned properties of polymeric NPs, CMC alleviates the side effects of drugs, enhances bioavailability, and improves anti-tumor activity. Furthermore, CMC can change its state from acid to base and vice versa depending on the pH. Hence, CMC is a pH-responsive polymer and a potentially smart nanocarrier. Since the extracellular environment of tumors has a lower pH than ordinary cells, the pH sensitivity of CMC can improve the targeted delivery of quercetin [20–23]. Polymeric compounds such as CMC can be cross-linked to form a hydrophilic network called hydrogel. Hydrogels can absorb large amounts of water without losing structural integrity. The porosity of these networks can be modified by controlling the cross-linking density. The large amount of water that can be absorbed by hydrogels, together with their stimulus-responsive properties, makes them suitable drug carriers. Owing to their pH and temperature sensitivity, their cargo can be released in a controlled way. In addition, their large water content makes them very biocompatible [24,25].

Despite the abovementioned advantages, hydrogels suffer from some drawbacks for usage as drug carriers. Due to their hydrophilic properties, encapsulating hydrophobic drugs in them can be challenging. Stimulus-responsive hydrogels demonstrate very low response time, and they are not homogeneous in general. Furthermore, they have low mechanical strength and a high rate of biodegradability [3,26–31]. One solution to circumvent the mentioned disadvantages is to incorporate inorganic nanoparticles within the hydrogel structure, developing nanocomposites. Introducing inorganic NPs can improve the mechanical properties of hydrogels, increase drug loading capacity, make them stimulus responsive, and enhance biocompatibility, thereby improving drug delivery efficacy [32]. Among inorganic nanoparticles, iron oxide-based ones and magnetite (Fe_3O_4) in particular have received considerable attention for drug delivery applications. Fe_3O_4 demonstrates outstanding magnetic and electronic properties and is sufficiently biocompatible [33,34]. Previous research conducted by Azizi [33] has proven that introducing magnetite NPs within a polymeric structure can enhance the swelling capacity, loading capacity, and thermal resistance of the nanocarriers. These improvements can be associated with the electrical properties and high surface area of magnetite. In another study, Mohammadi et al. [35] fabricated carboxymethylcellulose/polyacrylic acid/starch-modified Fe_3O_4 nanocomposite hydrogels for oral doxorubicin delivery. The swelling ratio of nanocomposites depended on both pH and the weight percentage of Fe_3O_4. This contributed to controlled drug release in the intestine (neutral environment) instead of the stomach (acidic environment). However, high surface area and instability at low pH can lead to the aggregation and oxidation of Fe_3O_4 molecules. Other inorganic nanoparticles with suitable characteristics for drug delivery are SiO_2. These NPs have high surface area and an amphiphilic surface and are biocompatible. Furthermore, various functional groups can be added to their surface thanks to the valence electrons of silicon atoms. Hence, they can be used for delivering cargos such as antineoplastic drugs [36,37]. In order to make up for the drawbacks of Fe_3O_4, core–shell NPs of Fe_3O_4 coated with SiO_2 can be used. Core–shell NPs have several advantages over normal NPs, including enhanced biodistribution, lower cytotoxicity, more controlled release of the drug, improved binding to biomolecules, and stimulus-responsive properties [38,39].

Herein, we prepared a hydrogel nanocomposite based on CMC that incorporated $Fe_3O_4@SiO_2$ core–shell nanoparticles (Scheme 1). QC was then loaded in the CMC/$Fe_3O_4@SiO_2$/QC nanocomposite. Since hydrogels have aqueous environments, loading hydrophobic drugs such as quercetin inside them can be challenging. However, the presence of core–shell NPs within the hydrogel structure provided a large surface area and made a slight contribution to improving the loading capacity of quercetin. The majority of the loaded quercetin was localized in the porous structure of SiO_2 (shell material). There have been reports of possible hydrogen bonding between CMC and quercetin, with the hydroxyl group of quercetin as the donor, in the literature. These interactions lead to the formation of a stable network comprising CMC and quercetin [40,41]. In addition, there have been reports of using $Fe_3O_4@SiO_2$ core–shell nanoparticles for adsorbing quercetin and other flavonoids [42,43]. These reports justify our choice of material for fabricating the drug delivery system. Furthermore, we examined the pH sensitivity of the prepared delivery system by analyzing the release behavior at two different pH values (5.4 and 7.4). The obtained results demonstrated a sustained but improved release of quercetin at the lowest pH. The sensitivity of CMC/$Fe_3O_4@SiO_2$/QC to pH could be associated with the protonation of carboxyl groups at low pH values, which leads to a reduction in electrostatic interactions between the drug and the carrier. These elucidations are in agreement with previous literature reports [44–46]. The preparation of hydrogel nanocomposites was followed by a water-in-oil-in-water (W-O-W) emulsification step. Loading the nanocomposites in W-O-W nanoemulsions has many advantages. Nanoemulsions provide benefits such as controlling the release of quercetin, inhibiting its degradation, biocompatibility, and capability for loading both hydrophilic and lipophilic drugs [47,48].

Therefore, the purpose of this research was to introduce a novel nanocarrier for treating cancer using QC. The pH sensitivity of the nanocarrier reduced the side effects of quercetin. In addition, the incorporation of core–shell nanoparticles slightly improved the drug loading in the polymeric carrier. Although the loading and entrapment efficiency did not improve significantly after introducing core–shell nanoparticles, the FTIR analysis revealed that in the presence of these nanoparticles, a share of loaded quercetin is entrapped inside them instead of the polymeric network. This entrapment contributes to the sustained release of quercetin, as the drug has to diffuse through extra layers before getting released. The effectiveness of the system was assessed against the A549 cell line. XRD and FTIR characterization was used to determine the crystalline structure, physical properties, and composition of the samples. A VSM test was employed to confirm the magnetic properties of Fe_3O_4. In addition, FESEM images were recorded to obtain insights into the morphology of the nanocarriers.

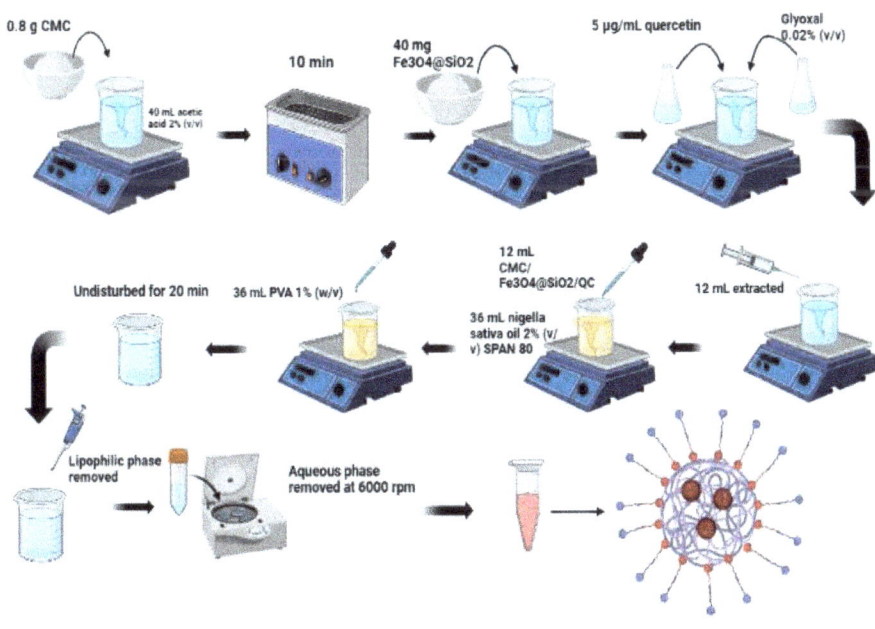

Scheme 1. Schematic representation of the procedure for the synthesis of double-nanoemulsion-entrapped nanocomposite of CMC/Fe$_3$O$_4$@SiO$_2$/QC.

2. Materials and Methods

2.1. Materials

Sodium carboxymethyl cellulose (M_w = 90,000) was purchased from Merck Co. (Darmstadt, Germany). FeCl$_3$.6H$_2$O (reagent grade, 97%) was purchased from Sigma Aldrich Co. (Burlington, MA, USA). Quercetin drug (>95% (HPLC), solid) was also obtained from Sigma Aldrich Co. SPAN 80 (molar mass = 428.60 g/mol) was purchased from Merck Co. Phosphate-buffered saline was obtained from Sigma Aldrich Co. Ammonium hydroxide solution and tetraethyl orthosilicate (M_w = 208.33) were obtained from Merck Co. (Darmstadt, Germany).

2.2. Preparation of Fe$_3$O$_4$ Nanoparticles

In order to synthesize Fe$_3$O$_4$ nanoparticles, 4.8 g of FeCl$_3$.6H$_2$O was added to 100 mL of deionized water. The solution was stirred at 700 rpm for 1 h under Ar atmosphere conditions, and the salts were completely dissolved in the water. Then, 10 mL of ammonium hydroxide (25%) was added to the solution dropwise over 10 min. The black precipitate of magnetite nanoparticles was formed instantly. This was followed by 1 h of mechanical stirring. The precipitate was then removed using an external magnet and washed five times using distilled water. Finally, the NPs were dried at 50 °C overnight [49].

2.3. Preparation of Fe$_3$O$_4$@SiO$_2$ Core–Shell Nanoparticles

The synthesis process began with dissolving 1 g of magnetite nanoparticles in a solution with 40 mL of ethanol and 10 mL of water using an ultrasonic bath. The solution was then transferred to a three-necked bottle. The pH of the solution was fixed at 10 using ammonia solution. A volume of 0.5 mL of tetraethyl orthosilicate was then added to the solution dropwise, and the resulting mixture was stirred for 6 h at 50 °C. Finally, core–shell nanoparticles of Fe$_3$O$_4$@SiO$_2$ were formed. The NPs were then thoroughly washed with ethanol and distilled water and dried at 60 °C for 24 h [49].

2.4. Preparation of Quercetin-Loaded CMC/Fe_3O_4@SiO_2 Hydrogel

The overall protocol for the preparation of the hydrogel was similar to those found in previous literature reports [50]. Firstly, 0.8 g of CMC was added to 40 mL of acetic acid 2% (v/v) solution. The solution was placed on a heater stirrer until complete dissolution of the polymer at room temperature was achieved and homogenous 2% (w/v) CMC solution was obtained. The homogenous solution was then placed in an ultrasonic bath for 10 min. Then, 40 mg of Fe_3O_4@SiO_2 NPs was added to the mixture, and the solution was placed on a heater stirrer until homogeneity was achieved. Then, 0.02% (v/v) glyoxal as the crosslinking agent was added to the mixture. The quantity of QC added was chosen so that the final concentration of the drug in the hydrogel was 5 µg/mL. After QC addition, the mixture was heated while stirring for another 30 min; finally, CMC/Fe_3O_4@SiO_2/QC hydrogel was obtained.

2.5. Preparation of Double-Emulsion-Encapsulated Hydrogel

The double-emulsion system was prepared using the method reported by Ahmadi et al. [51]. A volume of 12 mL of CMC/Fe_3O_4@SiO_2/QC hydrogel was extracted using a syringe and added to 36 mL of 2% (v/v) SPAN 80-containing nigella sativa oil dropwise (SPAN 80 as the hydrophobic surfactant). The hydrophobic phase was placed on a stirrer during this operation. Upon the addition of the hydrogel to the hydrophobic phase, spherical particles of nanocarriers were formed in the solution. After 10 min of stirring, 36 mL of PVA 1% (w/v) solution (PVA as the hydrophilic surfactant) was added to the mixture dropwise. Then, the stirring process was stopped, and the system was kept undisturbed for 20 min so that different layers could become separated. The lipophilic phase was then separated by means of a sampler. The aqueous phase was centrifuged at 6000 rpm, and the water was removed from the QC-loaded nanocarriers.

Prior to each test, the materials were powdered using a freeze-dryer. The samples were subjected to a temperature of $-20\ °C$ prior to freeze-drying.

2.6. Characterization of Nanoparticles

The morphology of nanoparticles was observed using a field emission scanning electron microscope (FE-SEM). Dynamic light scattering (DLS) was performed to determine the size distribution and zeta potential of the nanoparticles. X-ray diffraction (XRD) was employed to analyze the crystalline structure of the nanocomposite after the addition of each component. Fourier transform infrared (FTIR) spectroscopy was used to identify the different functional groups in the composite and to assess the interactions between the different components in the nanocomposite network. Finally, a vibrating sample magnetometer was used to verify the magnetic property of Fe_3O_4.

2.7. Drug Loading and Encapsulation Efficiency

The method applied for measuring loading and encapsulation efficiency is similar to the approach reported in the literature [52]. In order to determine the effect of incorporating core–shell nanoparticles within the polymeric network on the loading capacity and entrapment efficiency, CMC/QC and CMC/Fe_3O_4@SiO_2/QC NPs were added to phosphate-buffered saline. Next, ethyl acetate solvent was added to the mixture and stirred until homogeneity was obtained. The organic phase was then separated, and its quercetin content was measured using UV-Vis spectrophotometer. Equations (1) and (2) were used for calculating the loading efficiency and encapsulation efficiency, respectively.

$$\text{Loading Efficiency } (\%) = \frac{(\text{Total QC quantity}) - (\text{Free QC quantity})}{\text{Total nanocarrier quantity}} \quad (1)$$

$$\text{Entapment Efficiency } (\%) = \frac{(\text{Total QC quantity}) - (\text{Free QC quantity})}{\text{Total QC quantity}} \quad (2)$$

2.8. In Vitro Drug Release

The dialysis method was applied to study the release profile of the drug from nanoemulsions in vitro [53]. A beaker was filled with water at 37 °C to simulate the thermal conditions of the body. The beaker was placed on a magnetic stirrer to ensure uniform distribution of temperature throughout the vessel. Two Falcon test tubes were filled with phosphate-buffered saline containing 20% v/v of ethanol at pH values of 5.4 and 7.4 to represent the cancerous and healthy tissues' media, respectively. The Falcon test tubes were then immersed into the beaker. Then, two dialysis bags were filled with nanoemulsions and placed inside the tubes. Samples were taken from the tubes 0, 0.5, 1, 2, 3, 6, 12, 24, 48, 72, and 96 h after the beginning of the experiment. The extracted sample was replaced with a proportional amount of fresh PBS. A UV-Vis spectrophotometer was used to quantify the amount of released quercetin within the samples. The percentage of released drug was calculated using Equation (3).

$$\text{Released Drug percentage} = \frac{\text{Released drug}}{\text{Loaded drug}} \times 100 \qquad (3)$$

In Equation (3), "Loaded drug" is the amount of drug that was loaded in the nanoemulsions before placing them inside dialysis bags. "Released drug" is the amount of drug within the extracted samples, which was measured using a UV-Vis spectrophotometer. Since UV-Vis spectrophotometry was used to measure the amount of QC released from the different samples, the nanoparticles needed to have enough solubility in PBS. For this reason, 20% v/v ethanol was used to increase the solubility of the nanoparticles. The rationality of this method is consistent with previous literature reports on the dialysis bag technique [51,53,54].

2.9. Cell Culture

The A549 and L929 cell lines were cultured in RPMI 1640 and DMEM, respectively. The culturing procedure was performed with 100 µg/mL streptomycin, 100 U/mL penicillin, and 10% (v/v) fetal bovine serum. The cultivation media were kept in a humidified atmosphere containing 5% carbon dioxide.

2.10. MTT Assay

An MTT assay was employed to assess the cytotoxic effect of free QC, CMC, Fe_3O_4@SiO_2, CMC/Fe_3O_4@SiO_2, and CMC/Fe_3O_4@SiO_2/QC on the A549 and L929 cell lines. Each well of a 96-well plate was filled with the cultivation medium, which contained 10^4 cells. The plate was incubated for a day so that cell adhesion could happen. The cells were then treated with the abovementioned samples for 24 h. The reason for using various samples was to identify the effect of each component of the prepared nanocomposite on the antitumor activity of the whole system. In addition, the L929 cell line was also used to examine the cytotoxic effect of the prepared nanocarriers on a noncancerous cell line and obtain an estimate of the potential side effects of the delivery system. The control group was cultured in a similar medium (DMEM) without any treatment. After three days, the cells in each well were incubated with fresh DMEM and MTT solution. DMSO was also added to each well and stirred until formazan particles were dissolved. The control group was used as the reference for the reporting cell viability percentage of treated cells. An ELISA reader was used to determine optical density. All tests were performed in triplicate, and the standard error of the mean was calculated for each group using statistical data analysis methods.

2.11. Flow Cytometry Test

A flow cytometry test was employed to analyze the level of apoptosis and necrosis in A549 cells treated with CMC, Fe_3O_4@SiO_2, CMC/Fe_3O_4@SiO_2, and CMC/Fe_3O_4@SiO_2/QC. After 24 h of treatment, the cells were washed with phosphate-buffered saline. This was followed by creating a suspension of cells in a binding buffer and staining them using Annexin V-FITC. A flow cytometer was used to measure apoptosis and necrosis by analyzing the fluorescence intensity. Four quadrants, for necrotic death (Q1), late apoptotic death (Q2), early apoptotic death (Q3), and viability (Q4), were defined, and quadrant statistics were performed. The testing of all samples was performed three times.

3. Results and Discussion

3.1. DLS

The dynamic light scattering (DLS) technique was used to determine the size of CMC/Fe_3O_4@SiO_2/QC nanoparticles. The results indicated an average size of 151.6 nm of the nanoparticles, with a polydispersity index (PDI) of 0.14. It was evident that the obtained nanoparticles showed a uniform size distribution, since almost 70% of them had a size close to the average value. Moreover, the zeta potential values of the nanoparticles were measured to evaluate their stability. The average zeta potential value of the quercetin-loaded nanoparticles was 44.49 mV. According to literature reports, absolute zeta potential values higher than 30 mV are indicative of good colloidal stability [55]. Such high surface charge can prevent the aggregation of nanoparticles due to the repulsive electrostatic forces among them. Figure 1 shows the nanoparticles size distribution (left) and zeta potential distribution (right).

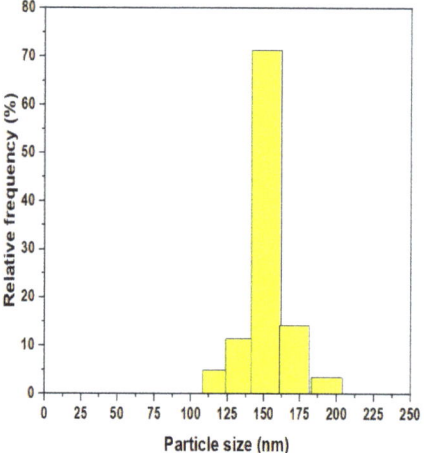
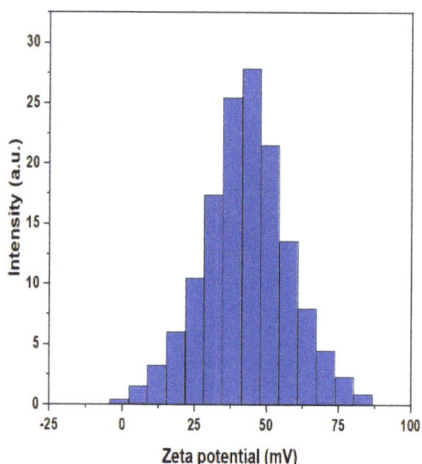

Figure 1. Particle size (**left**) and zeta potential (**right**) distributions for CMC/Fe_3O_4@SiO_2/QC nanocomposite.

3.2. FE-SEM

The powder of CMC/Fe_3O_4@SiO_2/QC was examined by means of FE-SEM to obtain insights into the morphology of the synthesized nanoparticles (Figure 2). It was evident from the images that the nanoparticles were spherical in shape, which is the most suitable geometry for drug delivery systems [56]. In addition, the nanoparticles displayed a smooth surface, which indicated good interphase adhesion between CMC and Fe_3O_4@SiO_2, likely attained via hydrogen bonding between the carboxyl groups of CMC and the hydroxyl groups of the inorganic nanoparticles, as evidenced in the FTIR analysis. It should be noted that the nanoparticles were dried using a freeze-dryer prior to SEM observation. As a result of this pretreatment, the particles were closely compacted together and appeared to show poor dispersity. Nevertheless, the DLS test, which measures the hydrodynamic size of the nanoparticles in solution, corroborated the uniform dispersity of the nanoparticles. As demonstrated in Figure 2, the nanoparticles had sizes well below 100 nm (e.g., 41.2 and 95.4 nm), which corroborates the nanoscale size of the delivery system. As expected, the obtained average size obtained with the DLS test was higher (around 150 nm), which was due to the agglomeration of some of the nanoparticles during the drying stage. A similar observation was made by Emami et al. [50] regarding the morphology of a chitosan/polyvinylpyrrolidone/α-Fe$_2$O$_3$ nanocomposite.

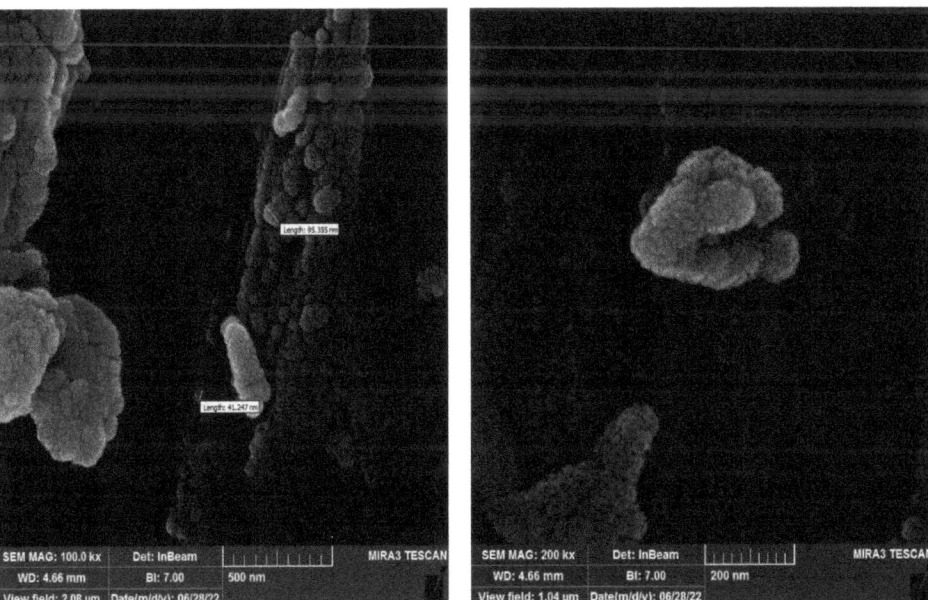

Figure 2. FE-SEM images of CMC/Fe$_3$O$_4$@SiO$_2$/QC nanocomposites. Scale bar: 500 nm (**left**) and 200 nm (**right**).

3.3. FTIR

Fourier transform infrared spectroscopy was used to assess the chemical interactions among the functional groups of the different composite component bonds in the samples and to corroborate the sample composition. Figure 3 compares the FTIR spectra of CMC, Fe$_3$O$_4$, Fe$_3$O$_4$@SiO$_2$, CMC/Fe$_3$O$_4$@SiO$_2$, and CMC/Fe$_3$O$_4$@SiO$_2$/QC. Regarding CMC, the peaks observed at 1407 and 1631 cm^{-1} were associated with symmetric and anti-symmetric stretching vibrations of the carboxyl group, respectively. Furthermore, the peak located around 3000 cm^{-1} indicated the stretching of the O-H bond. These results are in agreement with previous literature reports [21,57]. Regarding Fe$_3$O$_4$, the peak observed at 579 cm^{-1} indicated the stretching vibration of the metal–oxygen bond at the tetrahedral structure [58,59]. A new peak could be observed in Fe$_3$O$_4$@SiO$_2$ at around 1100 cm^{-1}, which could be assigned to the asymmetric vibration of bonds between oxygen and Si [60,61], which corroborated the formation of Fe$_3$O$_4$@SiO$_2$ NPs. The FTIR spectrum of CMC/Fe$_3$O$_4$@SiO$_2$ showed all the bands identified in both CMC and Fe$_3$O$_4$@SiO$_2$. A decrease in the intensity of some of the peaks was observed, which was indicative of the interactions between inorganic core–shell nanoparticles and the polymeric network. Based on the results reported by Hu et al. [62], the peak observed at 1682 cm^{-1} could be attributed to hydrogen bonds between the Fe$_3$O$_4$@SiO$_2$ nanoparticles and the carboxyl groups of CMC. Similarly, all of the characteristic peaks of CMC/Fe$_3$O$_4$@SiO$_2$ were preserved once QC was loaded in the carrier. However, a shift in the peak located around 3500 cm^{-1} and the decreased intensity of some of the other peaks indicated interactions between quercetin and CMC. In addition, the peak observed at 1080 cm^{-1} could be associated with the vibration of the bond between Si and the OH group of quercetin [62].

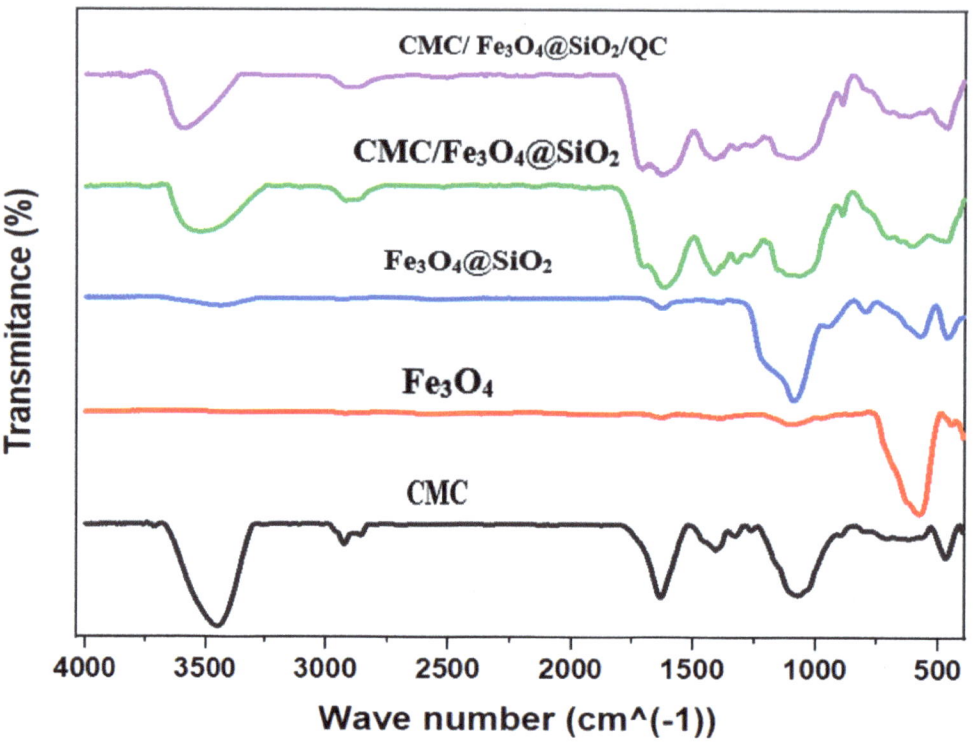

Figure 3. FTIR spectra of CMC, Fe_3O_4, $Fe_3O_4@SiO_2$, CMC/$Fe_3O_4@SiO_2$, and CMC/$Fe_3O_4@SiO_2$/QC.

3.4. XRD

The X-ray diffraction (XRD) technique was used to analyze the crystalline structures of CMC, Fe_3O_4, $Fe_3O_4@SiO_2$, CMC/$Fe_3O_4@SiO_2$, and CMC/$Fe_3O_4@SiO_2$/QC. Figure 4 shows the XRD patterns of the abovementioned samples. The six peaks observed for Fe_3O_4 at 2θ angles of 30, 35.4, 43, 53.35, 56.9, and 62.55° were in very good agreement with previous data reported in the literature [63]. These results proved the successful synthesis of magnetite. Regarding CMC, the peak at a diffraction angle of 21.84° was also consistent with former studies [64]. Regarding $Fe_3O_4@SiO_2$ nanoparticles, the XRD pattern depicted six peaks at diffraction angles of 30.34, 35.74, 43.34, 53.74, 57.24, and 62.89°, which were in agreement with a former XRD analysis of these nanoparticles [65]. The broad peak of CMC at 2θ = 21.84° indicated its amorphous structure. The XRD pattern of CMC/$Fe_3O_4@SiO_2$ was quite similar to that of CMC. However, a peak was observed at 2θ = 35.4°, which confirmed the successful incorporation of $Fe_3O_4@SiO_2$ nanoparticles within the CMC structure. The decreased intensity of this peak could be associated with the amorphous structure of CMC. Furthermore, it could be seen from the XRD results that no noticeable peaks were added to the pattern of CMC/$Fe_3O_4@SiO_2$/QC compared with that of unloaded nanocomposites. This result corroborated the entrapment of quercetin within the nanocomposite network. The encapsulation of quercetin within the nanocomposites inhibited crystal formation, which is a desirable result, as the crystalline form of quercetin is less soluble than its amorphous form [54].

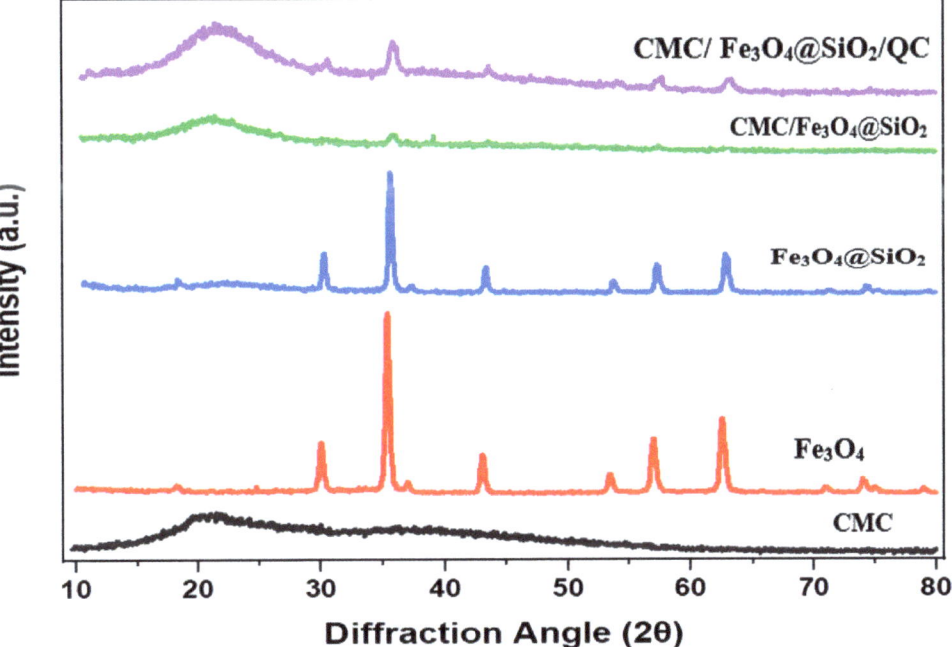

Figure 4. XRD patterns of CMC, Fe_3O_4, $Fe_3O_4@SiO_2$, CMC/$Fe_3O_4@SiO_2$, and CMC/$Fe_3O_4@SiO_2$/QC.

3.5. VSM

A vibrating sample magnetometer (VSM) was employed to determine hysteresis loops in Fe_3O_4, $Fe_3O_4@SiO_2$, CMC/$Fe_3O_4@SiO_2$, and CMC/$Fe_3O_4@SiO_2$/QC (Figure 5). The strength of the applied magnetic field was in the range of −15,000 to +15,000 kilo oersted. The curves of all samples were S-shaped and had zero coercivity, which revealed their super magnetic characteristic. Furthermore, it can be observed from Figure 5 that upon the addition of each component (SiO_2, CMC, and QC), the saturation magnetization value decreased. For instance, while this value was approximately −16 emu/g for Fe_3O_4, it fell to around −11 emu/g for CMC/$Fe_3O_4@SiO_2$/QC. This behavior could be attributed to the drop in the weight percentage of Fe_3O_4 upon the incorporation of the non-magnetic components.

3.6. Loading and Entrapment Efficiency

As mentioned before, a major challenge that inhibits the therapeutic effectiveness of QC is its low solubility, which leads to insufficient bioavailability [66]. In light of this fact, finding ways to improve the loading efficiency and entrapment efficiency of the drug on the developed nanocarriers is essential. Herein, we measured the loading efficiency and entrapment efficiency of the drug on both the CMC and CMC/$Fe_3O_4@SiO_2$ hydrogels to evaluate the impact of incorporating the core–shell nanoparticles in the polymeric structure. The loading efficiency and entrapment efficiency were calculated using Equations (1) and (2), respectively. Upon the incorporation of $Fe_3O_4@SiO_2$ NPs within the hydrogel network, the loading efficiency increased from 86.75% to 88.50%. In addition, the entrapment efficacy had a 2% rise from 45.00% to 47.25%. These data are summarized in Table 1. Based on the FTIR spectrum of the final formulation (CMC/$Fe_3O_4@SiO_2$/QC), the peak around 1080 cm^{-1} could be attributed to the vibration of the bond between Si and the hydroxyl group of quercetin. This characteristic could explain the slight improvement in the loading of quercetin and suggests that quercetin was mostly loaded in the SiO$_2$ particles. A similar result derived from an FTIR analysis was reported by Hu et al. [62].

Although the loading and entrapment efficiency did not improve significantly after introducing the core–shell nanoparticles, the FTIR analysis revealed that in the presence of these nanoparticles, a share of loaded quercetin is entrapped inside them instead of the polymeric network. This entrapment contributes to the sustained release of quercetin, as the drug has to diffuse through extra layers before getting released. In the following sections, it is shown that this sustained release profile improves anticancer efficiency of the drug delivery system.

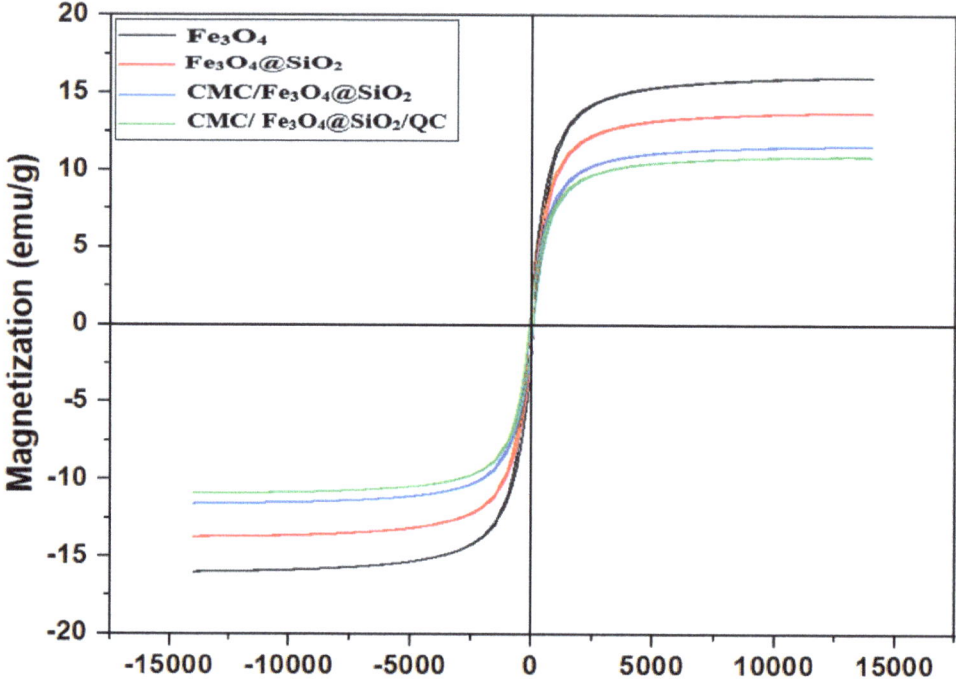

Figure 5. VMS results for Fe$_3$O$_4$, Fe$_3$O$_4$@SiO$_2$, CMC/Fe$_3$O$_4$@SiO$_2$, and CMC/Fe$_3$O$_4$@SiO$_2$/QC.

Table 1. Entrapment and loading efficiency of CMC and CMC/Fe$_3$O$_4$@SiO$_2$ hydrogels.

Hydrogel	Entrapment Efficiency (%)	Loading Efficiency (%)
CMC	45.00	86.75
CMC/Fe$_3$O$_4$@SiO$_2$	47.25	88.50

3.7. Quercetin Release Profile

The abovementioned dialysis method was employed to study the in vitro release of quercetin from the CMC/Fe$_3$O$_4$@SiO$_2$ nanocomposite. Figure 6 shows the cumulative release curve of quercetin throughout 96 h. The release profile of the drug was studied at pH values of 7.4 and 5.4. Buffer solutions with the mentioned pH values were used to simulate the microenvironment of normal and cancerous tissues, respectively. Both solutions were kept at 37 °C, which is the physiological temperature. The solutions were monitored for 96 h. Within the first 12 h, the cumulative amounts of the drug released from nanocarriers within neutral and acidic media were 33 and 43%, respectively. Wang et al. [67] examined the release profile of QC from F127, used as the carrier. Within 12 h, approximately 70% of quercetin was released in the neutral environment. Baksi and coworkers [68] studied release of quercetin from chitosan nanoparticles and obtained around 60% cumulative release within 12 h. In another study, Sunoqrot and colleagues [69] obtained about 60% release of quercetin from Eudragit S100 nanoparticles within 12 h at a pH of 7.2. Compared with the mentioned literature reports, the nanocomposite developed herein releases quercetin in a more sustained manner. This sustained release pattern in neutral environment is of paramount importance, since the administered nanoparticles have to pass through several normal tissues before reaching the tumor microenvironment. Hence, it is beneficial for the nanocarrier to retain most of the payload before reaching tumor sites. In light of this explanation, the sustained release pattern of quercetin can minimize the side effects of the drug. Likely, the stability of the quercetin molecules inside the nanocomposite network that makes this sustained release pattern feasible can be associated with the formation of hydrogen bonds between the hydroxyl groups of quercetin and the carboxylic acid groups of CMC [40].

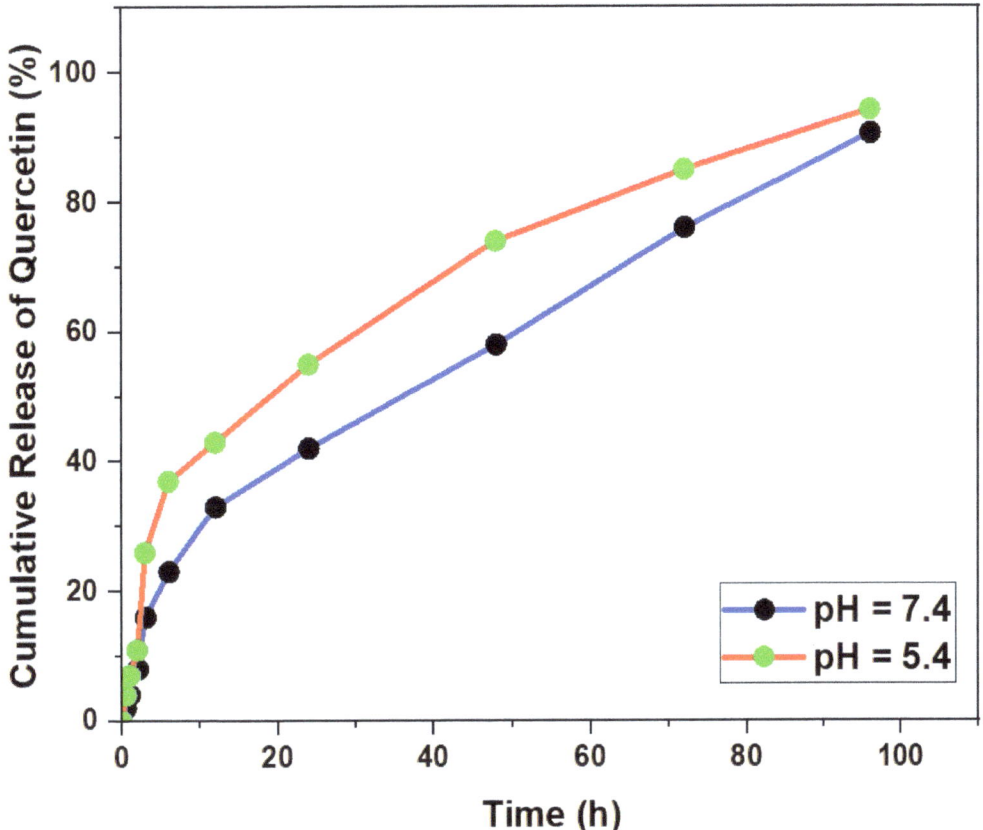

Figure 6. Cumulative curves of quercetin release from the nanocomposite carriers.

Furthermore, the difference in the quercetin release profile under acidic and neutral conditions is an evident indicator of the pH sensitivity of the CMC/Fe_3O_4@SiO_2 nanocomposite. Within 24 h, 42 and 58% of quercetin were released at pH values of 7.4 and 5.4, respectively. The increased amount of drug released under acidic pH conditions can be associated with the protonation of the carboxylic acid groups, which leads to the disruption of the hydrogen bonds between hydroxyl and carboxyl groups [62]. The pH sensitivity of the nanoparticles is a key factor in improving their therapeutic efficiency and decreasing the side effects. Many tissues of body and the blood itself have neutral pH values (close to 7.0), whereas the microenvironment of tumors has acidic pH. Provided that our nanoparticles are administered via intravenous injection, they would bypass the GI tract and reach systemic blood circulation directly. Hence, they would retain most of their therapeutic payload until they reach the tumor site, where the drug would be released. In this way, healthy tissues would be preserved from the toxicity of quercetin drug. In addition, the benefits of avoiding the GI tract include preventing the dissociation of the nanoparticle structure in the highly acidic environment of the stomach (pH < 2.0). Many literature reports have focused on developing similar pH-sensitive delivery systems for cancer therapy [70–79].

The incorporation of the nanocomposites within water-in-oil-in-water double emulsion also played a role in controlling the release of quercetin. The oil layer of the double-emulsion system acted as a membrane and inhibited the high burst release of quercetin after the disintegration of the nanocomposite network. In addition, the presence of SPAN 80 as a surfactant contributed to stabilize the nanoemulsions and to extend the drug release. Pourmadadi et al. [80] incorporated 5-fluorouacil/curcumin-loaded nanocomposites of agarose/chitosan within double nanoemulsions of W-O-W and obtained similar results in terms of release profile.

3.8. Drug Release Kinetic Modelling

The data obtained with in vitro drug release studies were used to develop a kinetic model for quercetin release. The data for both pH values of 7.4 and 5.4 were fitted to various models, including the first-order model, the zero-order model, the Korsmeyer–Peppas model, the Hixson–Crowell model, and the Higuchi model. By comparing the R-squared values of the different models, the Higuchi model was identified as the most accurate one for describing the profile of quercetin release data. Furthermore, hypothesis testing was performed to analyze the significance of the time variable in each model. All models had p-values lower than 0.0001, which indicated the significance of the chosen variable for modelling. Assuming that the null hypothesis claims the insignificance of the time variable on drug release, p-values lower than 0.0001 were proof for rejecting this hypothesis and validated the models. Figure 7 shows the fitting of the release data to different models.

The Higuchi model was the first mathematical model applied to study drug release from a porous media such as a porous polymer. Higuchi divides the porous medium in two regions. The first one is the inner region, which contains undissolved particles. The second one is the outer region, where drug particles are dissolved. The assumptions that need to be upheld upon using this model and that are thus valid in our system include the following: (i) drug diffusivity does not change; (ii) the size of the drug particles is negligible compared with the thickness of the medium walls; (iii) The drug diffuses in one dimension only; (iv) the initial drug concentration in the porous media is greater than its solubility; (v) the swelling and disintegration of the porous medium can be neglected. The formula of this model can be written as $Q = A\sqrt{D(2C - C_s)C_s t}$. In this formula, Q is the amount of released drug, t is the time, A is the area, C is the initial drug concentration, and C_s is the drug solubility in the porous media [81,82]. Table 2 shows the equations and R-squared values of the different kinetic models applied to quercetin release in both acidic and neutral environments.

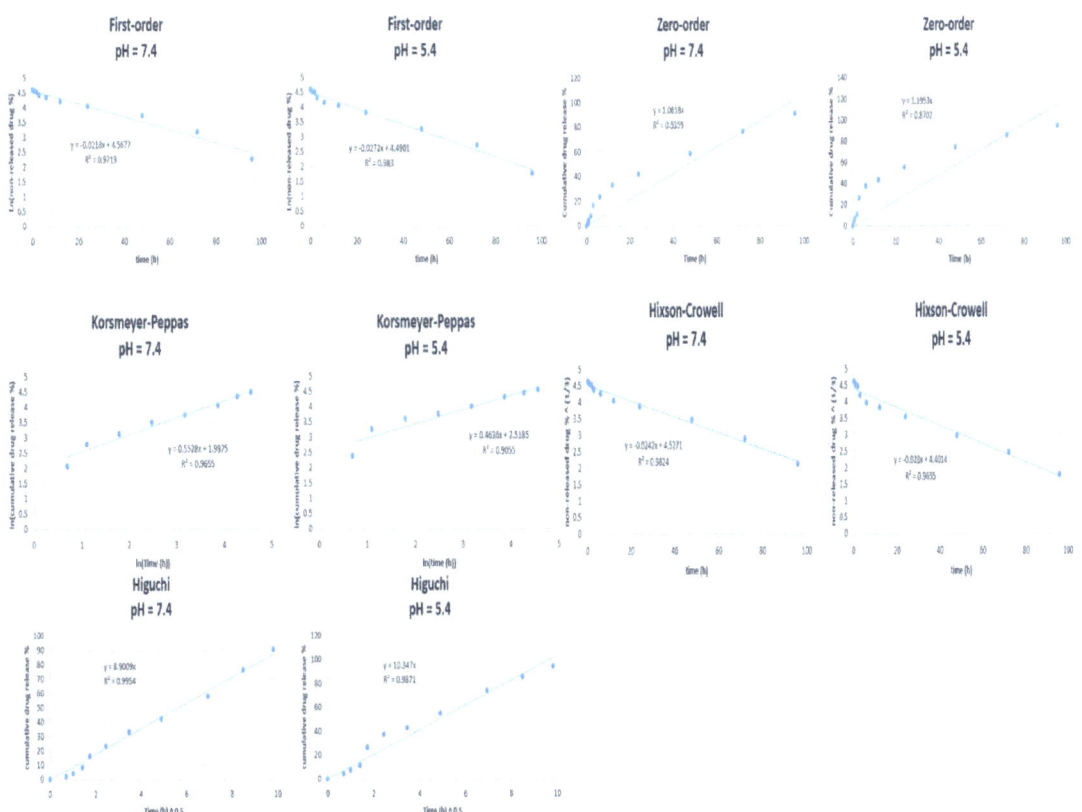

Figure 7. Fitting of the drug release data to different kinetic models.

Table 2. Equations and R-squared values for different kinetic models applied to drug release data.

Model	Equation	R^2
First-order (pH = 7.4)	$Ln\left(1 - \frac{M_t}{M_\infty}\right) = 4.5677 - 0.0218t$	0.9713
First-order (pH = 5.4)	$Ln\left(1 - \frac{M_t}{M_\infty}\right) = 4.4901 - 0.0272t$	0.9830
Zero-order (pH = 7.4)	$C_t = 1.0618t$	0.9359
Zero-order (pH = 5.4)	$C_t = 1.1953t$	0.8702
Korsmeyer–Peppas (pH = 7.4)	$Ln\left(\frac{M_t}{M_\infty}\right) = 1.9975 + 0.5528 Ln(t)$	0.9655
Korsmeyer–Peppas (pH = 5.4)	$Ln\left(\frac{M_t}{M_\infty}\right) = 2.5185 + 0.4636 Ln(t)$	0.9055
Hixson–Crowell (pH = 7.4)	$\left(1 - \frac{M_t}{M_\infty}\right)^{\frac{1}{3}} = 4.5271 - 0.0242t$	0.9824
Hixson–Crowell (pH = 5.4)	$\left(1 - \frac{M_t}{M_\infty}\right)^{\frac{1}{3}} = 4.4014 - 0.028t$	0.9655
Higuchi (pH = 7.4)	$Q = 8.9009 t^{0.5}$	0.9954
Higuchi (pH = 5.4)	$Q = 10.347 t^{0.5}$	0.9871

3.9. MTT Assay

An MTT assay was performed to assess the in vitro cytotoxicity of the prepared nanocarriers against the A549 lung cancer cell line and compare its antitumor activity with that of free quercetin. The same experiments were performed on the L929 fibroblast cell line to evaluate the potential side effects of the prepared nanocarriers. Both cell lines were incubated with $Fe_3O_4@SiO_2$, CMC, CMC/$Fe_3O_4@SiO_2$, CMC/$Fe_3O_4@SiO_2$/QC, and free quercetin. The concentration of each sample in each experiment was 5 μg/mL. A positive control group was also cultured to verify the cells' natural growth and proliferation. All experiments were performed three times to validate the accuracy of the results. The data in Figure 8 are reported as means ± SEMs.

The control group had 100% cell viability after 24 h, which corroborated the quality of the selected cells. Other than the control group, all the samples used in different experiments, including the final CMC/Fe_3O_4@SiO_2/QC nanocomposite, induced higher cytotoxicity in A549 cells compared with L929 cells. While CMC did not have any noticeable toxic impact on either of the cell lines (similar to the results presented in previous literature reports on CMC biological properties [20,83]), Fe_3O_4@SiO_2 nanoparticles eliminated around 15% of cancerous cells and exhibited the cytotoxic property. Shahabadi et al. [84] reported a similar biological behavior in Fe_3O_4@SiO_2 nanoparticles. However, their impact on the viability of L929 cells was still negligible, which is a desirable result. As expected, encapsulating these inorganic nanoparticles within the CMC hydrogel mitigated their cytotoxic effects on cancerous cells. Loading quercetin on the CMC/Fe_3O_4@SiO_2 hydrogel decreased the viability percentage of A549 cells to 65%, which was 27% less than the value of the raw nanocomposites. In addition, 97% of L929 cells survived 24 h of incubation with CMC/Fe_3O_4@SiO_2/QC nanocomposites. The viability percentages of A549 and L929 cells after 24 h of incubation with free quercetin were 71 and 89%, respectively. Sul and colleagues [85] performed an MTT assay using the A549 cell line and different concentrations of free quercetin. The lowest viability percentages of the cells hardly reached 70%, and the results were in agreement with this work. Nanomaterials loaded with Quercetin as an advanced tool for cancer treatment were reviewed by Caro and coworkers [86]. In addition, Milanezi et al. [87] performed an MTT assay using free quercetin and the L929 cell line and obtained cell viability values similar to those of this study. The selectivity of quercetin between normal cells and cancerous cells has been reported in the literature [88]. The comparison of these values with their counterparts in the CMC/Fe_3O_4@SiO_2/QC experiment is a clear indication of the superiority of the synthesized nanocomposites in terms of cytotoxicity and side effects. Quercetin-loaded nanocomposites eliminated more cancerous cells, whereas their effect on L929 fibroblast cells was less than that of the free drug. The entrapment of quercetin within core–shell nanoparticles and their consequent sustained release from the double-layer nanoemulsion led to the boosted apoptosis of A549 cells. This sustained release pattern of quercetin is of paramount importance, since this drug has low bioavailability and its sustained release can lead to prolonged exposure of cancerous cells to it.

It could be observed that incubating A549 cells with CMC/Fe_3O_4@SiO_2/QC nanoparticles did not offer a very low viability percentage (65%). This result can be attributed to the low concentration of the nanoparticles in the experiments (5 μg/mL). Here, we demonstrated that strategies such as using core–shell nanoparticles and a double-emulsion system could lead to a sustained release profile of the loaded drug, thereby decreasing the cell viability compared with the free quercetin. It is expected that increasing the initial dosage of the nanoparticles results in the elimination of more cancerous cells. It is worth mentioning that the potential low anticancer potency of quercetin could not have been responsible for this result, as multiple studies have already reported its efficient anticancer activity against various cell lines [3,51].

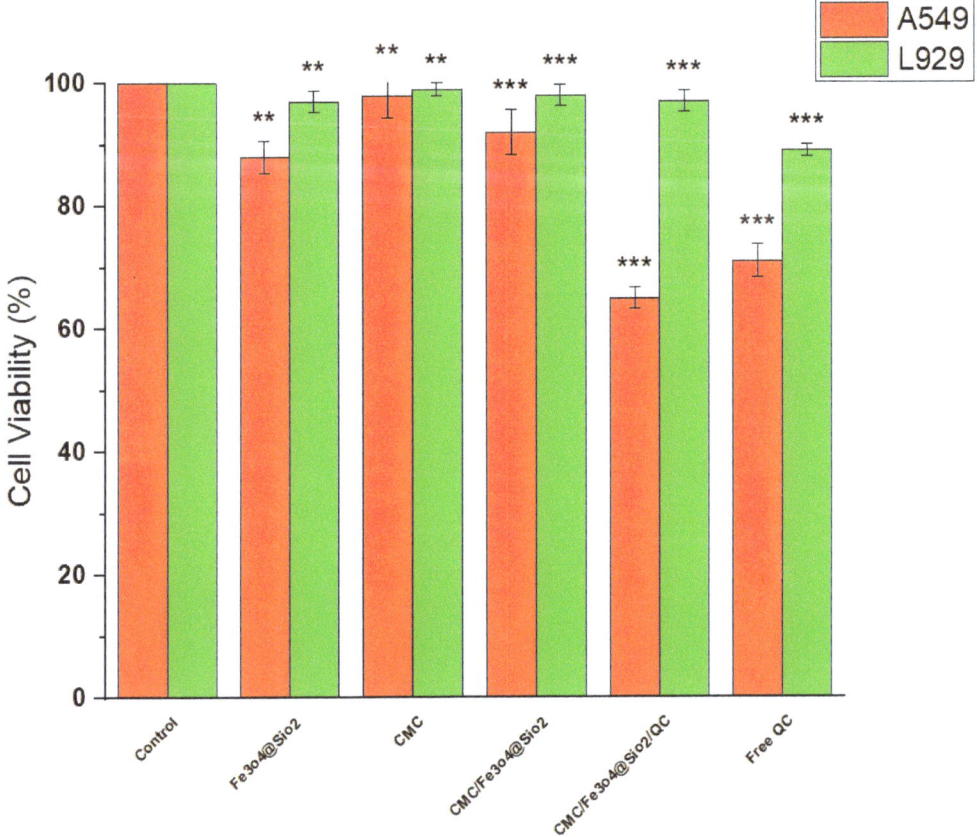

Figure 8. MTT assay results in a 24 h time frame. Incubation was performed in a 96-well plate, and the experiments were performed three times for each sample. The data are shown as means ± SEMs. (Difference between control group and the samples marked with ** was significant at $p < 0.01$; difference between control group and the samples marked with *** was significant at $p < 0.001$.)

3.10. Flow Cytometry Test

In order to analyze the cytotoxic impact of the designed system in more detail and verify the results obtained in the MTT test, a flow cytometry test was performed with free QC, CMC, $Fe_3O_4@SiO_2$, $CMC/Fe_3O_4@SiO_2$, and $CMC/Fe_3O_4@SiO_2/QC$ using the A549 cell line. The concentration of each sample for each experiment was 5 μg/mL. Figure 9 demonstrates the results of flow cytometry for each sample. As mentioned before, Q1, Q2, Q3, and Q4 quadrants represent necrotic death, late apoptotic death, early apoptotic death, and variability, respectively. It can be interpreted from Figure 9 that both CMC and $Fe_3O_4@SiO_2$ NPs had lower cell viability percentages than the nanocomposite form of the components ($CMC/Fe_3O_4@SiO_2$). Hence, the fabricated hydrogel that contained $Fe_3O_4@SiO_2$ NPs ameliorated the toxicity of both the polymeric component and the core–shell NPs. Nevertheless, the final formulation of the raw nanocarrier was noticeably cytotoxic itself, as 5.68%, 7.02%, and 3.69% of incubated cells were subjected to necrotic, late apoptotic, and early apoptotic death, respectively. Upon the encapsulation of QC within the nanocarrier network, a noticeable increase was observed in the apoptosis percentage with the increase in the share of cancerous cells subjected to either early or late apoptosis from 10.71% to 21.50% compared with the raw nanocarriers. The comparison of the cell viability values of the control group, the raw nanocarriers, and the loaded nanocarriers was consistent with the results obtained from MTT assay. Moreover, the QC-loaded nanoemulsion demonstrated superior performance in terms of apoptosis induction compared with free QC. While loaded nanoemulsions eliminated 21.50% of A549 cells through either early or late apoptosis, this figure, for free QC, was no more than 8.24%. This result was also in agreement with the MTT assay data. The significant difference between the apoptosis values of QC-loaded nanocarriers and free QC was likely associated with the very high value of late apoptosis (Q2) of the loaded nanocarriers. The strong interactions between entrapped QC and the polymeric and inorganic components of the nanocomposite, together with the encapsulation of the nanocomposites within the double-layer nanoemulsion, led to a very gradual release pattern. Such sustained release profile and the hindrance of burst release led to greater late apoptotic-mediated cell death. The share of the cells subjected to necrotic death hardly changed from the raw nanoemulsion to the loaded nanoemulsion or the free drug.

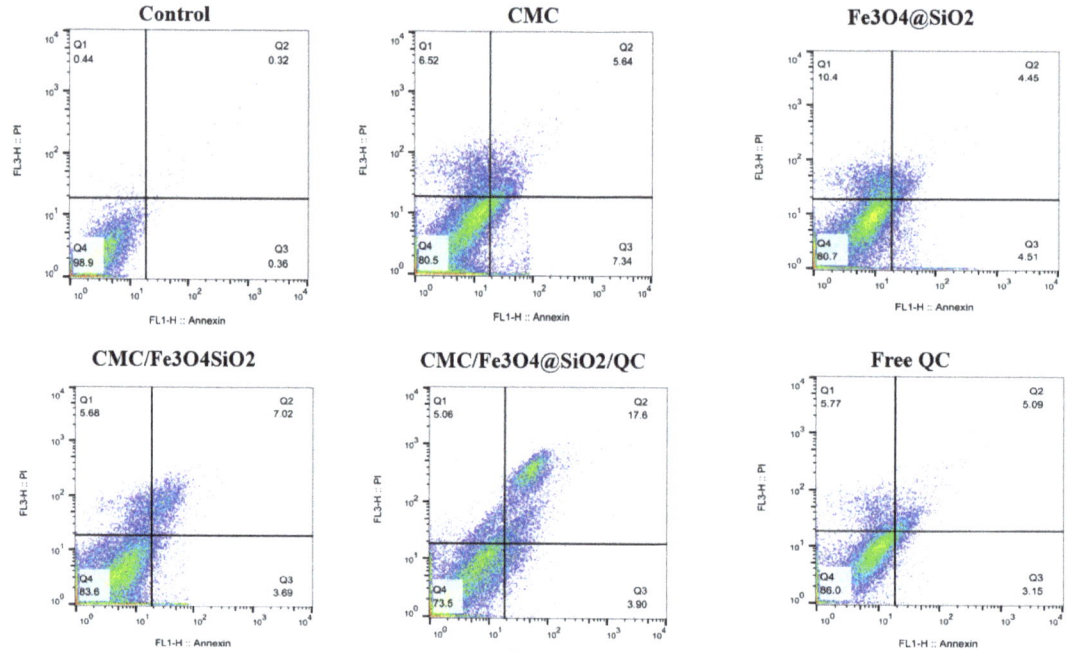

Figure 9. Results of flow cytometry test for CMC, $Fe_3O_4@SiO_2$, CMC/$Fe_3O_4@SiO_2$, CMC/$Fe_3O_4@SiO_2$/QC, and free QC using the A549 cell line.

4. Conclusions

Drawbacks such as ineffective biodistribution, insufficient biological half-life, low solubility, and instability have limited the therapeutic efficiency of quercetin drug against cancer. In this regard, the design of novel stimulus-responsive drug delivery systems that can extend the release time of QC, improve its loading efficiency, and trigger its release at specific target sites is necessary to circumvent these disadvantages. pH is an important factor that can markedly affect oral drug absorption and bioavailability, as the pH values of normal tissues and tumors are noticeably different. In this regard, pH-responsive nanocarriers can be developed for effective cancer therapy. In this study, a pH-responsive nanocomposite was synthesized based on CMC and $Fe_3O_4@SiO_2$ core–shell nanoparticles to improve the controlled release of QC at tumor sites. In addition, the nanocomposites were entrapped in a double-layer nanoemulsion to further extend the QC release time. In vitro drug release studies revealed that the developed nanoemulsion retained most of the payload in neutral medium and released a higher amount of drug in acidic environment due to the protonation of the carboxyl groups and the dissociation of hydrogen bonds. The prolonged drug release period and the pH-sensitiveness can account for the low bioavailability and inefficient biodistribution of QC, respectively. In addition, the incorporation of inorganic core–shell NPs with high surface area within the polymeric network slightly improved the loading efficiency of QC. Furthermore, an MTT assay and a flow cytometry test using the A549 cell line demonstrated the higher effectiveness of the QC-loaded nanoemulsion compared with free QC in killing cancerous cells and inducing apoptosis. An MTT assay was also performed using the L929 cell line, and the results demonstrated a low cytotoxic effect of the nanoemulsion on the fibroblast cell line compared with the free drug, which can be an indicator of low side effects of the developed nanocarrier. To sum up, a novel pH-responsive double-emulsion-entrapped hydrogel nanocomposite was prepared for effective cancer therapy.

Author Contributions: Methodology, A.R.; Data curation, M.M.E. and M.P.; Writing—original draft, M.M.E. and M.P.; Writing—review & editing, A.M.D.-P.; Supervision, A.R. and A.M.D.-P.; Funding acquisition, A.M.D.-P. All authors have read and agreed to the published version of the manuscript.

Funding: Financial support from Community of Madrid within the framework of the multi-year agreement with University of Alcalá in the line of action "Stimulus to Excellence for Permanent University Professors", Ref. EPU-INV/2020/012, is gratefully acknowledged.

Institutional Review Board Statement: Not applicable.

Informed Consent Statement: Not applicable.

Data Availability Statement: Data within this article are available upon request.

Conflicts of Interest: The authors declare that there are no conflict of interest regarding the publication of this article.

References

1. Siegel, R.L.; Miller, K.D.; Jemal, A. Cancer statistics, 2019. *CA Cancer J. Clin.* **2019**, *69*, 7–34. [CrossRef] [PubMed]
2. Bullo, S.; Buskaran, K.; Baby, R.; Dorniani, D.; Fakurazi, S.; Hussein, M.Z. Dual drugs anticancer nanoformulation using graphene oxide-PEG as nanocarrier for protocatechuic acid and chlorogenic acid. *Pharm. Res.* **2019**, *36*, 1–11. [CrossRef]
3. Samadi, A.; Pourmadadi, M.; Yazdian, F.; Rashedi, H.; Navaei-Nigjeh, M. Ameliorating quercetin constraints in cancer therapy with pH-responsive agarose-polyvinylpyrrolidone-hydroxyapatite nanocomposite encapsulated in double nanoemulsion. *Int. J. Biol. Macromol.* **2021**, *182*, 11–25. [CrossRef] [PubMed]
4. Klimaszewska-Wiśniewska, A.; Hałas-Wiśniewska, M.; Izdebska, M.; Gagat, M.; Grzanka, A.; Grzanka, D. Antiproliferative and antimetastatic action of quercetin on A549 non-small cell lung cancer cells through its effect on the cytoskeleton. *Acta Histochem.* **2017**, *119*, 99–112. [CrossRef] [PubMed]
5. Tang, S.-M.; Deng, X.-T.; Zhou, J.; Li, Q.-P.; Ge, X.-X.; Miao, L. Pharmacological basis and new insights of quercetin action in respect to its anti-cancer effects. *Biomed. Pharmacother.* **2019**, *121*, 109604. [CrossRef]
6. Reyes-Farias, M.; Carrasco-Pozo, C. The Anti-Cancer Effect of Quercetin: Molecular Implications in Cancer Metabolism. *Int. J. Mol. Sci.* **2019**, *20*, 3177. [CrossRef]
7. Vinayak, M.; Maurya, A.K. Quercetin Loaded Nanoparticles in Targeting Cancer: Recent Development. *Anti-Cancer Agents Med. Chem.* **2019**, *19*, 1560–1576. [CrossRef]
8. Xingyu, Z.; Peijie, M.; Dan, P.; Youg, W.; Daojun, W.; Xinzheng, C.; Xijun, Z.; Yangrong, S. Quercetin suppresses lung cancer growth by targeting Aurora B kinase. *Cancer Med.* **2016**, *5*, 3156–3165. [CrossRef]
9. Zhaorigetu; Farrag, I.M.; Belal, A.; Al Badawi, M.H.; Abdelhady, A.A.; Galala, F.M.A.A.; El-Sharkawy, A.; El-Dahshan, A.A.; Mehany, A.B.M. Antiproliferative, Apoptotic Effects and Suppression of Oxidative Stress of Quercetin against Induced Toxicity in Lung Cancer Cells of Rats: In vitro and In vivo Study. *J. Cancer* **2021**, *12*, 5249–5259. [CrossRef]
10. Huang, K.Y.; Wang, T.H.; Chen, C.C.; Leu, Y.L.; Li, H.J.; Jhong, C.L.; Chen, C.Y. Growth Suppression in Lung Cancer Cells Harboring EGFR-C797S Mutation by Quercetin. *Biomolecules* **2021**, *11*, 1271. [CrossRef]
11. Wang, Q.; Chen, Y.; Lu, H.; Wang, H.; Feng, H.; Xu, J.; Zhang, B. Quercetin radiosensitizes non-small cell lung cancer cells through the regulation of miR-16-5p/WEE1 axis. *IUBMB Life* **2020**, *72*, 1012–1022. [CrossRef] [PubMed]
12. Guo, H.; Ding, H.; Tang, X.; Liang, M.; Li, S.; Zhang, J.; Cao, J. Quercetin induces pro-apoptotic autophagy via SIRT1/AMPK signaling pathway in human lung cancer cell lines A549 and H1299 in vitro. *Thorac. Cancer* **2021**, *12*, 1415–1422. [CrossRef] [PubMed]

13. Li, Y.; Yao, J.; Han, C.; Yang, J.; Chaudhry, M.T.; Wang, S.; Liu, H.; Yin, Y. Quercetin, Inflammation and Immunity. *Nutrients* **2016**, *8*, 167. [CrossRef] [PubMed]
14. Cai, X.; Fang, Z.; Dou, J.; Yu, A.; Zhai, G. Bioavailability of quercetin: Problems and promises. *Curr. Med. Chem.* **2013**, *20*, 2572–2582. [CrossRef] [PubMed]
15. Chen, H.; Yao, Y. Phytoglycogen improves the water solubility and Caco-2 monolayer permeation of quercetin. *Food Chem.* **2017**, *221*, 248–257. [CrossRef]
16. Liu, L.; Tang, Y.; Gao, C.; Li, Y.; Chen, S.; Xiong, T.; Li, J.; Du, M.; Gong, Z.; Chen, H.; et al. Characterization and biodistribution in vivo of quercetin-loaded cationic nanostructured lipid carriers. *Coll. Surf. B Biointerfaces* **2014**, *115*, 125–131. [CrossRef]
17. Díez-Pascual, A.M. Inorganic-Nanoparticle Modified Polymers. *Polymers* **2022**, *14*, 1979. [CrossRef]
18. Zang, X.; Cheng, M.; Zhang, X.; Chen, X. Quercetin nanoformulations: A promising strategy for tumor therapy. *Food Funct.* **2021**, *12*, 6664–6681. [CrossRef]
19. Haseli, S.; Pourmadadi, M.; Samadi, A.; Yazdian, F.; Abdouss, M.; Rashedi, H.; Navaei-Nigjeh, M. A novel pH-responsive nanoniosomal emulsion for sustained release of curcumin from a chitosan-based nanocarrier: Emphasis on the concurrent improvement of loading, sustained release, and apoptosis induction. *Biotechnol. Prog.* **2022**, *38*, e3280. [CrossRef]
20. Rao, Z.; Ge, H.; Liu, L.; Zhu, C.; Min, L.; Liu, M.; Fan, L.; Li, D. Carboxymethyl cellulose modified graphene oxide as pH-sensitive drug delivery system. *Int. J. Biol. Macromol.* **2018**, *107*, 1184–1192. [CrossRef]
21. Wang, R.; Shou, D.; Lv, O.; Kong, Y.; Deng, L.; Shen, J. pH-Controlled drug delivery with hybrid aerogel of chitosan, carboxymethyl cellulose and graphene oxide as the carrier. *Int. J. Biol. Macromol.* **2017**, *103*, 248–253. [CrossRef] [PubMed]
22. Javanbakht, S.; Namazi, H. Doxorubicin loaded carboxymethyl cellulose/graphene quantum dot nanocomposite hydrogel films as a potential anticancer drug delivery system. *Mater. Sci. Eng. C* **2018**, *87*, 50–59. [CrossRef]
23. Rajabzadeh-Khosroshahi, M.; Pourmadadi, M.; Yazdian, F.; Rashedi, H.; Navaei-Nigjeh, M.; Rasekh, B. Chitosan/agarose/graphitic carbon nitride nanocomposite as an efficient pH-sensitive drug delivery system for anticancer curcumin releasing. *J. Drug Deliv. Sci. Technol.* **2022**, *74*, 103443. [CrossRef]
24. Sosnik, A.; Seremeta, K.P. Polymeric hydrogels as technology platform for drug delivery applications. *Gels* **2017**, *3*, 25. [CrossRef] [PubMed]
25. Liao, J.; Huang, H. Review on magnetic natural polymer constructed hydrogels as vehicles for drug delivery. *Biomacromolecules* **2020**, *21*, 2574–2594. [CrossRef]
26. Hamidi, M.; Azadi, A.; Rafiei, P. Hydrogel nanoparticles in drug delivery. *Adv. Drug Deliv. Rev.* **2008**, *60*, 1638–1649. [CrossRef]
27. Papagiannopoulos, A.; Vlassi, E.; Pispas, S.; Tsitsilianis, C.; Aurel, R. Polyethylene Oxide Hydrogels Crosslinked by Peroxide for the Controlled Release of Proteins. *Macromol.* **2002**, *1*, 37–48. [CrossRef]
28. Shin, Y.; Kim, D.; Hu, Y.; Kim, Y.; Hong, I.K.; Kim, M.S.; Jung, S. pH-Responsive Succinoglycan-Carboxymethyl Cellulose Hydrogels with Highly Improved Mechanical Strength for Controlled Drug Delivery Systems. *Polymers* **2021**, *13*, 3197. [CrossRef]
29. Rasoulzadeh, M.; Namazi, H. Carboxymethyl cellulose/graphene oxide bio-nanocomposite hydrogel beads as anticancer drug carrier agent. *Carbohydr. Polym.* **2017**, *168*, 320–326. [CrossRef]
30. Zhao, M.; Zhou, H.; Hao, L.; Chen, H.; Zhou, X. Natural rosin modified carboxymethyl cellulose delivery system with lowered toxicity for long-term pest control. *Carbohydr. Polym.* **2021**, *259*, 117749. [CrossRef]
31. He, X.; Tang, K.; Li, X.; Wang, F.; Liu, J.; Zou, F.; Yang, M.; Li, M. A porous collagen-carboxymethyl cellulose/hydroxyapatite composite for bone tissue engineering by bi-molecular template method. *Int. J. Biol. Macromol.* **2019**, *137*, 45–53. [CrossRef] [PubMed]
32. Xu, X.; Liu, Y.; Fu, W.; Yao, M.; Ding, Z.; Xuan, J.; Li, D.; Wang, S.; Xia, Y.; Cao, M. Poly (N-isopropylacrylamide)-based thermoresponsive composite hydrogels for biomedical applications. *Polymers* **2020**, *12*, 580. [CrossRef] [PubMed]
33. Azizi, A. Green synthesis of Fe_3O_4 nanoparticles and its application in preparation of Fe_3O_4/cellulose magnetic nanocomposite: A suitable proposal for drug delivery systems. *J. Inorg. Organomet. Polym. Mater.* **2020**, *30*, 3552–3561. [CrossRef]
34. Akhtar, H.; Pourmadadi, M.; Yazdian, F.; Rashedi, H. Kosmotropic and chaotropic effect of biocompatible Fe_3O_4 nanoparticles on egg white lysozyme; the key role of nanoparticle-protein corona formation. *J. Mol. Struct.* **2022**, *1253*, 132016. [CrossRef]
35. Mohammadi, R.; Saboury, A.; Javanbakht, S.; Foroutan, R.; Shaabani, A. Carboxymethylcellulose/polyacrylic acid/starch-modified Fe_3O_4 interpenetrating magnetic nanocomposite hydrogel beads as pH-sensitive carrier for oral anticancer drug delivery system. *Eur. Polym. J.* **2021**, *153*, 110500. [CrossRef]
36. Najafi, M.; Morsali, A.; Bozorgmehr, M.R. DFT study of SiO_2 nanoparticles as a drug delivery system: Structural and mechanistic aspects. *Struct. Chem.* **2019**, *30*, 715–726. [CrossRef]
37. Abolghasemzade, S.; Pourmadadi, M.; Rashedi, H.; Yazdian, F.; Kianbakht, S.; Navaei-Nigjeh, M. PVA based nanofiber containing CQDs modified with silica NPs and silk fibroin accelerates wound healing in a rat model. *J. Mater. Chem. B* **2021**, *9*, 658–676. [CrossRef]
38. Deshpande, S.; Sharma, S.; Koul, V.; Singh, N. Core–shell nanoparticles as an efficient, sustained, and triggered drug-delivery system. *ACS Omega* **2017**, *2*, 6455–6463. [CrossRef]
39. Chatterjee, K.; Sarkar, S.; Rao, K.J.; Paria, S. Core/shell nanoparticles in biomedical applications. *Adv. Colloid Interface Sci.* **2014**, *209*, 8–39. [CrossRef]

40. Tongdeesoontorn, W.; Mauer, L.J.; Wongruong, S.; Sriburi, P.; Rachtanapun, P. Physical and antioxidant properties of cassava starch–carboxymethyl cellulose incorporated with quercetin and TBHQ as active food packaging. *Polymers* **2020**, *12*, 366. [CrossRef]
41. Ezati, P.; Rhim, J.-W. Fabrication of quercetin-loaded biopolymer films as functional packaging materials. *ACS Appl. Polym. Mater.* **2021**, *3*, 2131–2137. [CrossRef]
42. He, H.; Yuan, D.; Gao, Z.; Xiao, D.; He, H.; Dai, H.; Peng, J.; Li, N. Mixed hemimicelles solid-phase extraction based on ionic liquid-coated Fe_3O_4/SiO_2 nanoparticles for the determination of flavonoids in bio-matrix samples coupled with high performance liquid chromatography. *J. Chromatogr. A* **2014**, *1324*, 78–85. [CrossRef] [PubMed]
43. Sani, T.H.; Hadjmohammadi, M.; Fatemi, M.H. Extraction and determination of flavonoids in fruit juices and vegetables using Fe_3O_4/SiO_2 magnetic nanoparticles modified with mixed hemi/ad-micelle cetyltrimethylammonium bromide and high performance liquid chromatography. *J. Sep. Sci.* **2020**, *43*, 1224–1231. [CrossRef] [PubMed]
44. Cai, W.; Guo, M.; Weng, X.; Zhang, W.; Owens, G.; Chen, Z. Modified green synthesis of Fe_3O_4@ SiO_2 nanoparticles for pH responsive drug release. *Mater. Sci. Eng. C* **2020**, *112*, 110900. [CrossRef]
45. Wang, W.; Wang, Q.; Wang, A. pH-responsive carboxymethylcellulose-g-poly (sodium acrylate)/polyvinylpyrrolidone semi-IPN hydrogels with enhanced responsive and swelling properties. *Macromol. Res.* **2011**, *19*, 57–65. [CrossRef]
46. Guo, X.; Xue, L.; Lv, W.; Liu, Q.; Li, R.; Li, Z.; Wang, J. Facile synthesis of magnetic carboxymethylcellulose nanocarriers for pH-responsive delivery of doxorubicin. *New J. Chem.* **2015**, *39*, 7340–7347. [CrossRef]
47. Chouaibi, M.; Mejri, J.; Rezig, L.; Abdelli, K.; Hamdi, S. Experimental study of quercetin microencapsulation using water-in-oil-in-water (W1/O/W2) double emulsion. *J. Mol. Liq.* **2019**, *273*, 183–191. [CrossRef]
48. Iqbal, M.; Zafar, N.; Fessi, H.; Elaissari, A. Double emulsion solvent evaporation techniques used for drug encapsulation. *Int. J. Pharm.* **2015**, *496*, 173–190. [CrossRef]
49. Bayat, A.; Shakourian-Fard, M.; Ehyaei, N.; Hashemi, M.M. A magnetic supported iron complex for selective oxidation of sulfides to sulfoxides using 30% hydrogen peroxide at room temperature. *RSC Adv.* **2014**, *4*, 44274–44281. [CrossRef]
50. Gerami, S.E.; Pourmadadi, M.; Fatoorehchi, H.; Yazdian, F.; Rashedi, H.; Nigjeh, M.N. Preparation of pH-sensitive chitosan/polyvinylpyrrolidone/α-Fe2O3 nanocomposite for drug delivery application: Emphasis on ameliorating restrictions. *Int. J. Biol. Macromol.* **2021**, *173*, 409–420. [CrossRef]
51. Ahmadi, M.; Pourmadadi, M.; Ghorbanian, S.A.; Yazdian, F.; Rashedi, H. Ultra pH-sensitive nanocarrier based on Fe2O3/chitosan/montmorillonite for quercetin delivery. *Int. J. Biol. Macromol.* **2021**, *191*, 738–745. [CrossRef] [PubMed]
52. Zavareh, H.S.; Pourmadadi, M.; Moradi, A.; Yazdian, F.; Omidi, M. Chitosan/carbon quantum dot/aptamer complex as a potential anticancer drug delivery system towards the release of 5-fluorouracil. *Int. J. Biol. Macromol.* **2020**, *165*, 1422–1430. [CrossRef] [PubMed]
53. Samadi, A.; Haseli, S.; Pourmadadi, M.; Rashedi, H.; Yazdian, F.; Navaei-Nigjeh, M. Curcumin-loaded Chitosan-Agarose-Montmorillonite Hydrogel Nanocomposite for the Treatment of Breast Cancer. In Proceedings of the 2020 27th National and 5th International Iranian Conference on Biomedical Engineering (ICBME), Tehran, Iran, 26–27 November 2020; pp. 148–153.
54. Nematollahi, E.; Pourmadadi, M.; Yazdian, F.; Fatoorehchi, H.; Rashedi, H.; Nigjeh, M.N. Synthesis and characterization of chitosan/polyvinylpyrrolidone coated nanoporous γ-Alumina as a pH-sensitive carrier for controlled release of quercetin. *Int. J. Biol. Macromol.* **2021**, *183*, 600–613. [CrossRef] [PubMed]
55. Júnior, J.A.A.; Baldo, J.B. The behavior of zeta potential of silica suspensions. *New J. Glass Ceram.* **2014**, *4*, 29. [CrossRef]
56. Prabha, G.; Raj, V. Preparation and characterization of chitosan—Polyethylene glycol-polyvinylpyrrolidone-coated superparamagnetic iron oxide nanoparticles as carrier system: Drug loading and in vitro drug release study. *J. Biomed. Mater. Res. Part B Appl. Biomater.* **2016**, *104*, 808–816. [CrossRef]
57. Javanbakht, S.; Pooresmaeil, M.; Namazi, H. Green one-pot synthesis of carboxymethylcellulose/Zn-based metal-organic framework/graphene oxide bio-nanocomposite as a nanocarrier for drug delivery system. *Carbohydr. Polym.* **2019**, *208*, 294–301. [CrossRef]
58. Yang, K.; Peng, H.; Wen, Y.; Li, N. Re-examination of characteristic FTIR spectrum of secondary layer in bilayer oleic acid-coated Fe_3O_4 nanoparticles. *Appl. Surf. Sci.* **2010**, *256*, 3093–3097. [CrossRef]
59. Ozkaya, T.; Toprak, M.S.; Baykal, A.; Kavas, H.; Köseoğlu, Y.; Aktaş, B. Synthesis of Fe_3O_4 nanoparticles at 100 °C and its magnetic characterization. *J. Alloys Compd.* **2009**, *472*, 18–23. [CrossRef]
60. Huang, S.; Li, C.; Cheng, Z.; Fan, Y.; Yang, P.; Zhang, C.; Yang, K.; Lin, J. Magnetic Fe_3O_4@ mesoporous silica composites for drug delivery and bioadsorption. *J. Colloid Interface Sci.* **2012**, *376*, 312–321. [CrossRef]
61. Beganskienė, A.; Sirutkaitis, V.; Kurtinaitienė, M.; Juškėnas, R.; Kareiva, A. FTIR, TEM and NMR investigations of Stöber silica nanoparticles. *Mater. Sci. (Medžiagotyra)* **2004**, *10*, 287–290.
62. Hu, X.; Wang, Y.; Zhang, L.; Xu, M.; Zhang, J.; Dong, W. Design of a pH-sensitive magnetic composite hydrogel based on salecan graft copolymer and Fe_3O_4@SiO_2 nanoparticles as drug carrier. *Int. J. Biol. Macromol.* **2018**, *107*, 1811–1820. [CrossRef]
63. Kazemi, S.; Pourmadadi, M.; Yazdian, F.; Ghadami, A. The synthesis and characterization of targeted delivery curcumin using chitosan-magnetite-reduced graphene oxide as nano-carrier. *Int. J. Biol. Macromol.* **2021**, *186*, 554–562. [CrossRef] [PubMed]
64. Anjali, T. Modification of carboxymethyl cellulose through oxidation. *Carbohydr. Polym.* **2012**, *87*, 457–460. [CrossRef] [PubMed]
65. Gao, M.; Li, W.; Dong, J.; Zhang, Z.; Yang, B. Synthesis and characterization of superparamagnetic Fe_3O_4@ SiO_2 core-shell composite nanoparticles. *World J. Condens. Matter Phys.* **2011**, *1*, 49–54. [CrossRef]

66. Tran, T.H.; Guo, Y.; Song, D.; Bruno, R.S.; Lu, X. Quercetin-Containing Self-Nanoemulsifying Drug Delivery System for Improving Oral Bioavailability. *J. Pharm. Sci.* **2014**, *103*, 840–852. [CrossRef]
67. Wang, T.; Wu, C.; Li, T.; Fan, G.; Gong, H.; Liu, P.; Yang, Y.; Sun, L. Comparison of two nanocarriers for quercetin in morphology, loading behavior, release kinetics and cell inhibitory activity. *Mater. Express* **2020**, *10*, 1589–1598. [CrossRef]
68. Baksi, R.; Singh, D.P.; Borse, S.P.; Rana, R.; Sharma, V.; Nivsarkar, M. In vitro and in vivo anticancer efficacy potential of Quercetin loaded polymeric nanoparticles. *Biomed. Pharmacother.* **2018**, *106*, 1513–1526. [CrossRef]
69. Sunoqrot, S.; Abujamous, L. pH-sensitive polymeric nanoparticles of quercetin as a potential colon cancer-targeted nanomedicine. *J. Drug Deliv. Sci. Technol.* **2019**, *52*, 670–676. [CrossRef]
70. Chen, M.; Chen, C.; Shen, Z.; Zhang, X.; Chen, Y.; Lin, F.; Ma, X.; Zhuang, C.; Mao, Y.; Gan, H.; et al. Extracellular pH is a biomarker enabling detection of breast cancer and liver cancer using CEST MRI. *Oncotarget* **2017**, *8*, 45759–45767. [CrossRef]
71. Li, Z.; Huang, J.; Wu, J. pH-Sensitive nanogels for drug delivery in cancer therapy. *Biomater. Sci.* **2021**, *9*, 574–589. [CrossRef]
72. Norouzi, M.; Nazari, B.; Miller, D.W. Injectable hydrogel-based drug delivery systems for local cancer therapy. *Drug Discov. Today* **2016**, *21*, 1835–1849. [CrossRef] [PubMed]
73. He, X.; Li, J.; An, S.; Jiang, C. pH-sensitive drug-delivery systems for tumor targeting. *Ther. Deliv.* **2013**, *4*, 1499–1510. [CrossRef] [PubMed]
74. Sun, X.; Zhang, G.; Wu, Z. Nanostructures for pH-sensitive Drug Delivery and Magnetic Resonance Contrast Enhancement Systems. *Curr. Med. Chem.* **2018**, *25*, 3036–3057. [CrossRef] [PubMed]
75. Zhai, Y.; Wang, J.; Qiu, L. Drug -driven self-assembly of pH-sensitive nano-vesicles with high loading capacity and anti-tumor efficacy. *Biomater. Sci.* **2021**, *9*, 3348–3361. [CrossRef]
76. Liu, J.; Huang, Y.; Kumar, A.; Tan, A.; Jin, S.; Mozhi, A.; Liang, X.J. pH-sensitive nano-systems for drug delivery in cancer therapy. *Biotechnol. Adv.* **2014**, *32*, 693–710. [CrossRef]
77. Sethuraman, V.; Janakiraman, K.; Krishnaswami, V.; Kandasamy, R. Recent Progress in Stimuli-Responsive Intelligent Nano Scale Drug Delivery Systems: A Special Focus Towards pH-Sensitive Systems. *Curr. Drug Targets* **2021**, *22*, 947–966. [CrossRef]
78. Pfeiffer, J.K. Chapter 4—Innate Host Barriers to Viral Trafficking and Population Diversity: Lessons Learned from Poliovirus. In *Advances in Virus Research*; Maramorosch, K., Shatkin, A.J., Murphy, F.A., Eds.; Academic Press: Cambridge, MA, USA, 2010; Volume 77, pp. 85–118.
79. Constable, P. CHAPTER 111—Clinical Acid-Base Chemistry. In *Critical Care Nephrology*, 2nd ed.; Ronco, C., Bellomo, R., Kellum, J.A., Eds.; W.B. Saunders: Philadelphia, PA, USA, 2009; pp. 581–586. [CrossRef]
80. Pourmadadi, M.; Ahmadi, M.; Abdouss, M.; Yazdian, F.; Rashedi, H.; Navaei-Nigjeh, M.; Hesari, Y. The synthesis and characterization of double nanoemulsion for targeted Co-Delivery of 5-fluorouracil and curcumin using pH-sensitive agarose/chitosan nanocarrier. *J. Drug Deliv. Sci. Technol.* **2022**, *70*, 102849. [CrossRef]
81. Dash, S.; Murthy, P.N.; Nath, L.; Chowdhury, P. Kinetic modeling on drug release from controlled drug delivery systems. *Acta Pol. Pharm.* **2010**, *67*, 217–223.
82. Paul, D.R. Elaborations on the Higuchi model for drug delivery. *Int. J. Pharm.* **2011**, *418*, 13–17. [CrossRef]
83. Javanbakht, S.; Nabi, M.; Shadi, M.; Amini, M.M.; Shaabani, A. Carboxymethyl cellulose/tetracycline@UiO-66 nanocomposite hydrogel films as a potential antibacterial wound dressing. *Int. J. Biol. Macromol.* **2021**, *188*, 811–819. [CrossRef]
84. Shahabadi, N.; Falsafi, M.; Mansouri, K. Improving antiproliferative effect of the anticancer drug cytarabine on human promyelocytic leukemia cells by coating on Fe_3O_4@SiO_2 nanoparticles. *Colloids Surfaces. B Biointerfaces* **2016**, *141*, 213–222. [CrossRef] [PubMed]
85. Sul, O.J.; Ra, S.W. Quercetin Prevents LPS-Induced Oxidative Stress and Inflammation by Modulating NOX2/ROS/NF-kB in Lung Epithelial Cells. *Molecules* **2021**, *26*, 6949. [CrossRef] [PubMed]
86. Caro, C.; Pourmadadi, M.; Eshaghi, M.M.; Rahmani, E.; Shojaei, S.; Paiva-Santos, A.C.; Rahdar, A.; Behzadmehr, R.; García-Martín, M.L.; Díez-Pascual, A.M. Nanomaterials loaded with Quercetin as an advanced tool for cancer treatment. *J. Drug. Del. Sci. Technol.* **2022**, *78*, 103938. [CrossRef]
87. Milanezi, F.G.; Meireles, L.M.; de Christo Scherer, M.M.; de Oliveira, J.P.; da Silva, A.R.; de Araujo, M.L.; Endringer, D.C.; Fronza, M.; Guimarães, M.C.C.; Scherer, R. Antioxidant, antimicrobial and cytotoxic activities of gold nanoparticles capped with quercetin. *Saudi Pharm. J.* **2019**, *27*, 968–974. [CrossRef] [PubMed]
88. Shafabakhsh, R.; Asemi, Z. Quercetin: A natural compound for ovarian cancer treatment. *J. Ovarian Res.* **2019**, *12*, 55. [CrossRef]

Article

Reduced Cardiotoxicity of Ponatinib-Loaded PLGA-PEG-PLGA Nanoparticles in Zebrafish Xenograft Model

Hissa F. Al-Thani [1,2], Samar Shurbaji [1], Zain Zaki Zakaria [1], Maram H. Hasan [1], Katerina Goracinova [3,4], Hesham M. Korashy [3] and Huseyin C. Yalcin [1,*]

1. Biomedical Research Center, Qatar University, Doha P.O. Box 2713, Qatar; hissa.althani@qu.edu.qa (H.F.A.-T.); ss1104227@student.qu.edu.qa (S.S.); zain.zakaria@qu.edu.qa (Z.Z.Z.); mhasan@qu.edu.qa (M.H.H.)
2. Department of Biomedical Science, College of Health Sciences, QU Health, Qatar University, Doha P.O. Box 2713, Qatar
3. Department of Pharmaceutical Sciences, College of Pharmacy, QU Health, Qatar University, Doha P.O. Box 2713, Qatar; kago@ff.ukim.edu.mk (K.G.); hkorashy@qu.edu.qa (H.M.K.)
4. Faculty of Pharmacy, Ss. Cyril and Methodius University in Skopje, Mother Theresa 47, 1000 Skopje, North Macedonia
* Correspondence: hyalcin@qu.edu.qa

Abstract: Tyrosine kinase inhibitors (TKIs) are the new generation of anti-cancer drugs with high potential against cancer cells' proliferation and growth. However, TKIs are associated with severe cardiotoxicity, limiting their clinical value. One TKI that has been developed recently but not explored much is Ponatinib. The use of nanoparticles (NPs) as a better therapeutic agent to deliver anti-cancer drugs and reduce their cardiotoxicity has been recently considered. In this study, with the aim to reduce Ponatinib cardiotoxicity, Poly(D,L-lactide-co-glycolide)-b-poly(ethyleneoxide)-b-poly(D,L-lactide-co-glycolide) (PLGA-PEG-PLGA) triblock copolymer was used to synthesize Ponatinib in loaded PLGA-PEG-PLGA NPs for chronic myeloid leukemia (CML) treatment. In addition to physicochemical NPs characterization (NPs shape, size, size distribution, surface charge, dissolution rate, drug content, and efficacy of encapsulation) the efficacy and safety of these drug-delivery systems were assessed in vivo using zebrafish. Zebrafish are a powerful animal model for investigating the cardiotoxicity associated with anti-cancer drugs such as TKIs, to determine the optimum concentration of smart NPs with the least side effects, and to generate a xenograft model of several cancer types. Therefore, the cardiotoxicity of unloaded and drug-loaded PLGA-PEG-PLGA NPs was studied using the zebrafish model by measuring the survival rate and cardiac function parameters, and therapeutic concentration for in vivo efficacy studies was optimized in an in vivo setting. Further, the efficacy of drug-loaded PLGA-PEG-PLGA NPs was tested on the zebrafish cancer xenograft model, in which human myelogenous leukemia cell line K562 was transplanted into zebrafish embryos. Our results demonstrated that the Ponatinib-loaded PLGA-PEG-PLGA NPs at a concentration of 0.001 mg/mL are non-toxic/non-cardio-toxic in the studied zebrafish xenograft model.

Keywords: zebrafish; leukemia; nanomedicine; nanoparticle; pre-clinical; cardiotoxicity; cancer; Ponatinib; xenograft; PLGA

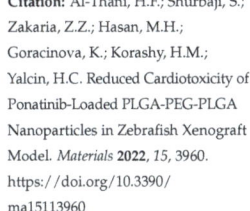

Citation: Al-Thani, H.F.; Shurbaji, S.; Zakaria, Z.Z.; Hasan, M.H.; Goracinova, K.; Korashy, H.M.; Yalcin, H.C. Reduced Cardiotoxicity of Ponatinib-Loaded PLGA-PEG-PLGA Nanoparticles in Zebrafish Xenograft Model. *Materials* **2022**, *15*, 3960. https://doi.org/10.3390/ma15113960

Academic Editor: Abdelwahab Omri

Received: 5 April 2022
Accepted: 13 May 2022
Published: 2 June 2022

Publisher's Note: MDPI stays neutral with regard to jurisdictional claims in published maps and institutional affiliations.

Copyright: © 2022 by the authors. Licensee MDPI, Basel, Switzerland. This article is an open access article distributed under the terms and conditions of the Creative Commons Attribution (CC BY) license (https://creativecommons.org/licenses/by/4.0/).

1. Introduction

Cancer is the second leading cause of death worldwide with a high number of incidents [1]. Cancer evolves from mutations that cause the activation of oncogenes or/and inactivation of the tumor suppressor genes leading to uncontrolled cell growth and proliferation, which further trigger other complications in the body that eventually might lead to death [2]. Leukemia is a type of cancer that is characterized by the uncontrolled growth of the hematopoietic stem cells from the bone marrow [3]. There are several subtypes of leukemia and the most encountered subtype among adults is CML [3]. CML is generally diagnosed by the presence of the Philadelphia chromosome that harbors the *BCR-ABL* oncogene, which would cause abnormal cell proliferation and complications in the patients [4].

Therefore, the demand for successful anti-cancer therapeutics and the development of effective tools for early cancer detection and screening have increased. For example, the introduction of TKIs [5] such as imatinib, nilotinib, Ponatinib, and dasatinib as anti-cancer drugs particularly for CML has aided in improving the overall outcomes of the patients and increasing their survival rates [6,7]. However, due to some encountered toxicity of these drugs, especially to the heart [8], the necessity to employ nanotechnology in anti-cancer treatments has increasingly progressed [9,10]. This is due to the higher efficiency and precision of nanoparticles (NPs) in targeting cancer cells and reducing toxicity associated with anti-cancer drugs [11].

Due to the biocompatibility and biodegradability of the PLGA and PEG triblock copolymers, they have been extensively used in drug delivery [12,13]. Moreover, PLGA and PEG polymers have gained the approval of the US Food and Drug Administration (FDA) and the European Medicine Agency (EMA) to be utilized in many therapeutic applications due to their low toxicities and solubilization effects [14,15]. Moreover, PEG polymer can prolong the half-life and the circulation period of the NPs in the body by permitting further reticuloendothelial system recognition. This is because PEG polymers have steric stability and they can selectively evade the attachment of the opsonin proteins on the NPs surface [12,16]. Furthermore, the PLGA and PEG triblock copolymer has shown their ability to deliver several different anti-cancer drugs such as cisplatin, methotrexate, doxorubicin [17], and irinotecan [18].

Zebrafish have been used as a research model in many applications, such as cancer and neuronal disorders studies, due to their numerous unique characteristics [19–21]. For example, they have a high genetic resemblance to humans with about 70% orthologue genes, making them a useful model for genetic manipulation [22]. Moreover, they are easy to maintain, have short maturation and developing time, and their transparent embryos have made imaging and studying internal organs such as the heart much easier [23]. In addition, due to their lack of adaptive immunity during the first months of development, zebrafish are a good model for the xenotransplantation of human tumor cells to develop a cancer model to study human cancers and testing of anti-cancer drugs [24]. Zebrafish are also considered a useful animal model for investigating and screening the toxicity of several agents such as anti-cancer drugs [25].

In this study, we first developed and characterized PLGA-PEG-PLGA NPs and then loaded the generated PLGA-PEG-PLGA NPs with different concentrations of Ponatinib. Next, we tested these NPs efficacy in reducing the cardiotoxic effect of Ponatinib in the zebrafish xenograft model. According to our results, PLGA-PEG-PLGA NPs demonstrated their efficacy in reducing the well-known cardiotoxic side effects of Ponatinib in our studied model.

2. Materials and Methods

2.1. Cell Culture

Human CML K-562 cell line was obtained from the American-type culture collection (ATCC) (ATCC, Manassas, VA, USA) and as a kind gift from Shahab Uddin Khan from the interim Translational Research Institute (iTRI) at Hamad Medical Corporation (HMC), Doha, Qatar. Cells have been cultured according to the optimum conditions described by the manufacturer. The cells were cultured in RPMI 1640 supplemented with 10% FBS, 10,000 U/mL Penicillin-Streptomycin, and 100× GlutaMAX at 37 °C in a humidified 5% CO_2 incubator. All reagents were obtained from Gibco (ThermoScientific, Waltham, MA, USA). The cells' culture medium was changed every other day to obtain the optimum cell count and maintain their viability at 90% following this equation: No. of viable cells/total No. of cells × 100. The cell counting was performed by obtaining all the cell suspension from the T75 flasks into sterile (15 mL or 50 mL) tubes, then centrifuged at 1300 rpm for 5 min using centrifuge 5804 (Eppendorf, Hamburg, Germany), the supernatant was discarded and then the pellet was re-suspended in 3–2 mL RPMI 1640 media. Then the cell count was performed manually using a KOVA™ Glasstic™ Slide 10 with Grids (Fisher Scientific, Waltham, MA, USA) by taking 20 µL of the cell suspension mixed with 20 µL of the trypan blue stain (ThermoScientific, Waltham, MA, USA) and then 20 µL of the mixture was loaded in the hemocytometer. Only the cells in the large 4 squares at the edges were counted under a light microscope. After that, the cell count in an ml was done following the equation: cell count × dilution factor (2) × the hemocytometer constant (10^4). After that to determine how much media were required to add into each T-75 flask for passaging the cells the following was followed:

$$\frac{\text{No. of cell count} \times \text{how much media was added to the pellet} \times 2}{8 \times 10^5}$$

2.2. Fluorescent Labeling of CML Cells before Xenotransplantation

Once the K-562 cells have reached confluency (1×10^6 cells/mL), they have been harvested by pelleting using a centrifuge at 1200 rpm for 5 min, the supernatant was then discarded then re-suspended in 3 mL PBS mixed with 6 µL of 5 µg/mL CM-Dil fluorescent dye Invitrogen (ThermoScientific, Waltham, MA, USA). Then, the labeled cells were incubated for 5 min at 37 °C followed by a 15–20 min incubation at 4 °C. After that, the cells were checked under the fluorescence microscope Olympus IX73 (Olympus, Hamburg, Germany) using fluorescent filters with excitation/emission spectra of 553/570 nm maxima.

2.3. Zebrafish Husbandry

Wild-type zebrafish embryos (AB strain) were used for this experiment. All animal experiments were carried out according to national and international guidelines for the use of zebrafish in experimental settings [26] and following the animal protocol guidelines required by Qatar University and the policy on zebrafish research established by the department of research in the Ministry of Public Health, Qatar (Ministry of Public Health, 2017). This study has been approved by the Institutional Animal Care and Use Committee (IACUC) (QU-IACUC 019/2020).

2.4. Observation of Embryos

Zebrafish embryos were counted and investigated for survival and morphological changes in 24 h intervals for 3 days using Zeiss SteREO Discovery V8 microscope with a Hamamatsu Orca Flash high-speed camera and images were analyzed using HCImage software. Dead embryos were scored according to the opaque color which they exhibit. Observed abnormalities have been investigated and recorded. Non-viable embryos were eliminated as soon as they were observed, whereas embryos with abnormal development were kept till the endpoint of the experiment.

After the incubation period, 72 hpf, six embryos from each experimental group were stabilized using 3% methylcellulose and visualized under the microscope. A 10 s bright field video of the beating heart and the body was recorded for each embryo at 100 frames per second (fps). The same region in the dorsal aorta (DA) and the posterior cardinal vein (PCV) was localized to measure the flow velocity, arterial pulse, and vessel diameter using Viewpoints MicroZebralab version 3.6 application [20,21].

2.5. Xenograft's Injection Procedure

The zebrafish embryos were exposed to Pronase for 10 min at 24 h post-fertilization to remove the chorion. After that, they were incubated till 2 or 3 days-post-fertilization (dpf) at 28 °C. Dechorionated embryos were transferred to an injection slide, and they were anesthetized with 1% Tricane solution (Western Chemical Inc, Ferndale, WA, USA) for destabilization. The fluorescently labeled K562 cells were then injected into the yolk sac to allow the cells to enter the blood circulation using a fashioned glass capillary needle (World Precision Instruments, Florida, FL, USA). About 300 K562 cells were injected into each embryo, using the Harvard Apparatus PLI 90A picolitre injector (Harvard Apparatus, Holliston, MA, USA) at the BRC zebrafish facility. The embryos were first anesthetized with 200 mg/L Tricane for 5 min and were aligned properly to have their body on one side to allow easier access to their yolk sacs. Then, a capillary needle was prepared using borosilicate glass microcapillaries following the setting on the Narishige PC-100 puller. A total of 10 µL of the cells' solution was then loaded into the needle and the needle was placed into a manipulator. The manipulator was then adjusted until holding the needle at a 45° angle to an embryo. The needle tip was broken slightly with tweezers, and the cells' solution was gently injected into the zebrafish embryos' yolk sacs. After that, the xenotransplanted embryos were transferred into new plates and fresh egg water and kept at 34 °C till the endpoint at 7 dpf. The zebrafish larvae were imaged under the fluorescence microscope using the ZEISS ZEN Microscope software (Carl-Zeiss, GmbH, Munich, Germany) each day after injection to check the cancer cell spread and to measure the tumor size.

2.6. Preparation of the NPs

The Poly(D,L-lactide-co-glycolide)-b-poly(ethylene oxide)-b-poly(D,L-lactide-co-glycolide) PLGA- PEG- PLGA polymer used to generate the NPs was purchased from Akina Inc., USA. A total of 25 mg of PLGA- PEG- PLGA polymer (Mw 6000:10,000:6000), along with 5 mg of the fluorescently labeled (DLLA: GA 50:50; Mn 10,000–20,000, Sigma-Aldrich, St. Louis, MO, USA) were dissolved in 10 mL of Tetrahydrofuran (THF) (VWR, USA). To induce nanoprecipitation and embryonic NPs formation, the organic solution was transferred drop-by-drop into 20 mL Milli Q water containing 5 mg Pluronic F127 (Sartorius, Germany). The dispersion was kept overnight with a magnetic stirrer to evaporate the organic solvent. The next day, the NPs dispersion was filtered through a 0.45 micron filter, and the filtrate was placed in the ultrafiltration tube (Vivaspin®® 20 Ultrafiltration Unit) (Sartorius Stedim Biotech, Germany) to wash and concentrate the NPs (three washing cycles with Milli Q water at 4500 RPM for 10 min).

Samples of drug-loaded NPs with increasing concentrations of the active substance were prepared using the previously described procedure for unloaded NPs, with the addition of 5 mg, 10 mg, and 15 mg Ponatinib in the organic solution. Moreover, three washing cycles were performed to remove the free unencapsulated drug and the excess of the surface agent and concentrate the NPs. NPs dispersions with known concentrations were prepared by redispersion of the concentrated NPs in a certain volume of Milli Q water.

2.7. NPs Characterization

2.7.1. Transmission Electron Microscope (TEM)

TF20: Tecnai G2 200kV TEM (FEI, Hillsboro, OR, USA) has been used to characterize the PLGA-PEG-PLGA NPs. The procedure was carried out by the Central Laboratories Unit (CLU) at Qatar University, by depositing a large droplet (around 10 µL) from each NPs sample onto a TF20 holder, and images were then obtained using a voltage of 200 kV.

2.7.2. Scanning Electron Microscope (SEM)

The particles' surface morphology was assessed using NOVA NANOSEM 450 (N-SEM) (FEI, Hillsboro, OR, USA) by the Central Laboratories Unit (CLU) at Qatar University. SEM uses a field emission gun as a source of electrons. The electron beam then travels through the column while being adjusted by different lenses till reaching the sample. The electrons interact with the sample producing secondary electrons and characteristic X-rays that can be detected by a special detector to produce electron images and elemental spectra correspondingly.

2.7.3. Nanoparticles' Size

The size of PLGA-PEG-PLGA NPs has been measured by Zetasizer Nano ZS (Malvern Instruments, Malvern, UK). The cuvette was filled with the NPs dispersion and inserted into the machine after selecting the corresponding refractive index of the NP.

2.7.4. Zeta Potential Measurement

The surface charges of the loaded and unloaded PLGA-PEG-PLGA NPs were determined by the Zetasizer Nano ZS (Malvern Instruments, UK). The machine measures the Zeta potential by using electrophoretic light scattering. The PLGA-PEG-PLGA refractive index was obtained from the literature [27] and the NPs solution was then placed in a disposable folded capillary cell to be processed by the machine.

2.7.5. Ponatinib Dissolution Rate

To determine the dissolution rate of the loaded drug in the NPs, a dialysis membrane method was performed. This was performed using the Float-A-Lyzer G2 membrane (MWCO 20 kDa), which traps the particles inside and allows the loaded drug to be released into the surrounding media. The NPs solution of 1 or 0.5 mL has been loaded inside the dialysis tube which was placed inside a beaker filled with PBS buffer (pH 7.4) with a magnetic stirrer at 37 °C for 24 h. After that, samples were taken for HPLC analysis from the same spot of the PBS buffer at regular intervals (1 h, 3 h, 5 h, and 24 h) results are shown in (Supplementary Figure S6).

2.7.6. High-Performance Liquid Chromatography (HPLC)

For the efficacy of and quantitative assessment of loading and drug content of encapsulated Ponatinib, the HPLC method was used. Adequate volume of Ponatinib loaded PLGA-PEG-PLGA NP dispersion was diluted with a mobile phase and the quantification procedure was performed as described below. The efficacy of encapsulation was determined using the following equation:

EE (%) = Amount of active substance in NP/Total amount of active substance × 100

Degree content was calculated using the equation:

DC (%) = Amount of active substance in NP/Total amount of NPs × 100

Quantitative analysis was performed on WATERS ACQUITY UPLC system HPLC system, using a C18 column, the flow rate of 1.2 mL/min, and injection volume of 5 µL. The mobile phase was composed of: (A) KH_2PO_4 0.0037 mol/L (40%), pH 3.5 adjusted by H_3PO_4; and: (B) Acetonitrile (60%). The quantification was performed using PDA/UV detector at 250 nm wavelength.

2.8. Unloaded NPs Toxicity

The zebrafish embryos at 24 h post-fertilization (hpf) were exposed to 200 µL Pronase solution (1 mg/mL) (Sigma-Aldrich, MO, USA) for 10 min to remove the chorion. Dechorioned embryos were then evaluated under the stereomicroscope (Zeiss, GmbH, Germany) and segregated into 6-wells plates equally (about 20 or 24 embryos in each well). After that, different concentrations of the unloaded NPs were prepared to determine the optimum concentration that will not cause any toxicity to the zebrafish embryos. Different concentrations of NPs have been prepared by diluting the proper amount of the NPs in (0.3 mg/mL) 1-phenyl-2-thiourea (PTU) in the egg water, the concentrations were: 1, 0.75, 0.5, 0.25, 0.1 mg/mL. The embryos then were incubated with different concentrations of NPs at 30 °C and the survival rate was then measured at 48 and 72 hpf.

2.9. Loaded NPs Toxicity

The zebrafish embryos at 24 h post-fertilization (hpf) were exposed to 200 µL Pronase (1 mg/mL) solution for 10 min to remove the chorion. Dechorionated embryos were then evaluated under the stereomicroscope and segregated into 6-well plates equally (about 20 or 24 embryos in each well). After that, three different concentrations of loaded NPs with Ponatinib (5 mg, 10 mg, and 15 mg) were prepared to select the least toxic concentration for the zebrafish embryos. Different concentrations of NPs in PTU containing egg water were prepared as follows: 1, 0.75, 0.5, 0.25, 0.1, 0.05, 0.01, 0.005, 0.0025 mg/mL to choose for loading the Ponatinib drug. The negative control group was untreated zebrafish embryos kept in egg water. The embryos were then incubated at 30 °C and the survival rate was measured at 48 and 72 hpf. The survival rate was calculated by dividing the number of viable embryos by the total number of embryos multiplied by 100.

2.10. Xenograft Exposure to Loaded PLGA-PEG-PLGA NPs Assay

Injected 2-dpf zebrafish embryos were allowed to recover for half an hour after injecting of K562 cells before exposing them to 0.001 mg/mL loaded PLGA-PEG-PLGA NPs with 10 mg and 15 mg Ponatinib. The embryos were separated and placed into 6-well plates: two wells for each group (control, 10 mg, and 15 mg) with 10 embryos in each. The 0.001 mg/mL concentration of loaded PLGA-PEG-PLGA NPs was prepared by diluting it in egg water. For a total volume of 3 mL, the required amount for one well, i.e., 1.440 µL of the 15 and 10 mg NPs were diluted in egg water. Then the embryos were incubated at 34 °C and started to be imaged at 4-dpf.

2.11. Survival Rate Analysis

On day 2 pf, the dead embryos were removed from the 6-well cell culture plates to avoid influencing the surviving embryos during the toxicity experiments. The numbers of the dead, surviving, and abnormal embryos of each NPs concentration group were recorded until 3 dpf. The survival rate was calculated by dividing the number of viable embryos by the total number of embryos, multiplied by 100.

2.12. Cardiovascular Structure/Function Analysis

To assess the cardiovascular toxicity side effects of both unloaded and loaded NPs, the analysis was carried out at 3-dpf for the embryos in all the treated groups to see the influence of interference on cardiac function, structure, and blood flow. The treated embryos were first placed in a concave slide for imaging using 3% methylcellulose for immobilization. Under the Hamamatsu Orca high-speed camera and Zeiss Lumar V12 stereomicroscope (Carl-Zeiss, GmbH, Germany), images and high-speed time-lapse movies were recorded at about 100 fps for the heart and tail of each embryo through the HCImage software (Hamamatsu, Japan). Then to assess for heart failure due to the toxicity of the NPs, tail videos have been analyzed for the Red Blood Cells (RBCs) movement within the blood flow using the MicroZebraLab (Viewpoint, Lyon, France). Tracking the RBCs aids in measuring the blood velocity by following an in-house algorithm from Viewpoint for tracking RBCs. This algorithm has also been used to measure heart rate in beats per minute. Heartbeat and blood flow velocity parameters are widely used to assess cardiac function in zebrafish. Lower heartbeat and/or blood flow velocity indicates deteriorated heart function.

2.13. Statistical Analysis

Statistical analysis was performed using GraphPad Prism version 9.0.0 software (GraphPad Software, San Diego, CA, USA). Data were analyzed using one way-ANOVA with Dunnet's multiple comparison test. A p-value of less than 0.05 was considered statistically significant. One asterisk (*) indicates $p < 0.05$, two asterisks (**) indicate $p < 0.01$, three asterisks (***) $p < 0.001$, and four asterisks (****) indicate $p < 0.0001$.

3. Results

3.1. Fluorescent K562

Olympus fluorescent microscope was used to image the fluorescent K562 CML cells stained with CM-Dil fluorescent dye. The mCherry fluorescent filter with excitation/emission spectra of 587/610 has been chosen to examine the fluorescent K562 CML cells as the CM-Dil fluorescent dye has an excitation/emission of 553/570 nm maxima. Supplementary Figure S2 represents an image of the fluorescent K562 cells at 60× magnification. As seen from the figure, most of the K562 cells were successfully fluorescently stained with CM-Dil dye.

3.2. PLGA-PEG-PLGA NPs Preparation and Characterization

3.2.1. PLGA-PEG-PLGA NPs Preparation

In the course of the nanoprecipitation process, particles are generated by simultaneous polymer/drug nucleation, molecular growth, and aggregation during the micromixing of water and the organic solvent phase. Supersaturation is the force behind all these processes influencing the size and distribution of the NP population. The growth of the nanoparticles will terminate due to the combined effect of the polymer/drug dilution and steric hindrance of Pluronic F127 which deposits at the polymer core/water interface, affecting the aggregation dynamic, particle size, and the drug content of the NPs. Therefore, during the nanoprecipitation process, the type and ratio of solvent to non-solvent, as well as the polymer/drug/stabilizing agent concentrations, have to be carefully selected to achieve high efficacy of drug encapsulation, adequate particle size, and low polydispersity index [28–30]. Our preliminary experiments pointed to the 5 mg Pluronic F 127/20 mL water phase, among the tested 2.5, 5, and 10 mg Pluronic F127/20 mL, as the most favorable concentration leading to the highest drug loading. No significant improvement in drug loading or influence on targeted particle size and distribution with further increase of concentration from 5 to 10 mg was noticed. The addition of surfactant in the organic phase increased drug loading, however, and the particle size and particle size distribution were also significantly increased. Two types of organic solvents, THF, with lower density and surface tension, and DMSO showing higher density and surface tension compared to water were also tested. The improved micromixing of the water phase with the lower density and lower surface tension organic solvent contributed to the generation of high-uniformity batches with significantly smaller NPs without any compromise on the drug-loading efficacy. Further, polymer concentration was adjusted to 25 mg or 30 mg/10mL THF to avoid the slow-down effect on the micro-mixing due to increasing viscosity of higher concentrations of the polymer solution which might lead to increased particle size and polydispersity index. Finally, three increasing concentrations, 5 mg, 10 mg, and 15 mg Ponatinib in the polymer solution were also selected to test the influence of drug concentration on the efficacy of loading, particle size, and distribution. Samples with increasing concentration of Ponatinib showed improved efficacy of loading and drug content as well as acceptable particle size for passive tumor targeting. Considering the results presented above, the final selected formula from our preliminary design studies was 30 mg PLGA-PEG-PLGA polymer with increasing concentrations of Ponatinib (1:6, 1:3, and 1:2 drug/polymer ratio) and 5 mg Pluronic F125 in 20 mL of water to prepare formulation A, B, and C for further physicochemical, morphological, in vitro, and in vivo characterization.

3.2.2. PLGA-PEG-PLGA NPs Morphology

Transmission and Scanning Electron microscopes (TEM and SEM), respectively, were used to characterize the shape of the PLGA-PEG-PLGA NPs. Supplementary Figure S3A represents the shape of PLGA-PEG-PLGA NP, by TEM micrograph and it shows the NPs with their characteristic round shape. Supplementary Figure S3B represents the shape of PLGA-PEG-PLGA NPs using SEM showing the 3D spherical shape of the NPs.

3.2.3. PLGA-PEG-PLGA NPs Size

The size of loaded PLGA-PEG-PLGA NPs prepared with 5 mg, 10 mg, and 15 mg Ponatinib has been measured in water dispersion using the Zetasizer Nano ZS (Malvern Instruments, UK). The particle size range of the NPs with increasing loading of Ponatinib was from 80 to 100 nm. The Z-average hydrodynamic diameter (Dh (nm) ± SD, n = 6) of each NPs' group was as follows: 74.55 nm ± 28.74 for the PLGA-PEG-PLGA NPs loaded with 5 mg Ponatinib (Supplementary Figure S4A), 125 nm ± 26.91 for the PLGA-PEG-PLGA NPs loaded with 10 mg Ponatinib (Supplementary Figure S4B) and 116.9 nm ± 42.92 for PLGA-PEG-PLGA NPs loaded with 15 mg Ponatinib (Supplementary Figure S4C).

The Z-average hydrodynamic diameter (Dh (nm) ± SD) of the unloaded PLGA-PEG-PLGA NPs was 84.33 nm ± 13.83 (Supplementary Figure S5), indicating that PLGA-PEG-PLGA NPs size increased with the loading of Ponatinib.

3.2.4. PLGA-PEG-PLGA NPs Surface Charge

The surface charge of the PLGA-PEG-PLGA NPs (water dispersion) had been assessed to characterize more of its material properties; thus, its interaction properties with the biological system can be predicted. For that, the Zeta potential for the loaded PLGA-PEG-PLGA NPs had been measured by the Zetasizer Nano ZS (Malvern Instruments, UK), and the surface charge ((mV) ± SD, n = 5) of the particles showed a net positive charge of 12.3 mV ± 5.5; 15.2 mV ± 3.4 and 16.7 mV ± 2.5 for 15, 10 and 5 mg Ponatinib loaded PLGA-PEG-PLGA NPs.

The Zeta potential for the unloaded PLGA-PEG-PLGA NPs demonstrated a net negative surface charge with an average of −2.66 ± 0.185 (mV) ± STD Table 1. The surface charge of the Ponatinib is positive (protonation of its terminal methylpiperazinyl nitrogen) with an average of 30.86 mV+/−2.744 (n = 5) Table 2, indicating that the positive zeta potential of loaded NPs is an additional confirmation for successful drug loading into the drug-delivery system.

Table 1. Unloaded PLGA-PEG-PLGA NPs surface charge.

Sample Name	Zeta Potential (mV)
Unloaded Nano particles 1	−2.48
Unloaded Nano particles 2	−2.85
Unloaded Nano particles 3	−2.65
Mean	−2.66
STD	0.185

Table 2. Ponatinib drug surface charge.

Sample Name	Zeta Potential (mV)
Ponatinib 1	32.5
Ponatinib 2	29.7
Ponatinib 3	33.7
Ponatinib 4	31.7
Ponatinib 5	26.7
Mean	30.86
STD	2.744

3.2.5. Ponatinib Dissolution Rate from PLGA-PEG-PLGA NPs

Dissolution rate experiments were performed using the dialysis method. In a phosphate buffer pH 7.4 pointed to a very slow dissolution rate from the prepared Ponatinib PLGA-PEG-PLGA NPs with no burst release except for the (sample A) prepared using 5 mg Ponatinib or 1:6 drug/polymer ratio (25% of the drug was released within the first 3 hours).

For samples B and C, prepared using 10 and 15 mg Ponatinib (1:3 and 1:2 drug/polymer ratio), respectively, less than 10% released drug was determined within 24 h from all the samples (HPLC analysis). Results are demonstrated in Supplementary Figure S6. These release pattern favors the accumulation of the drug at the site of action incorporated within the NPs at the same time decreasing the off-site effects and toxicity.

3.2.6. Efficacy of Loading and Drug Content

The efficacy of loading and drug content increased with the increasing concentration of Ponatinib during preparation. Calculated values were $18 \pm 3.3\%$ (n = 6), $20.8 \pm 2.1\%$ (n = 6) and $21.8 \pm 2.7\%$ (n = 6) for 5 mg, 10 mg, and 15 mg Ponatinib; or 1:6; 1:3, and 1:2 drug to polymer ratio during the preparation of the NPs. Calculated drug content was 3%, 6.5%, and 10.5% for samples prepared with 1:6, 1:3, and 1:2 drug-to-polymer ratios, respectively.

3.3. Unloaded PLGA-PEG-PLGA NPs Toxicity

3.3.1. Survival Rate

The survival rate of the zebrafish embryos at 72 h post-fertilizing (hpf) was calculated for the negative control (NC) which was untreated embryos kept in egg water and the treated groups of unloaded PLGA-PEG-PLGA NPs. Figure 1 indicates that there was a significant decrease in the survival rate of the 1.0 mg/mL group when compared to the negative control group. Meanwhile, the experimental groups with the lowest concentrations (0.75, 0.5, 0.25, 0.1, and 0.05 mg/mL) of unloaded PLGA-PEG-PLGA NPs did not show any significant difference when compared to the control group.

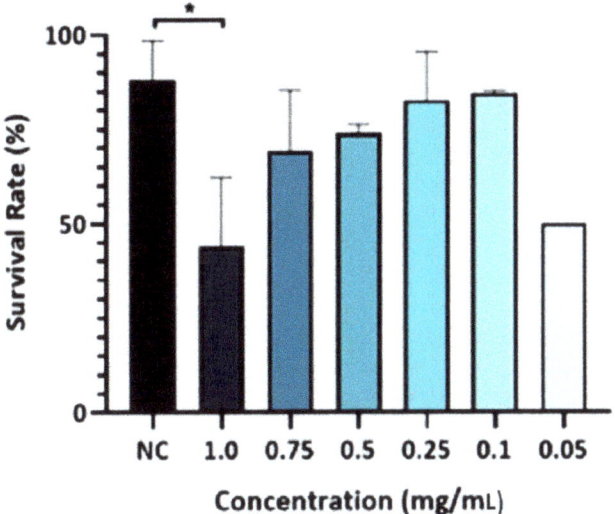

Figure 1. Survival rate of zebrafish embryos exposed to unloaded PLGA-PEG-PLGA NPs. The survival rate of the zebrafish embryos at 72 h post-fertilizing (hpf) was calculated for the negative control (NC) and the treated groups of different concentrations of unloaded PLGA-PEG-PLGA NPs. The survival rate of embryos exposed to unloaded PLGA-PEG-PLGA NPs (1 mg/mL) significantly decreased the survival rate of embryos compared to the NC, at 72 hpf. * = $p < 0.05$.

3.3.2. Cardiac Function Assessment

First, we had investigated the effect of treating 72 hpf zebrafish embryos with Ponatinib (2.5 µm) on different cardiac function parameters. Cardiac function parameters measurements were examined on the zebrafish embryos' two main blood arteries—(DA) dorsal aorta and (PCV) posterior cardinal vein—to examine the effect of Ponatinib treatment on the cardiovascular system of the zebrafish. Velocity, diameter, and pulse were measured using ZebraLab software. Cardiac parameters measurement was only possible on Ponatinib (2.5 µM) exposed embryos because of the severe effect of higher concentrations of Ponatinib on embryos' viability. As demonstrated in Figure 2, Ponatinib reduced blood flow velocity (almost no flow) in both the DA and the PCV. Other tested parameters did not show a significant difference when compared to both the control untreated group, (embryos were kept in egg water only) and the vehicle control (embryos were exposed to 0.1% DMSO in egg water). Our findings showed that high Ponatinib concentrations may alter cardiomyogenesis in zebrafish.

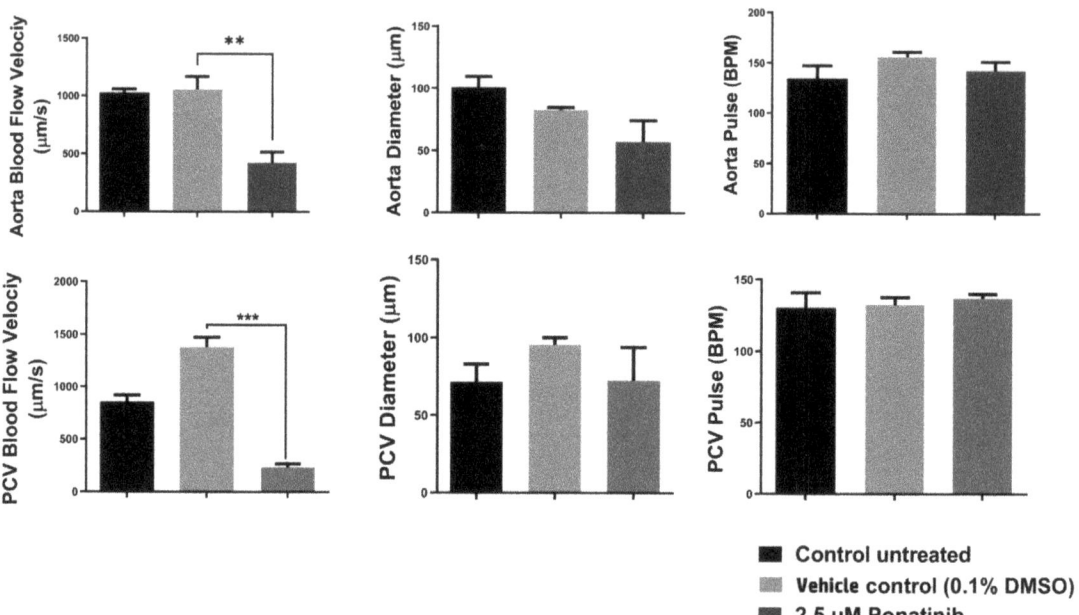

Figure 2. Cardiac function assessment and blood flow analysis for Ponatinib-tested embryos. Measurement of cardiac function for treated groups with Ponatinib at 72 hpf in comparison to the control untreated group (embryos were kept in egg water only) and the vehicle control group (0.1% DMSO in egg water); DMSO is the vehicle in which the Ponatinib drug was dissolved. All data are presented as mean ± SEM (6 embryos were used in each group; the experiment was performed in triplicate). D'Agostino and Pearson omnibus normality test was performed on all data to determine distribution. All parameters passed the test. Accordingly, a one-way-ANOVA with Sidak posthoc test was performed to compare pair: control vs. negative control and negative control and 2.5 µM Ponatinib. (**) = $p < 0.01$, (***) = $p < 0.001$.

We followed this with the investigation of (DA) and (PCV) vessel diameter, and blood flow velocity in the zebrafish embryos at 72 h -post fertilizing (hpf) treated with (0.75, 0.5, 0.25, 0.1, and 0.05 mg/mL) unloaded PLGA-PEG-PLGA NPs. It was demonstrated that there was a significant reduction in the heartbeat of the group which was treated with unloaded PLGA-PEG-PLGA NPs (0.75 mg/mL) when compared to the negative control Figure 3A.

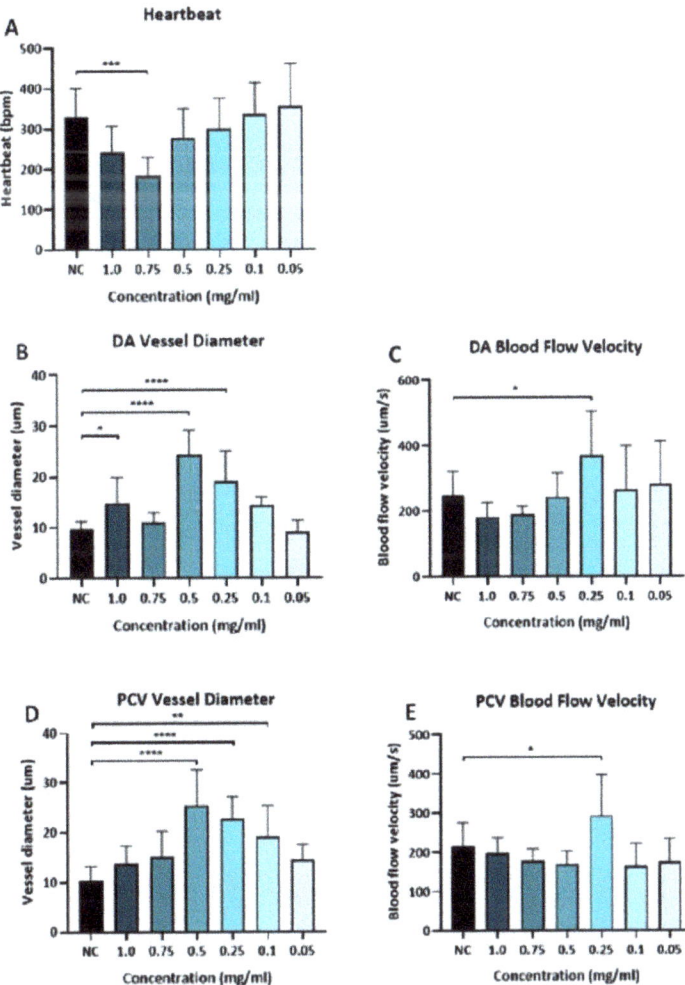

Figure 3. Cardiac function assessment of unloaded PLGA-PEG-PLGA. Cardiac function assessment of heart heartbeat, the dorsal aorta (DA) and posterior cardinal vein (PCV) vessel diameter, and blood flow velocity of the zebrafish embryos at 72hr post-fertilizing (hpf) treated with (0.75, 0.5, 0.25, 0.1, and 0.05 mg/mL) unloaded PLGA-PEG-PLGA NPs. (**A**) Heartbeat of embryos exposed to different concentrations of unloaded PLGA-PEG-PLGA NPs, a significant reduction in the heartbeat of the group which was treated with unloaded PLGA-PEG-PLGA NPs (0.75 mg/mL). (**B**) DA vessel diameter of embryos exposed to different concentrations of unloaded PLGA-PEG-PLGA NPs, vessel diameter showed to be enlarged significantly in groups of (1.0, 0.5, and 0.25 mg/mL) (**C**) DA blood flow velocity of embryos exposed to different concentrations of unloaded PLGA-PEG-PLGA NPs, blood velocity was increased significantly in the unloaded PLGA-PEG-PLGA NPs (0.25 mg/mL) group. (**D**) PCV vessel diameter of embryos exposed to different concentrations of unloaded PLGA-PEG-PLGA NPs, the vessel diameter showed to be enlarged in groups treated with unloaded PLGA-PEG-PLGA NPs (0.5, 0.25, 0.1 mg/mL). (**E**) PCV blood flow velocity of embryos exposed to different concentrations of unloaded PLGA-PEG-PLGA NPs, the blood velocity showed to be increased significantly in the unloaded PLGA-PEG-PLGA NPs (0.25 mg/mL) group. * = $p < 0.05$, (**) = $p < 0.01$, (***) = $p < 0.001$, (****) = $p < 0.0001$.

The dorsal aorta (DA) vessel diameter was seen to be enlarged in groups of (1.0, 0.5, and 0.25 mg/mL) and the blood velocity was increased significantly in the unloaded PLGA-PEG-PLGA NPs (0.25 mg/mL) group compared to the negative control Figure 3B,C.

In the posterior cardinal vein (PCV) the vessel diameter showed to be enlarged in groups treated with unloaded PLGA-PEG-PLGA NPs (0.5, 0.25, 0.1 mg/mL), and also the blood velocity increased significantly in the unloaded PLGA-PEG-PLGA NPs (0.25 mg/mL) group Figure 3D,E. Based on these results, we concluded that only a high concentration of unloaded NPs seems to be toxic to the animals.

3.4. Mortality and Visible Morphological Changes in Zebrafish

A simple visual comparison of treated zebrafish embryos to controls at 24, 48, and 72 hpf was undertaken to explore Ponatinib's teratogenic potential. Aristolochic acid (AA) (1 μM) was used as a positive control (PC) that induces cardiac failure per prior research [31,32]. Ponatinib treatment drastically reduces embryos' survival and tail flicking in a concentration-dependent manner, as demonstrated in Figure 4 panels A and B, respectively. The hatching rate of 48 h-treated zebrafish embryos was significantly different from that of untreated embryos. At low concentrations, Ponatinib increased the hatching rate significantly, while decreasing the hatching of embryos at a greater concentration (10 μM). At 72 hpf, survival was significantly reduced in the PC-treated group (AA, 1 μM).

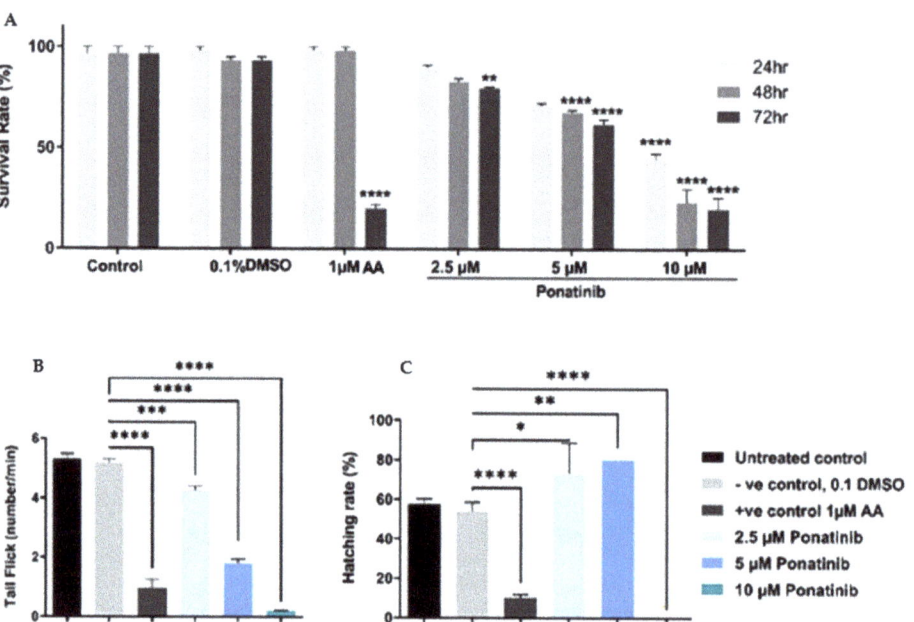

Figure 4. Mortality and visible morphological changes in Ponatinib-treated zebrafish embryos. (A) The survival rate of embryos exposed to different concentrations of Ponatinib compared to the Positive control (Aristolochic acid -AA) and NC (0.1% DMSO), at different timepoints: 24, 48, and 72 hpf, $n = 20$. (B) Assessment of potential neuro/muscular toxicity at 24 hpf by locomotion/tail-coiling assay. The plot represents the average tail coiling (burst/min) measured by DanioScope software. $n = 20$. (C) Ponatinib effect on the zebrafish embryos hatching rate $n = 20$. D'Agostino and Pearson omnibus normality test was performed on all data to determine distribution. All parameters passed the test. Accordingly, a one-way-ANOVA with Sidak posthoc test was performed to compare pair: Control vs. negative control and negative control and 2.5 mm Ponatinib. * = $p < 0.05$, (**) = $p < 0.01$, (***) = $p < 0.001$, (****) = $p < 0.0001$.

3.5. Loaded PLGA-PEG-PLGA NPs Toxicity

3.5.1. Survival Rate

The survival rate of the zebrafish embryos at 72 h post-fertilizing (hpf) was calculated for the negative control and the treated groups with three concentrations of loaded PLGA-PEG-PLGA NPs with Ponatinib: 3% of the drug content (sample A), 6.5% of drug content (sample B), and 10.5% of drug content (sample C). Figure 5A demonstrates the survival rate of treated embryos with different concentrations of the PLGA-PEG-PLGA NPs of Sample A. Data indicate that higher concentrations of the loaded PLGA-PEG-PLGA NPs (1 and 0.75 mg/mL) had the lowest survival rate compared to the other groups. Figure 5B demonstrates the survival rate of treated embryos with different concentrations of the PLGA-PEG-PLGA NPs loaded in sample B. Data indicate groups that were treated with loaded PLGA-PEG-PLGA NPs (1 and 0.5 mg/mL) showed the lowest survival rate. In Figure 5C, data demonstrates the survival rate of treated embryos with different concentrations of the PLGA-PEG-PLGA NPs of sample C, and it is indicated that the treated groups with loaded PLGA-PEG-PLGA NPs (1 and 0.75 mg/mL) showed the lowest survival rate when compared to the other groups.

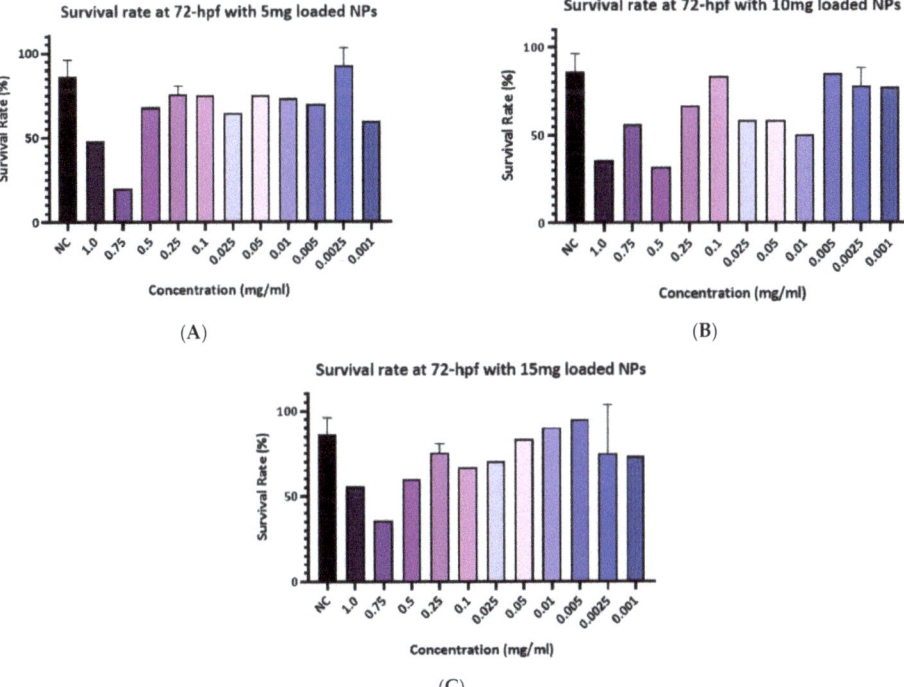

Figure 5. The survival rate of embryos exposed to different concentrations of Ponatinib-loaded PLGA-PEG-PLGA NPs. (**A**) The survival rate of embryos exposed to Ponatinib (5 mg) PLGA-PEG-PLGA NPs compared to the NC, at 72 hpf, higher concentrations of the loaded PLGA-PEG-PLGA NPs (1 and 0.75 mg/mL) had the lowest survival rate compared to the other groups. (**B**) The survival rate of embryos exposed to Ponatinib (10 mg) PLGA-PEG-PLGA NPs compared to the NC, at 72 hpf, groups that were treated with loaded PLGA-PEG-PLGA NPs (1 and 0.5 mg/mL) showed the lowest survival rate. (**C**) The survival rate of embryos exposed to different concentrations of Ponatinib (15 mg) loaded PLGA-PEG-PLGA NPs compared to the NC, at 72 hpf. The treated groups with loaded PLGA-PEG-PLGA NPs (1 and 0.75 mg/mL) showed the lowest survival rate when compared to the other groups.

Based on these results, only high concentrations of loaded NPs were toxic to the embryos and the concentration of PLGA-PEG-PLGA (0.001 mg/mL) of samples (B and C) showed a similar survival rate to the negative control. Thus, this concentration is considered to be the optimum for performing the next experiments.

3.5.2. Cardiac Function Assessment

The cardiac function was assessed by analyzing the heartbeat, the dorsal aorta (DA) and posterior cardinal vein (PCV) vessel diameter, and blood flow velocity.

The heartbeat of groups treated with sample A NPs (0.005 and 0.0025 mg/mL), sample B NPs (0.005, 0.0025 and 0.001 mg/mL) sample C NPs (0.0025 and 0.001 mg/mL) were significantly reduced compared to the negative control, as shown in Figure 6A.

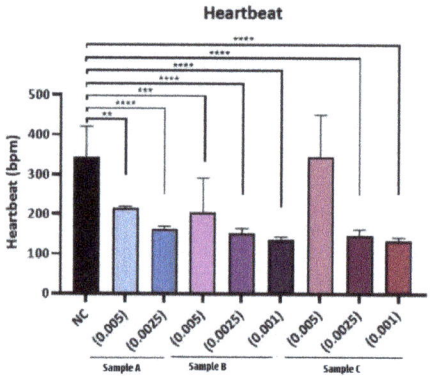

(A)

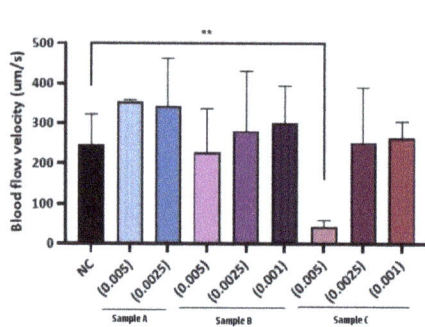

(B)

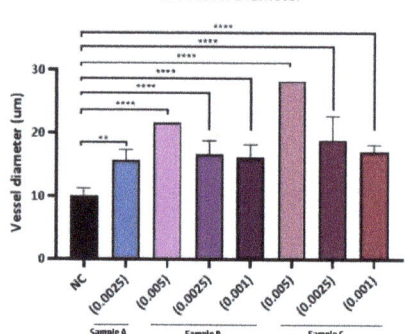

(C)

Figure 6. Cont.

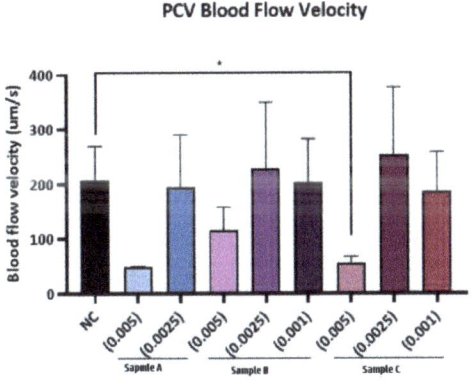
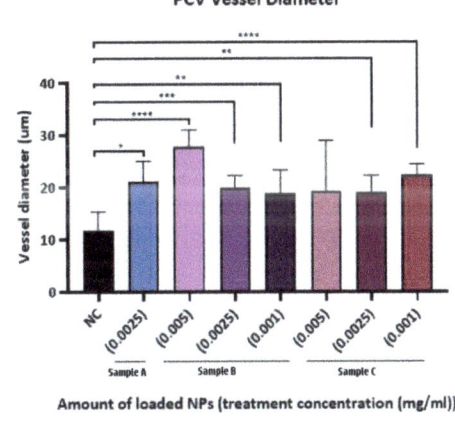

(D) (E)

Figure 6. Cardiac function assessment of loaded PLGA-PEG-PLGA NPs. (**A**) Heartbeats of groups treated with sample A NPs (0.005 and 0.0025 mg/mL), sample B NPs (0.005, 0.0025, and 0.001 mg/mL), and sample C NPs (0.0025 and 0.001 mg/mL) were significantly reduced compared to (NC) (**B**) DA blood-flow velocity, blood velocity was significantly reduced in treated groups with sample C NPs in the concentration of (0.005 mg/mL) (**C**) The DA vessel diameter, an enlargement in the groups treated with sample A NPs at a concentration of (0.005 and 0.0025 mg/mL), sample B NPs at a concentration of (0.0025 mg/mL) and sample C NPs at a concentration of (0.0025 and 0.001 mg/mL) (**D**) PCV blood flow velocity was significantly reduced in the group treated with sample C NPs (0.005 mg/mL), and (**E**) PCV vessel diameter of embryos exposed to different concentrations of samples A, B, and C PLGA-PEG-PLGA NPs, was significantly enlarged in all treated groups when compared to control. * = $p < 0.05$, (**) = $p < 0.01$, (***) = $p < 0.001$, (****) = $p < 0.0001$.

Blood velocity was significantly reduced in treated groups with sample C NPs at a concentration of (0.005 mg/mL) and slightly high in the treated group with sample C NPs at a concentration of (0.0025 mg/mL). The dorsal aorta (DA) vessel diameter demonstrated an enlargement in the groups treated with sample A NPs at a concentration of (0.005 and 0.0025 mg/mL), sample B NPs at a concentration of (0.0025 mg/mL), and sample C NPs at a concentration of (0.0025 and 0.001 mg/mL) as shown in Figure 6B,C.

In the posterior cardinal vein (PCV), the measured vessel diameters were significantly enlarged in all drug and NPs combinations, except for the group treated with sample C NPs (0.005 mg/mL), where the blood velocity was significantly reduced in Figure 6D,E.

Relying on these data, the concentration of NPs (0.001 mg/mL) of samples B and C manifested a non-toxic effect on treated groups; therefore, these concentrations would be used further in the xenograft experiments.

3.6. Zebrafish Xenograft Model

K562 CML cell line was successfully transplanted into the 72 hpf zebrafish embryos. Figure 7 represents a xenografted embryo from 1 day-post-injection to 3 days-post-injection (dpi) compared to a negative control embryo to differentiate between the autofluorescence of the embryos. The fluorescently labeled cancer cells using CM-Dil red fluorescent dye demonstrated an increase in proliferation leading to an increase in the tumor size and the migration of cancer cells to distant sites of the embryo over time as indicated by the white arrows. The yolk sac area (white X), which was the injection site of the cells, showed to contain concentrated tumor cells.

K562 CML cell line was also successfully transplanted when injected into the 48 hpf zebrafish embryos. Figure 8 represents a xenografted embryo from 1 day-post-injection until 5 days-post-injection (dpi) compared to a negative control embryo to differentiate between the autofluorescence of embryos. The fluorescently labeled cancer cells proliferated and resulted in an increase in tumor size and migration to distal sites of the embryo over time.

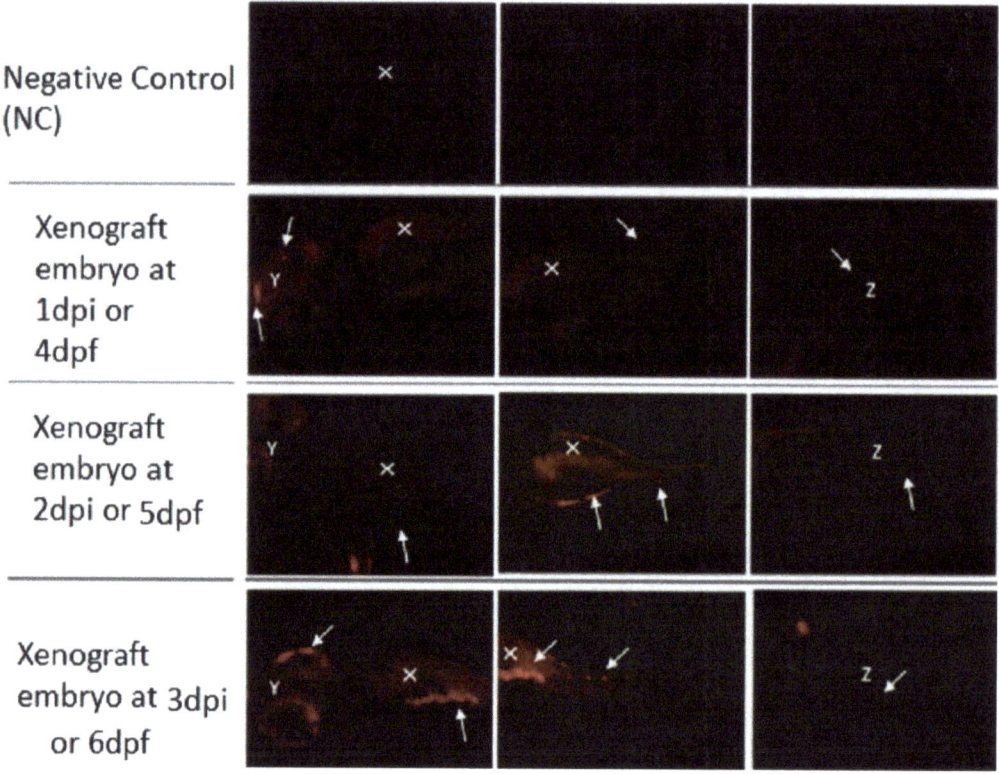

Figure 7. Zebrafish Xenograft model injected at 3 dpf. Representative fluorescence images of zebrafish screening at 4 dpf to 6 dpf using fluorescent microscopy and investigation of fluorescent K562 cells proliferation (white solid arrows) all over the animal body (Y—eyes; X—yolk sac; Z—tail) using mCherry fluorescence filter. Original magnification 100×.

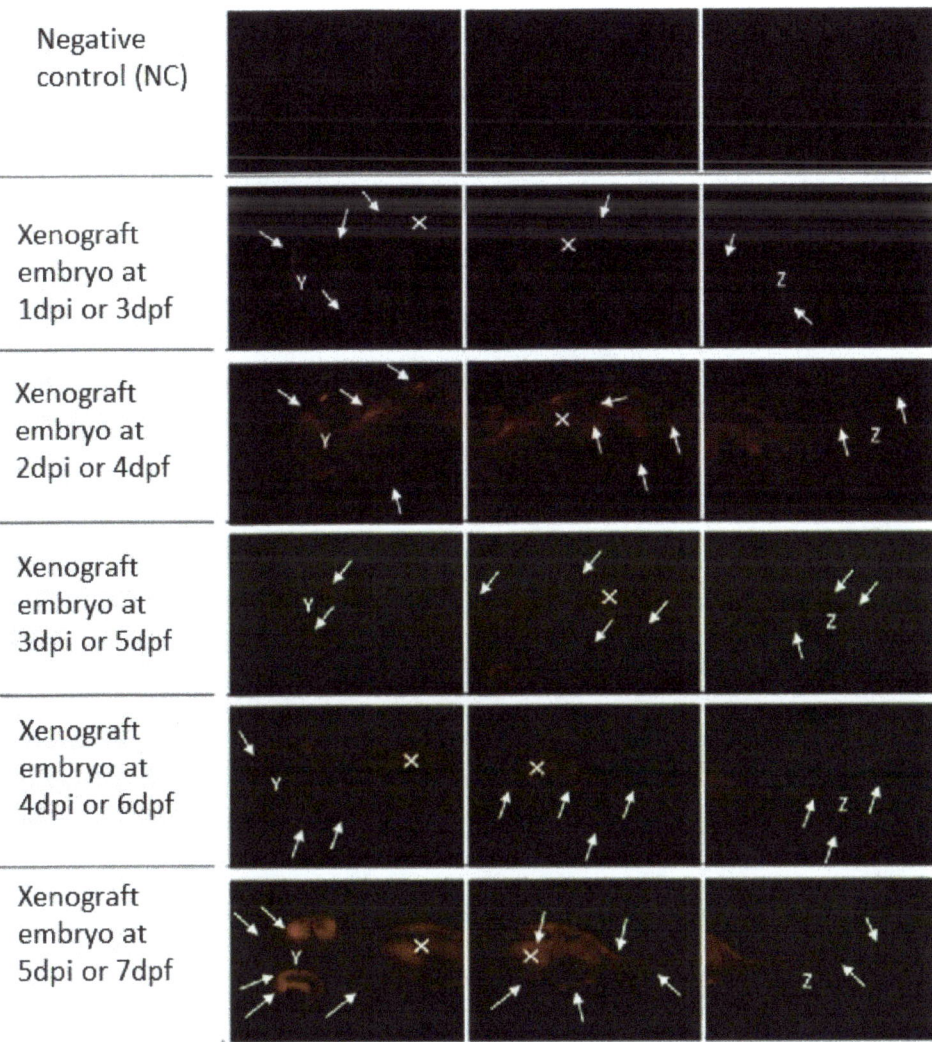

Figure 8. Zebrafish Xenograft model injected at 2 dpf. Representative fluorescence images of zebrafish screening at 3 dpf to 7 dpf using fluorescent microscopy and investigation of fluorescent K562 cells proliferation (white solid arrows) throughout the animal body (Y—eyes; X—yolk sac; Z—tail) using mCherry fluorescence filter. Original magnification 100×.

3.7. Xenograft Model Exposed to Loaded PLGA-PEG-PLGA NPs

On the same day of K562 cells injection, the 2 dpf xenograft embryos were exposed to PLGA-PEG-PLGA NPs (0.001 mg/mL) loaded with 10 mg Ponatinib. Figure 9 demonstrates the effect of exposing xenograft embryos to sample B NPs (0.001 mg/mL) from day 2 post-injection until day 5 post-injection (dpi) compared to a negative control embryo.

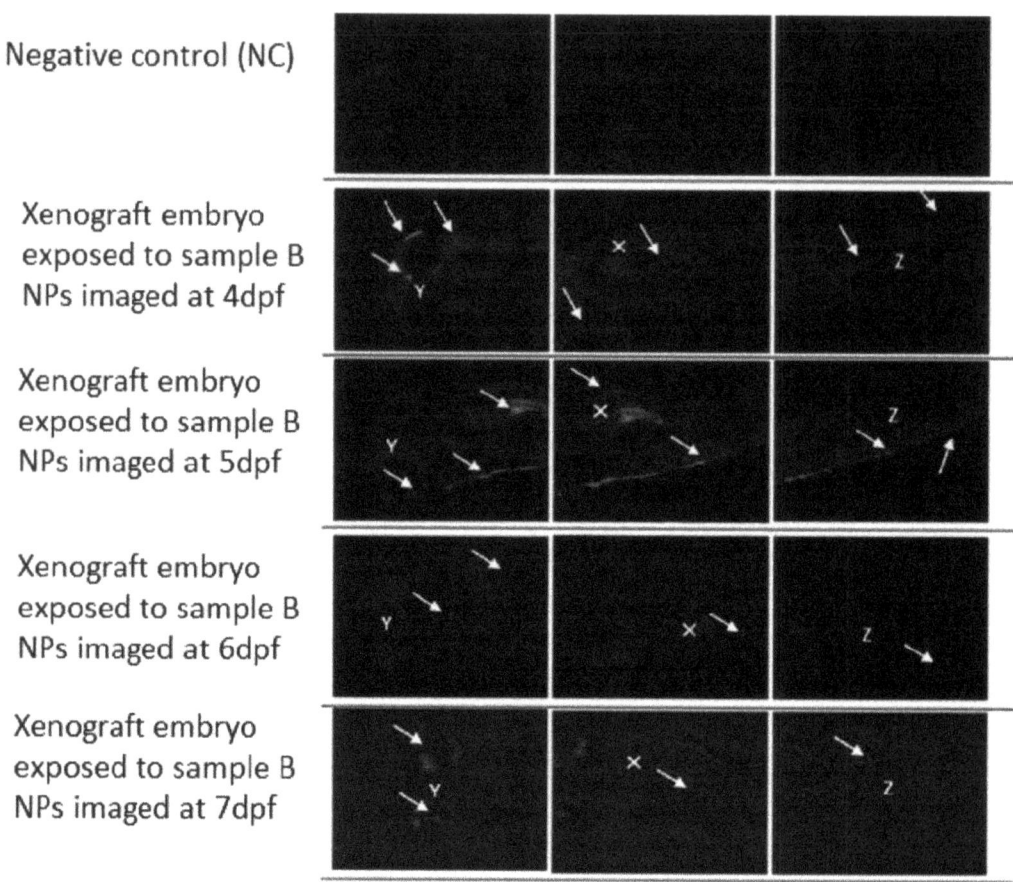

Figure 9. Xenograft model exposed sample B PLGA-PEG-PLGA NPs with 10 mg Ponatinib. Representative fluorescence images of zebrafish screening at 4 dpf until 7 dpf using fluorescent microscopy and investigation of fluorescent K562 cells proliferation (White solid arrows) throughout the animal body (Y—eyes; X—yolk sac; Z—tail) using mCherry fluorescence filter after exposing zebrafish embryos to of sample B PLGA-PEG-PLGA NPs (0.001 mg/mL). Original magnification 100×.

As shown in Figure 10 for the group of xenografted 2 dpf embryos which were treated with sample C PLGA-PEG-PLGA NPs (0.001 mg/mL). The fluorescently labeled K562 cells increased in number and migrated to the distal location from the original site of injection over time as indicated by the white arrows. The yolk sac area (white X) (injection site of the K562 cells) showed the largest tumor cells mass. K562 cells had also circulated through the blood as shown in the embryo's eyes (white Y) and tail (white Z). As seen in all Figures 8–10, the loaded NPs took a long time to release Ponatinib, with no obvious decrease in tumor size.

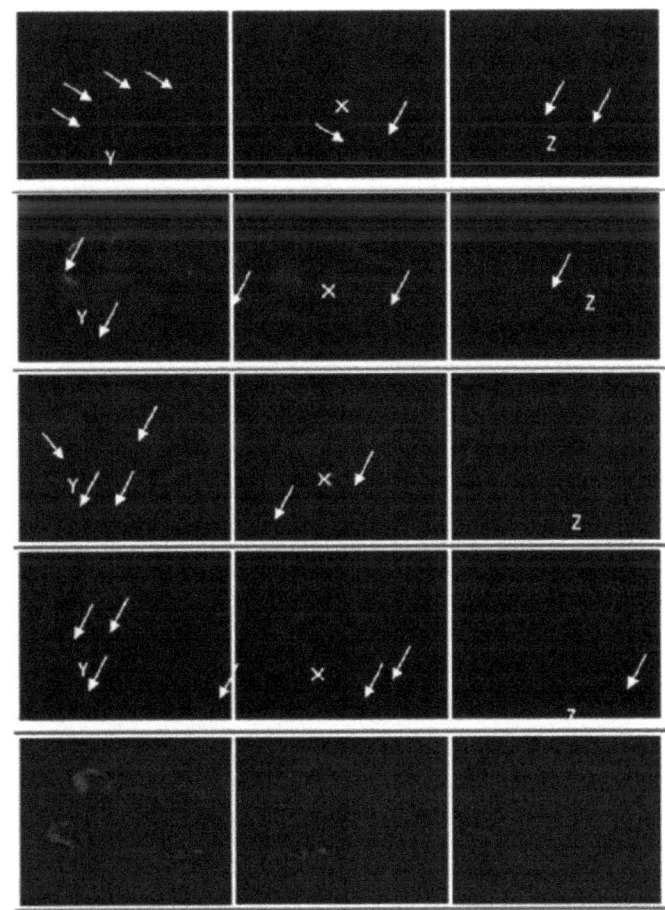

Figure 10. Xenograft model exposed to sample C PLGA-PEG-PLGA NPs. Representative fluorescence images of zebrafish screening at 4 dpf to 7 dpf using fluorescent microscopy and investigation of fluorescent K562 cells proliferation (white solid arrows) throughout the animal body (Y—eyes; X—yolk sac; Z—tail) using mCherry fluorescence filter after treating zebrafish embryos with sample C PLGA-PEG-PLGA NPs (0.001 mg/mL). Original magnification 100×.

3.8. The Uptake of Drug-Loaded PLGA-PEG-PLGA NPs

As shown in Figure 11 the sample B and C PLGA-PEG-PLGA NPs were successfully uptaken after 2 dpi, and NPs were stained using 5DTAF green fluorescent dye. White arrows are showing the fluorescence-labeled NPs through the embryo's eye (white Y), yolk sac (white X), and tail (white Z).

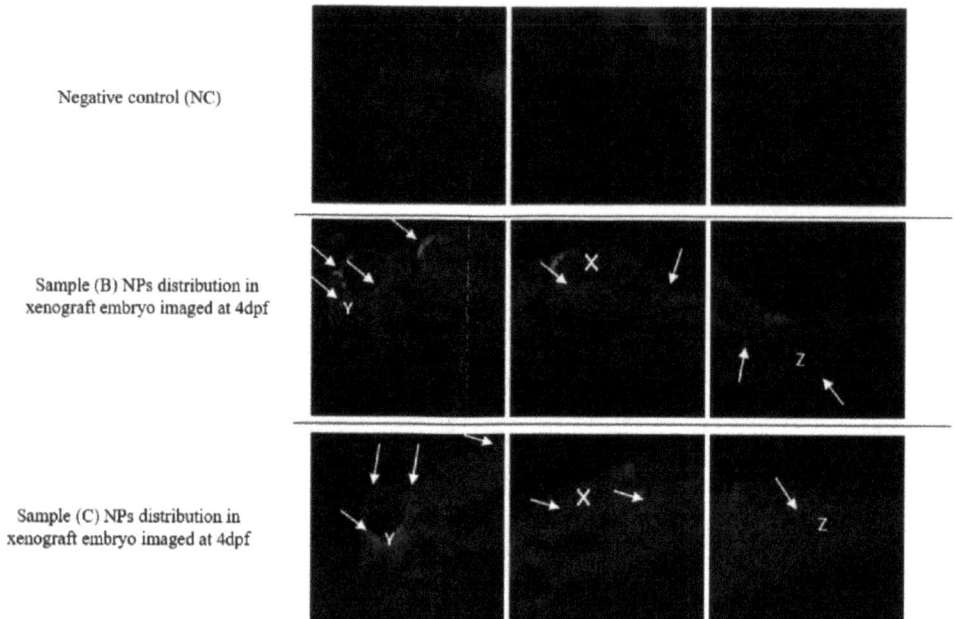

Figure 11. Loaded PLGA-PEG-PLGA NPs uptake. Representative fluorescence images of zebrafish screening at 4 dpf using fluorescent microscopy and investigation of fluorescent PLGA-PEG-PLGA NPs distribution (white solid arrows) throughout the animal body (Y—eyes; X—yolk sac; Z—tail) using GFP fluorescence filter. Original magnification 100×.

4. Discussion

Since cancer is one of the main causes of death worldwide, many research studies are currently focusing on finding novel and efficient therapeutic tools to reduce side effects associated with conventional therapies for cancer [33]. Nanomedicine is one of the new approaches that overcome some of the related issues of conventional cancer therapies including their low bioavailability and low specificity. [34]. Thus, encapsulating the anti-cancer drugs or related active agents in NPs would increase their biocompatibility, solubility, stability in body fluids, and their retention time in tumor vasculature which would enhance the efficacy of the treatment [35–37].

Moreover, nanomedicine could also support the cardio-oncology field which is an inter-disciplinary field of studying, detecting, and treating cardiovascular adverse effects associated with cancer therapies [38]. Although TKIs are the effective and preferred choice of therapy in several types of cancers including CML, their toxicity remains a major concern, particularly their cardiotoxic effects in cancer patients [39]. This prompted us to investigate the efficacy of loading the Ponatinib drug as a member of the TKI family into the PLGA-PEG-PLGA NPs in the enhancement of the anti-cancer activity and the reduction of cardiotoxic effects related to this.

In the current study, the smart PLGA-PEG-PLGA NPs were synthesized by the nano-precipitation method. The PLGA-PEG-PLGA NPs characteristics (size, shape, efficacy of loading, drug content, surface charge, and dissolution rate) had been investigated. The size of the unloaded PLGA-PEG-PLGA NPs was approximately 84.33 nm. Sulaiman et al. (2019) has shown that the PLGA-PEG-PLGA NPs size is in the range of 206 to 402 nm, Dimchevska et al., (2017) demonstrated that the size will vary depending upon the experimental conditions and polymer characteristics, and the most efficient way to optimize the formulation is to use experimental design for preparing PLGA-PEG-PLGA NPs [30,40].

SEM has shown that NPs are spherical in three dimensions with a smooth surface. This is because SEM is used to examine material surfaces and is based on scattered electrons [41]. While TEM showed that NPs are round in shape, TEM was used to show the NPs in a higher magnification as it is based on transmitted electrons; also, it has higher electron energy than SEM, which allows them to penetrate through the particles to define any internal elements in the particles [41]. The surface charge of the empty PLGA-PEG-PLGA NPs has been determined by measuring the Zeta potential that would aid in determining more about the particle properties and their interaction with the biological system. The unloaded PLGA-PEG-PLGA NPs revealed a negatively charged surface (−2.66 mV), as the polymer is affected by the PLGA copolymer end-group [40].

The cardiotoxicity of Ponatinib (2.5 µM) and the unloaded PLGA-PEG-PLGA NPs have been investigated to determine if this type of polymers would cause any cardiotoxicity or adverse effects. Our results demonstrated a significant effect of Ponatinib on reducing the velocity of aortic and PCV blood flow; these results came in agreement with a previous study conducted by Singh and coworkers [42]. The zebrafish model has been used in this study due to the transparency of the zebrafish embryos bodies which allows a non-invasive examination of the organ development [43]. The high concentrations of PLGA-PEG-PLGA NPs inside the studied model (1 mg/mL and 0.75 mg/mL) showed some toxicity as confirmed by the low survival rate, decreased heartbeat, low DA diameter, and decreased blood flow velocity of the treated groups when compared to the control group l. While the other groups treated with lower concentrations (0.5, 0.25, 0.1, and 0.05 mg/mL) demonstrated no significant difference in the survival rate and other aforementioned measured parameters when compared to the control group. This supports the fact that PLGA-PEG-PLGA NPs have low toxic effects and may be used to improve bioavailability and antitumor targeting [44–46].

Before Ponatinib loading to NPs cardiotoxicity experiments we investigated the effect of treating zebrafish embryos with different concentrations of Ponatinib on their viability, tail flicking, and hatching rate at different time points, it has been shown that Ponatinib had the minimum effect on all measured parameters at a concentration of (2.5 µM). The dosing rationale that was tested for Ponatinib was based on the literature [47]. Increased concentrations of Ponatinib significantly exerted a neuro/muscular toxic effect on embryos at 24 hpf reflected by the reduced tail flicking, these results can be explained by the toxic effect of Ponatinib at higher doses on angiogenesis as an important step in embryo organs development [48].

The smart NPs PLGA-PEG-PLGA loaded with Ponatinib (samples A—3%, B—6.05%, and C—10.5% drug loading) were used to reduce the drug's toxic side effects, especially on the cardiovascular system. The loading of the drug was successfully performed, and this was indicated by the change in the surface charge of the PLGA-PEG-PLGA NPs from negative (−2.66 mV) to positive (12.3, 15.2, and 16.7 mV) charge for C, B, and A samples, respectively during the drug incorporation. Ponatinib surface charge is positive (30.86 mV) due to protonation of its on its terminal methylpiperazinyl nitrogen across water [49]. This phenomenon is often seen in nanoparticles. One example is described by Ku et al. (2010), who disclosed the change in the FMSNs surface charge from negative (−22.43 mV) to positive (18.93 mV) due to the conjugation of PAMAM of a positive charge, and eventually, the charged change almost to neutral (1.49 mV) revealing an additional modification of PEG [50]. HPLC analysis has confirmed the presence of Ponatinib in the PLGA-PEG-PLGA NPs with drug content of 3% for sample A, 6.05% for sample B, and 10.5% for sample C. Increasing concentrations of Ponatinib during the NPs preparation resulted in increased efficacy of encapsulation, probably due to electrostatic interaction between the protonated drug and the hydroxyl-terminated polymer which improved packaging and chain entanglement during the nanoprecipitation. For sample A, prepared using the lowest quantity of Ponatinib (5 mg; 1:6 drug/polymer ratio), resulting in the lowest efficacy of encapsulation and drug content, there was an initial burst release of the drug within a short time after the immersion in the dissolution medium which is additional confirmation of improved packaging of the drug and polymer chains with a higher concentration of Ponatinib during the sample preparation. Burst release is undesirable, as it would shorten the drug's overall therapeutic duration and increase the drug's toxic potential due to excessive burst release [51]. on the other hand, PLGA-PEG-PLGA NPs prepared with higher concentrations of Ponatinib during the nanoprecipitation process (10mg Ponatinib, a drug-to-polymer ratio of 1:3 for sample B; and 15 mg Ponatinib, a drug-to-polymer ratio of 1:2 for sample C) did not show the same pattern of burst release over a 48 h period, marking these concentrations and samples out to be better candidates for in vivo testing.

The cardiotoxicity of different concentrations of sample B and sample C PLGA-PEG-PLGA NPs was tested in zebrafish embryos before the investigation of their efficacy, as a less cardiotoxic therapeutic tool to treat CML. The concentration of 1, 0.75, 0.5, 0.25, 0.1, 0.025, 0.05, and 0.01 mg/mL NPs of all groups (samples A, B, and C) had shown very clear toxicity based on demonstrated embryo survival rates which were the lowest associated with abnormal morphology.This was similar to the toxicity results of the treatment of Ponatinib drug only, as the embryos were deformed with heart edema and abnormal heart structure as well as for the absence of blood flow in the PCV and DA. This could be due to the burst release of Ponatinib which has been determined to cause cardiotoxicity. However, the lowest concentrations (0.005 and 0.0025 mg/mL) had shown a better effect but still, there were some observed abnormalities in the embryos, thus sample A PLGA-PEG-PLGA NPs were excluded and a lower concentration (0.001 mg/mL) from samples B and C PLGA-PEG-PLGA NPs were tested, and they demonstrated the best results in terms of survival rate, normal morphology, and cardiac output.

Successfully, a zebrafish xenograft model has been generated to investigate the efficacy of those loaded with sample B and C PLGA-PEG-PLGA NPs, in reducing cardiotoxicity and as effective anti-cancer therapy to treat CML. This was achieved by transplanting the human K562 cell line into 2 dpf zebrafish embryos. This xenograft model has also been successfully generated and confirmed by the spread of the tumor cells to distal sites from the yolk sac throughout 6 days-post-injection, which is consistent with previous studies. Corkery et.al. (2011) has also used the K562 cells that were stained by the CM-Dil dye to give a red fluorescence color. These cells were then transplanted into the zebrafish embryos and the embryos were then kept for 1 h at 28 °C for a recovery period and this aided in enhancing the embryos' survival rate [52]. However, in this study, the embryos have been immediately incubated at 34 °C without a recovery period and this might be the reason behind the low survival rate of the injected embryos after 1 day-post-injection. Moreover, Pruvot et al. (2011) has shown successful transplantation of the K562 cell line into the zebrafish embryos [53].

Finally, samples B and C PLGA-PEG-PLGA NPs (0.001 mg/mL) concentration were introduced to the injected zebrafish embryos (2 dpf) half an hour after the injection. The tumor cells are not reduced clearly over the 6 days after injection, this could be due to the long release time of Ponatinib from the PLGA-PEG-PLGA NPs.

Possible limitations of this study includes that it being mostly dependent on zebrafish embryos that needed proper care and training for handling and that the xenograft model required an even a higher level of handling as the embryos are injured. The treated groups of the zebrafish embryos with the loaded PLGA-PEG-PLGA NPs were only observed till the ethical endpoint of 7-dpf,;thus, the effect of the loaded PLGA-PEG-PLGA NPs in reducing tumor cells was only observed for a few days despite the fact that Ponatinib release that could happen after a few days of the endpoint. For that, xenografted embryos need to be observed for a longer time, e.g., 10-dpf. Moreover, due to the lack of FTIR < X-ray and DSC studies, the paradox of the presence of burst release at the lowest drug concentration cannot be explained. Moreover, deducting background fluorescence per unit area of the fluorescence images would give better quantitative measurements.

5. Conclusions

In summary, the zebrafish is a suitable animal model for investigating the cardiotoxicity associated with anti-cancer drugs such as TKIs, determining the optimum concentration of smart NPs with the least side effects, and generating a xenograft model of several cancer types.

In this study, PLGA-PEG-PLGA NPs were synthesized to carry the TKIs drugs. These NPs have been shown to carry Ponatinib drugs (Samples B and C for a long time, allowing for longer circulation in the zebrafish body. Zebrafish animal model was used for testing the cardiotoxicity of a range of different concentrations of loaded and unloaded PLGA-PEG-PLGA NPs and the least concentrations were shown to be of low toxicity and enhanced survival rate. The concentrations of 0.1 and 0.05 mg/mL of the unloaded PLGA-PEG-PLGA NPs are the best in terms of low cardiotoxicity and high survival rate, while 0.001 mg/mL concentration of samples B or C PLGA-PEG-PLGA NPs has been shown to be the optimum concentration among the rest of the concentrations. Lastly, these loaded NPs have been exposed to the successfully generated CML xenograft zebrafish model, however, no obvious reduction in the tumor mass was seen, indicating the slow release of Ponatinib from PLGA-PEG-PLGA NPs.

Generally, PLGA-PEG-PLGA NPs could be a good candidate for CML treatment, but their cellular internalization should be enhanced. This could be achieved by coating and labeling the surface of PLGA-PEG-PLGA NPs with specific ligands that are unique to CML cells.

Supplementary Materials: The following supporting information can be downloaded at: https://www.mdpi.com/article/10.3390/ma15113960/s1. Figure S1: QU-IACUC approval; Figure S2: Representative fluorescence images for K562 cells stained with CM-Dil dye (Red). Fluorescently labeled K562 cells at magnification 60×; Scale bar, 0.03 mm; Figure S3: TEM and SEM micrographs of PLGA-PEG-PLGA NPs. (A) TEM image of PLGA-PEG-PLGA Np on scale bar, 50 nm. (B) SEM image of PLGA-PEG-PLGA NPS on scale bar, 1 μm; Figure S4: Ponatinib loaded PLGA-PEG-PLGA NPs intensity-based particles size distribution. Representative graphs of Nanosizer 2000-Malvern for loaded PLGA-PEG-PLGA NPs size. (A) The size of Sample A PLGA-PEG-PLGA NPs is 74.55+/−28.74 (d.nm) ± SD. (B) The size of Sample B PLGA-PEG-PLGA NPs is 125+/−26.91 (d.nm) ± SD. (C) The size of Sample C PLGA-PEG-PLGA NPs is 116.9+/−42.92 (d.nm) ± SD; Figure S5: Unloaded PLGA-PEG-PLGA NPs intensity-based particles size distribution. Representative graph of Nanosizer 2000-Malvern for unloaded PLGA-PEG-PLGA NPs size. The size of unloaded PLGA-PEG-PLGA NPs is 84.33+/−13.83 (d.nm) ± SD. Figure S6: Ponatinib Dissolution Rate from PLGA-PEG-PLGA NPs. Representative graphs of HPLC for Ponatinib Dissolution Rate from PLGA-PEG-PLGA NPs (A) Standard graph of Ponatinib drug peak at 1.678 RT. (B) sample A PLGA-PEG-PLGA NPs at 1 h. (C) sample A PLGA-PEG-PLGA NPs at 3 h. (D) sample A PLGA-PEG-PLGA NPs at 5 h. (E) sample A PLGA-PEG-PLGA NPs at 24 h. (F) sample B PLGA-PEG-PLGA NPs at 1 h. (G) PLGA-PEG-PLGA NPs at 3 h. (H) sample B PLGA-PEG-PLGA NPs at 5 h. (I) sample B PLGA-PEG-PLGA NPs at 24 h. (J) sample C PLGA-PEG-PLGA NPs at 1 h. (K) sample C PLGA-PEG-PLGA NPs at 3 h. (L) sample C PLGA-PEG-PLGA NPs at 5 h. (M) PLGA-PEG-PLGA NPs at 24 h. (N) sample B PLGA-PEG-PLGA NPs at 48 h. (O) sample C PLGA-PEG-PLGA NPs at 48 h.

Author Contributions: Conceptualization, H.C.Y.; data curation, H.F.A.-T.; formal analysis, H.F.A.-T.; funding acquisition, H.C.Y.; investigation, H.C.Y.; methodology, H.F.A.-T., S.S., Z.Z.Z. and K.G.; project administration, H.C.Y.; resources, H.C.Y.; supervision, H.C.Y.; validation, H.C.Y.; visualization, H.F.A.-T.; writing—original draft, H.F.A.-T. and S.S.; writing—review and editing, M.H.H., K.G., H.M.K. and H.C.Y. All authors have read and agreed to the published version of the manuscript.

Funding: Qatar University Student Grant (QUST-2-CHS-2020-11). Also, the authors would like to thank QU Scholarship, Qatar University for the financial support to publish this article.

Institutional Review Board Statement: The Qatar University Institutional and Use Committee (QU-IACUC) had approved the protocol in the current work, number QU-IACUC 019/2020.

Conflicts of Interest: The authors disclose no potential conflict of interest.

References

1. Nagai, H.; Kim, Y.H. Cancer prevention from the perspective of global cancer burden patterns. *J. Thorac. Dis.* **2017**, *9*, 448. [CrossRef]
2. Sarkar, S.; Horn, G.; Moulton, K.; Oza, A.; Byler, S.; Kokolus, S.; Longacre, M. Cancer development, progression, and therapy: An epigenetic overview. *Int. J. Mol. Sci.* **2013**, *14*, 21087–21113. [CrossRef]
3. Davis, A.S.; Viera, A.J.; Mead, M.D. Leukemia: An overview for primary care. *Am. Fam. Physician* **2014**, *89*, 731–738.
4. Hanlon, K.; Copland, M. Chronic myeloid leukaemia. *Medicine* **2017**, *45*, 287–291. [CrossRef]
5. Winkler, G.C.; Barle, E.L.; Galati, G.; Kluwe, W.M. Functional differentiation of cytotoxic cancer drugs and targeted cancer therapeutics. *Regul. Toxicol. Pharmacol.* **2014**, *70*, 46–53. [CrossRef]
6. O'Brien, S.G.; Guilhot, F.; Larson, R.A.; Gathmann, I.; Baccarani, M.; Cervantes, F.; Cornelissen, J.J.; Fischer, T.; Hochhaus, A.; Hughes, T. Imatinib compared with interferon and low-dose cytarabine for newly diagnosed chronic-phase chronic myeloid leukemia. *N. Engl. J. Med.* **2003**, *348*, 994–1004. [CrossRef]
7. Geskovski, N.; Matevska-Geshkovska, N.; Dimchevska Sazdovska, S.; Glavas Dodov, M.; Mladenovska, K.; Goracinova, K. The impact of molecular tumor profiling on the design strategies for targeting myeloid leukemia and EGFR/CD44-positive solid tumors. *Beilstein J. Nanotechnol.* **2021**, *12*, 375–401. [CrossRef]
8. Skubitz, K.M. Cardiotoxicity monitoring in patients treated with tyrosine kinase inhibitors. *Oncologist* **2019**, *24*, e600. [CrossRef]
9. Shurbaji, S.; Anlar, G.G.; Hussein, E.A.; Elzatahry, A.; Yalcin, H.C. Effect of flow-induced shear stress in nanomaterial uptake by cells: Focus on targeted anti-cancer therapy. *Cancers* **2020**, *12*, 1916. [CrossRef]
10. Shurbaji, S.; Manaph, N.P.A.; Ltaief, S.M.; Al-Shammari, A.R.; Elzatahry, A.; Yalcin, H.C. Characterization of MXene as a cancer photothermal agent under physiological conditions. *Front. Nanotechnol.* **2021**, *63*, 689718. [CrossRef]
11. Hua, S.; De Matos, M.B.; Metselaar, J.M.; Storm, G. Current trends and challenges in the clinical translation of nanoparticulate nanomedicines: Pathways for translational development and commercialization. *Front. Pharmacol.* **2018**, *9*, 790. [CrossRef]
12. Biswas, S.; Kumari, P.; Lakhani, P.M.; Ghosh, B. Recent advances in polymeric micelles for anti-cancer drug delivery. *Eur. J. Pharm. Sci.* **2016**, *83*, 184–202. [CrossRef]
13. Dimchevska, S.; Geskovski, N.; Koliqi, R.; Matevska-Geskovska, N.; Gomez Vallejo, V.; Szczupak, B.; Sebastian, E.S.; Llop, J.; Hristov, D.R.; Monopoli, M.P.; et al. Efficacy assessment of self-assembled PLGA-PEG-PLGA nanoparticles: Correlation of nano-bio interface interactions, biodistribution, internalization and gene expression studies. *Int. J. Pharm.* **2017**, *533*, 389–401. [CrossRef]
14. Tabatabaei Mirakabad, F.S.; Nejati-Koshki, K.; Akbarzadeh, A.; Yamchi, M.R.; Milani, M.; Zarghami, N.; Zeighamian, V.; Rahimzadeh, A.; Alimohammadi, S.; Hanifehpour, Y. PLGA-based nanoparticles as cancer drug delivery systems. *Asian Pac. J. Cancer Prev.* **2014**, *15*, 517–535. [CrossRef]
15. Zhang, K.; Tang, X.; Zhang, J.; Lu, W.; Lin, X.; Zhang, Y.; Tian, B.; Yang, H.; He, H. PEG–PLGA copolymers: Their structure and structure-influenced drug delivery applications. *J. Control. Release* **2014**, *183*, 77–86. [CrossRef]
16. Oerlemans, C.; Bult, W.; Bos, M.; Storm, G.; Nijsen, J.F.W.; Hennink, W.E. Polymeric micelles in anticancer therapy: Targeting, imaging and triggered release. *Pharm. Res.* **2010**, *27*, 2569–2589. [CrossRef]
17. Ma, H.; He, C.; Cheng, Y.; Yang, Z.; Zang, J.; Liu, J.; Chen, X. Localized co-delivery of doxorubicin, cisplatin, and methotrexate by thermosensitive hydrogels for enhanced osteosarcoma treatment. *ACS Appl. Mater. Interfaces* **2015**, *7*, 27040–27048. [CrossRef]
18. Ci, T.; Chen, L.; Yu, L.; Ding, J. Tumor regression achieved by encapsulating a moderately soluble drug into a polymeric thermogel. *Sci. Rep.* **2014**, *4*, 5473. [CrossRef]
19. Al-Thani, H.F.; Shurbaji, S.; Yalcin, H.C. Zebrafish as a Model for Anticancer Nanomedicine Studies. *Pharmaceuticals* **2021**, *14*, 625. [CrossRef]
20. Benslimane, F.M.; Zakaria, Z.Z.; Shurbaji, S.; Abdelrasool, M.K.A.; Al-Badr, M.A.H.; Al Absi, E.S.K.; Yalcin, H.C. Cardiac function and blood flow hemodynamics assessment of zebrafish (*Danio rerio*) using high-speed video microscopy. *Micron* **2020**, *136*, 102876. [CrossRef]
21. Yalcin, H.C.; Amindari, A.; Butcher, J.T.; Althani, A.; Yacoub, M. Heart function and hemodynamic analysis for zebrafish embryos. *Dev. Dyn.* **2017**, *246*, 868–880. [CrossRef]
22. Howe, K.; Clark, M.D.; Torroja, C.F.; Torrance, J.; Berthelot, C.; Muffato, M.; Collins, J.E.; Humphray, S.; McLaren, K.; Matthews, L. The zebrafish reference genome sequence and its relationship to the human genome. *Nature* **2013**, *496*, 498–503. [CrossRef]
23. Teame, T.; Zhang, Z.; Ran, C.; Zhang, H.; Yang, Y.; Ding, Q.; Xie, M.; Gao, C.; Ye, Y.; Duan, M.; et al. The use of zebrafish (*Danio rerio*) as biomedical models. *Anim. Front.* **2019**, *9*, 68–77. [CrossRef]
24. Lam, S.; Chua, H.; Gong, Z.; Lam, T.; Sin, Y. Development and maturation of the immune system in zebrafish, *Danio rerio*: A gene expression profiling, in situ hybridization and immunological study. *Dev. Comp. Immunol.* **2004**, *28*, 9–28. [CrossRef]
25. Zakaria, Z.Z.; Benslimane, F.M.; Nasrallah, G.K.; Shurbaji, S.; Younes, N.N.; Mraiche, F.; Da'as, S.I.; Yalcin, H.C. Using zebrafish for investigating the molecular mechanisms of drug-induced cardiotoxicity. *BioMed Res. Int.* **2018**, *2018*, 1642684. [CrossRef]
26. Reed, B.; Jennings, M. *Guidance on the Housing and Care of Zebrafish Danio Rerio*; Royal Society for the Prevention of Cruelty to Animals (RSPCA): Horsham, UK, 2011.
27. Huang, W.; Zhang, C. Tuning the Size of Poly(lactic-co-glycolic Acid) (PLGA) Nanoparticles Fabricated by Nanoprecipitation. *Biotechnol. J.* **2018**, *13*, 1700203. [CrossRef]

28. Lince, F.; Marchisio, D.L.; Barresi, A.A. Strategies to control the particle size distribution of poly-epsilon-caprolactone nanoparticles for pharmaceutical applications. *J. Colloid Interface Sci.* **2008**, *322*, 505–515. [CrossRef]
29. Shen, H.; Hong, S.; Prud'homme, R.K.; Liu, Y. Self-assembling process of flash nanoprecipitation in a multi-inlet vortex mixer to produce drug-loaded polymeric nanoparticles. *J. Nanopart. Res.* **2011**, *13*, 4109–4120. [CrossRef]
30. Dimchevska, S.; Geskovski, N.; Petruševski, G.; Chacorovska, M.; Popeski-Dimovski, R.; Ugarkovic, S.; Goracinova, K. SN-38 loading capacity of hydrophobic polymer blend nanoparticles: Formulation, optimization and efficacy evaluation. *Drug Dev. Ind. Pharm.* **2017**, *43*, 502–510. [CrossRef]
31. Narumanchi, S.; Wang, H.; Perttunen, S.; Tikkanen, I.; Lakkisto, P.; Paavola, J. Zebrafish Heart Failure Models. *Front. Cell Dev. Biol.* **2021**, *9*, 1061. [CrossRef]
32. Huang, C.-C.; Chen, P.-C.; Huang, C.-W.; Yu, J. Aristolochic acid induces heart failure in zebrafish embryos that is mediated by inflammation. *Toxicol. Sci.* **2007**, *100*, 486–494. [CrossRef]
33. Pucci, C.; Martinelli, C.; Ciofani, G. Innovative approaches for cancer treatment: Current perspectives and new challenges. *Ecancermedicalscience* **2019**, *13*, 961. [CrossRef] [PubMed]
34. Martinelli, C.; Pucci, C.; Ciofani, G. Nanostructured carriers as innovative tools for cancer diagnosis and therapy. *APL Bioeng.* **2019**, *3*, 011502. [CrossRef] [PubMed]
35. Albanese, A.; Tang, P.S.; Chan, W.C. The effect of nanoparticle size, shape, and surface chemistry on biological systems. *Annu. Rev. Biomed. Eng.* **2012**, *14*, 1–16. [CrossRef]
36. Maeda, H. Toward a full understanding of the EPR effect in primary and metastatic tumors as well as issues related to its heterogeneity. *Adv. Drug Deliv. Rev.* **2015**, *91*, 3–6. [CrossRef]
37. Gerlowski, L.E.; Jain, R.K. Microvascular permeability of normal and neoplastic tissues. *Microvasc. Res.* **1986**, *31*, 288–305. [CrossRef]
38. Wickramasinghe, C.D.; Nguyen, K.-L.; Watson, K.E.; Vorobiof, G.; Yang, E.H. Concepts in cardio-oncology: Definitions, mechanisms, diagnosis and treatment strategies of cancer therapy-induced cardiotoxicity. *Future Oncol.* **2016**, *12*, 855–870. [CrossRef]
39. Moslehi, J.J. Cardiovascular Toxic Effects of Targeted Cancer Therapies. *N. Engl. J. Med.* **2016**, *375*, 1457–1467. [CrossRef]
40. Sulaiman, T.N.S.; Larasati, D.; Nugroho, A.K.; Choiri, S. Assessment of the Effect of PLGA Co-polymers and PEG on the Formation and Characteristics of PLGA-PEG-PLGA Co-block Polymer Using Statistical Approach. *Adv. Pharm. Bull.* **2019**, *9*, 382. [CrossRef]
41. Inkson, B. Scanning electron microscopy (SEM) and transmission electron microscopy (TEM) for materials characterization. In *Materials Characterization Using Nondestructive Evaluation (NDE) Methods*; Elsevier: Amsterdam, The Netherlands, 2016; pp. 17–43.
42. Singh, A.P.; Glennon, M.S.; Umbarkar, P.; Gupte, M.; Galindo, C.L.; Zhang, Q.; Force, T.; Becker, J.R.; Lal, H. Ponatinib-induced cardiotoxicity: Delineating the signalling mechanisms and potential rescue strategies. *Cardiovasc. Res.* **2019**, *115*, 966–977. [CrossRef]
43. Cassar, S.; Adatto, I.; Freeman, J.L.; Gamse, J.T.; Iturria, I.; Lawrence, C.; Muriana, A.; Peterson, R.T.; Van Cruchten, S.; Zon, L.I. Use of Zebrafish in Drug Discovery Toxicology. *Chem. Res. Toxicol.* **2020**, *33*, 95–118. [CrossRef] [PubMed]
44. Devulapally, R.; Foygel, K.; Sekar, T.V.; Willmann, J.K.; Paulmurugan, R. Gemcitabine and antisense-microRNA co-encapsulated PLGA–PEG polymer nanoparticles for hepatocellular carcinoma therapy. *ACS Appl. Mater. Interfaces* **2016**, *8*, 33412–33422. [CrossRef] [PubMed]
45. Devulapally, R.; Paulmurugan, R. Polymer nanoparticles for drug and small silencing RNA delivery to treat cancers of different phenotypes. *Wiley Interdiscip. Rev. Nanomed. Nanobiotechnol.* **2014**, *6*, 40–60. [CrossRef]
46. Wang, T.-Y.; Choe, J.W.; Pu, K.; Devulapally, R.; Bachawal, S.; Machtaler, S.; Chowdhury, S.M.; Luong, R.; Tian, L.; Khuri-Yakub, B. Ultrasound-guided delivery of microRNA loaded nanoparticles into cancer. *J. Control. Release* **2015**, *203*, 99–108. [CrossRef]
47. Sharma, A.; Burridge, P.W.; McKeithan, W.L.; Serrano, R.; Shukla, P.; Sayed, N.; Churko, J.M.; Kitani, T.; Wu, H.; Holmström, A.; et al. High-throughput screening of tyrosine kinase inhibitor cardiotoxicity with human induced pluripotent stem cells. *Sci. Transl. Med.* **2017**, *9*, eaaf2584. [CrossRef]
48. Ai, N.; Chong, C.-M.; Chen, W.; Hu, Z.; Su, H.; Chen, G.; Lei Wong, Q.W.; Ge, W. Ponatinib exerts anti-angiogenic effects in the zebrafish and human umbilical vein endothelial cells via blocking VEGFR signaling pathway. *Oncotarget* **2018**, *9*, 31958–31970. [CrossRef]
49. Klein, T.; Vajpai, N.; Phillips, J.J.; Davies, G.; Holdgate, G.A.; Phillips, C.; Tucker, J.A.; Norman, R.A.; Scott, A.D.; Higazi, D.R.; et al. Structural and dynamic insights into the energetics of activation loop rearrangement in FGFR1 kinase. *Nat. Commun.* **2015**, *6*, 7877. [CrossRef]
50. Ku, S.; Yan, F.; Wang, Y.; Sun, Y.; Yang, N.; Ye, L. The blood–brain barrier penetration and distribution of PEGylated fluorescein-doped magnetic silica nanoparticles in rat brain. *Biochem. Biophys. Res. Commun.* **2010**, *394*, 871–876. [CrossRef]
51. Paolini, M.S.; Fenton, O.S.; Bhattacharya, C.; Andresen, J.L.; Langer, R. Polymers for extended-release administration. *Biomed. Microdevices* **2019**, *21*, 45. [CrossRef]
52. Corkery, D.P.; Dellaire, G.; Berman, J.N. Leukaemia xenotransplantation in zebrafish–chemotherapy response assay in vivo. *Br. J. Haematol.* **2011**, *153*, 786–789. [CrossRef]
53. Pruvot, B.; Jacquel, A.; Droin, N.; Auberger, P.; Bouscary, D.; Tamburini, J.; Muller, M.; Fontenay, M.; Chluba, J.; Solary, E. Leukemic cell xenograft in zebrafish embryo for investigating drug efficacy. *Haematologica* **2011**, *96*, 612–616. [CrossRef]

Article

Rubus ellipticus Sm. Fruit Extract Mediated Zinc Oxide Nanoparticles: A Green Approach for Dye Degradation and Biomedical Applications

Jyoti Dhatwalia [1], Amita Kumari [1,*], Ankush Chauhan [2], Kumari Mansi [3], Shabnam Thakur [1], Reena V. Saini [4], Ishita Guleria [1], Sohan Lal [1], Ashwani Kumar [5], Khalid Mujasam Batoo [6], Byung Hyune Choi [7], Amanda-Lee E. Manicum [8] and Rajesh Kumar [9,*]

1. School of Biological and Environmental Sciences, Faculty of Sciences, Shoolini University of Biotechnology & Management Sciences, Solan 173212, Himachal Pradesh, India; dhatwaliajyoti3096@gmail.com (J.D.); shabnamthakur780@gmail.com (S.T.); ishita.thakur93@gmail.com (I.G.); sohanlal4810@gmail.com (S.L.)
2. Chettinad Hospital and Research Institute, Chettinad Academy of Research and Education, Kanchipuram 603103, Tamil Nadu, India; ankushchauhan18@gmail.com
3. Advanced School of Chemical Sciences, Shoolini University of Biotechnology & Management Sciences, Solan 173212, Himachal Pradesh, India; mansi528sharma@gmail.com
4. Central Research Laboratory MMIMSR, Department of Biotechnology MMEC, Maharishi Markandeshwar (Deemed to be University), Mullana 133207, Haryana, India; reenavohra10@gmail.com
5. Patanjali Research Institute, Haridwar 249405, Uttarakhand, India; ashu5157@gmail.com
6. King Abdullah Institute for Nanotechnology, College of Science, King Saud University, Building No. 04, Riyadh 11451, Saudi Arabia; khalid.mujasam@gmail.com
7. Department of Biomedical Sciences, Inha University College of Medicine, 100 Inha-ro, Incheon 22212, Korea; bryan@inha.ac.kr
8. Department of Chemistry, Faculty of Science, Arcadia Campus, Tshwane University of Technology, Pretoria 0183, South Africa; manicumae@tut.ac.za
9. Department of Physics, Faculty of Physical Sciences, Sardar Vallabhbhai Patel Cluster University, Mandi 175001, Himachal Pradesh, India
* Correspondence: amitabot@gmail.com (A.K.); rajesh.shoolini@gmail.com (R.K.)

Abstract: *Rubus ellipticus* fruits aqueous extract derived ZnO-nanoparticles (NPs) were synthesized through a green synthesis method. The structural, optical, and morphological properties of ZnO-NPs were investigated using XRD, FTIR, UV-vis spectrophotometer, XPS, FESEM, and TEM. The Rietveld refinement confirmed the phase purity of ZnO-NPs with hexagonal wurtzite crystalline structure and p-63-mc space group with an average crystallite size of 20 nm. XPS revealed the presence of an oxygen chemisorbed species on the surface of ZnO-NPs. In addition, the nanoparticles exhibited significant in vitro antioxidant activity due to the attachment of the hydroxyl group of the phenols on the surface of the nanoparticles. Among all microbial strains, nanoparticles' maximum antibacterial and antifungal activity in terms of MIC was observed against *Bacillus subtilis* (31.2 µg/mL) and *Rosellinia necatrix* (15.62 µg/mL), respectively. The anticancer activity revealed 52.41% of A549 cells death (IC$_{50}$: 158.1 ± 1.14 µg/mL) at 200 µg/mL concentration of nanoparticles, whereas photocatalytic activity showed about 17.5% degradation of the methylene blue within 60 min, with a final dye degradation efficiency of 72.7%. All these results suggest the medicinal potential of the synthesized ZnO-NPs and therefore can be recommended for use in wastewater treatment and medicinal purposes by pharmacological industries.

Keywords: ZnO-NPs; photocatalyst; antioxidant; antimicrobial; anticancer activity

1. Introduction

Nanotechnology is an important field of modern technology that deals with the study of particles having a size of 1–100 nm [1]. The nanoparticles (NPs), which are synthesized through this technology, have various applications such as detectors, surface coating agents, catalysts, and antimicrobial agents [2]. Different types of metal oxides NPs have been synthesized through this technique like silver oxide (AgO), silicon dioxide (SiO_2), titanium dioxide (TiO_2), indium (III) oxide (In_2O_3), tin (IV) oxide (SnO_2), zinc oxide (ZnO) and copper oxide (CuO), etc. After $SiOTiO_2$, ZnO is one of the most abundantly produced metal oxides [3]. The Zinc oxide nanoparticles (ZnO-NPs) are easy to synthesize, showed less toxicity, therefore, have been reported as the most popular metal oxide nanoparticle [4]. In addition to these, their roles are reported in the literature on drug delivery, wound healing, bioimaging, antimicrobial, antidiabetic, and anticancer activities [4]. Due to zero toxicity, they are widely utilized in the cosmetic industry (in sunscreen and facial creams) [4,5], and food industries as additives [6].

The various methods that have been used for the synthesis of ZnO-NPs are the physical, chemical, hybrid, microwave, and biological methods [7–12]. Among all these methods, the most abundantly used method currently is green synthesis, i.e., plant-mediated synthesis of nanoparticles because of its simplicity, cost-effectiveness, and environmentally friendly behavior [13].

Due to the presence of phytochemicals such as phenols, flavonoids, and terpenoids in the plants, they are used for the synthesis of nanoparticles and act as non-toxic reducing, capping, and stabilizing agents [14]. The role of ZnO-NPs is also reported in various biological applications such as antibacterial, antifungal, and anticancer activities. Additionally, due to their catalytic activity, ZnO-NPs are also used for wastewater treatment [15].

The ZnO-NPs synthesized from the extracts of *Rosa indica*, *Justicia adhatoda*, *Vitex negundo*, *Mentha pulegium*, and *Cassia fistula* were reported to inhibit the growth of broad-spectrum pathogens [16–20]. Currently, plant-mediated synthesized nanoparticles are observed to possess photocatalytic activity [15,20–22] and they have shown more effective results than chemically prepared nanoparticles [23]. *Rubus ellipticus* Sm., commonly known as yellow Himalayan raspberry, Aiselu, or Aekhae, is an evergreen shrub belonging to the family Apocynaceae. It is native to China, Indonesia, the Indian subcontinent, and Sri Lanka [24]. Its fruits are rich in glycosides, flavonoids, phenols, resin, pectin, and tannins [25,26]. They are commonly used for the treatment of fever, cough, dysentery, constipation, diarrhea, curing bone fracture, and relieving stomach worms in children [25]. In addition, the fruit extract was also reported to possess good antioxidant, antimicrobial, and anticancer activity [26]. Due to the various biological applications of plant-mediated nanoparticles, the present study focuses on the synthesis of environmentally friendly, non-toxic ZnO-NPs using the fruit extract of *R. ellipticus*. These particles are further screened for antimicrobial, anticancer, and photocatalytic activity.

2. Material and Methods

2.1. Plant Material Collection

R. ellipticus fruits were collected from the Shimla (2000 masl) district, Himachal Pradesh, India. The collected fruits were dried in an oven at 40 °C, crushed using a grinder, and the coarse powder was stored in an airtight container for further use. The authentication of the plants was done in the Botanical Survey of India (BSI), Dehradun, India with accession number 117.

Extract Preparation

The fruit extract was prepared by dissolving 10 g of powdered sample in 100 mL of distilled water in a water bath at 60 °C for 8 h. The extract was filtered through Whatman filter paper number 41 and dried in the oven at 37 °C. The dried extract was collected and stored at 5 °C in a refrigerator for further analysis [27].

2.2. Synthesis of ZnO-NPs

ZnO-NPs were prepared from the fruit's aqueous extract of *R. ellipticus* by following the method of Chauhan et al. [27]. For the preparation of ZnO-NPs, the 10 mL of fruits aqueous extract (100 mg of fruits' aqueous extract in 10 mL of distilled water) was mixed with 50 mL of zinc acetate dihydrate (0.5 M; Loba Chemie, Mumbai, India; 98% purity) in a beaker (Figure 1). To the solution, 5 mL of sodium hydroxide (0.2 M) was added, and the whole solution was magnetically stirred for 20 min. at 40 °C temperature. The whole solution was magnetically stirred for 20 min. at a temperature of 40 °C. The resulting solution in the flask was stirred for two hours until white precipitates formed. The precipitates were collected, centrifuged, and washed with deionized water for the removal of impurities. The sample was collected and then dried at 60 °C for 24 h.

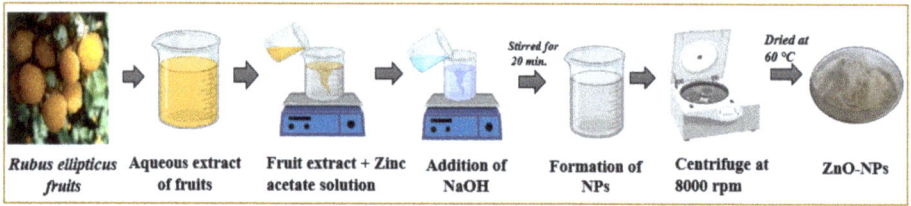

Figure 1. Schematic representation of biosynthesis of ZnO-NPs using fruit extract.

2.3. Characterization of ZnO-NPs

The characterization of the synthesized ZnO-NPs was done by Fourier transform infrared spectroscopy (FTIR; Perkin Elmer, Waltham, MA, USA) at the resolution of 1 cm^{-1} and a scan range of 4000 cm^{-1} to 250 cm^{-1}) for functional group identification, X-ray photoelectron spectroscopy (XPS; Prevac, Rogów, Poland) for determination of elemental composition, chemical state, and electronic state of the elements in the material, X-ray diffraction spectroscopy (XRD; PANalytical, Almelo, Netherlands) for the determination of amorphous and crystalline nature of the material, scanning electron microscope [(SEM-Mapping; SU8010 Series) (Hitachi, Tokyo, Japan)] for checking texture and elemental composition of nanoparticles, and Transmission electron microscope [TEM; JEM 2100 Plus (JEOL, New Delhi, India)] used for size and morphology of nanoparticles at Sophisticated Analytical Instrumentation Facility (SAIF), Punjab University, Chandigarh, India. The absorption spectrum of green synthesized nanoparticles was analyzed by Ultraviolet-visible spectrophotometry (UV 2450, Shimadzu corporation, Kyoto, Japan) in Advanced Materials Research Centre, Indian Institute of Technology (IIT), Mandi, Himachal Pradesh, India.

2.4. Antioxidant Assays

2.4.1. DPPH Free Radical Scavenging Assay

The antioxidant potential of ZnO-NPs synthesized from aqueous fruit extract of *R. ellipticus* was analyzed by following the methodology with slight modifications [28]. In this method, 2,2-diphenyl-1-picrylhydrazyl (DPPH: 0.1 M) and ZnONPs (20–100 µg/mL) were prepared in methanol just before the experiment. Firstly, the 1 mL of each concentration was separately mixed with 1 mL of DPPH into the test tube. After this, the test tubes were incubated in the dark for 30 min. In the last step, a 200 µL reaction mixture was taken from each test tube and poured into a 96-well microtiter plate. The absorbance (Abs) was taken through a multileader (VARIOSKAN-LUX, Thermo Fisher Scientific, Waltham, MA, USA) at 517 nm. This assay was performed in triplicate and repeated for aqueous fruit extract. DPPH was used as a control and ascorbic acid was used as a positive control. Methanol was used as a blank for the assay. The radical scavenging activity was calculated as follows:

$$\text{DPPH activity (\%)} = \frac{[\text{Control (Abs)} - \text{Test sample (Abs)}]}{\text{Control (Abs)}} \times 100$$

2.4.2. ABTS Assay

The antioxidant effect of the fruit extract was studied using ABTS (2,2-azino-bis-3-ethyl benzthiazoline-6-sulphonic acid) radical cation decolorization assay [28]. ABTS radical cations (ABTS$^+$) were produced by reacting 7 mM ABTS solution with 2.45 mM ammonium persulphate; the mixture was prepared by mixing the two stock solutions in equal quantities and allowed to react for 16 h at room temperature in the dark. After that, 0.1 mL of extract of each sample (at four different concentrations 25 µg/mL, 50 µg/mL, 75 µg/mL, 100 µg/mL) were mixed with 0.9 mL ABTS solution in the test tubes. The reaction mixture was incubated in dark conditions for 30 min. The absorbance was read at 745 nm in a UV-vis spectrophotometer (VARIOSKAN-LUX, Thermo Fisher Scientific, Waltham, MA, USA) and the percentage inhibition was calculated by using the formula:

$$\text{ABTS assay (\%)} = \frac{[\text{Control (Abs)} - \text{Test sample (Abs)}]}{\text{Control (Abs)}} \times 100$$

2.4.3. FRAP Free Radical Scavenging Assay

The free radical scavenging activity of ZnO-NPs synthesized from aqueous fruit extract at different concentrations (20–100 µg/mL) was analyzed through ferric ion reducing antioxidant power (FRAP) assay with slight modifications [29]. In this method, 1 mL of each concentration was separately mixed with 1 mL of sodium phosphate buffer (0.2 M in distilled water; pH 6.6) into the test tube. After this, 1 mL of 1% potassium ferricyanide was added into all the test tubes and incubated at 50 °C in a water bath for 20 min. After this, 1 mL of trichloroacetic acid (10%) was added to all test tubes and the mixtures were centrifuged for 10 min. at 5000 rpm. In the next step, 1 mL of the reaction mixture was taken from each sample and 1 mL of distilled water was added to it. After this, 0.2 mL of ferric chloride (0.1% in distilled water) was added to each test tube. In the last step, 200 µL solution was taken from each test tube and poured into the 96-well microtiter plate. The blank was prepared similarly except that potassium ferricyanide (1%) was replaced by distilled water. The absorbance was measured at 700 nm using a multileader (VARIOSKAN-LUX, Thermo Fisher Scientific, Waltham, MA, USA)) [30]. The results were recorded in triplicate and ascorbic acid (20–100 µg/mL) was used as a positive control. The antioxidant capacity of each sample was calculated from the linear calibration curve of ferrous sulfhate (y = 0.004x + 0.0377; R^2) and expressed as µM FeSO$_4$ equivalents.

2.5. Antimicrobial Assay

2.5.1. Selection of Strains

For the antibacterial assay, two strains of Gram-positive bacteria [*Staphylococcus aureus* (MTCC 731), *Bacillus subtilis* (MTCC 441)], and Gram-negative bacteria [*Pseudomonas aeruginosa* (MTCC 424), *Escherichia coli* (MTCC 739)] were selected, whereas for antifungal assay two pathogenic strains of fungi [*Fusarium oxysporum* (SR266-9) and *Rosellinia necatrix*; (HG964402.1.)] were selected. The bacterial and fungal strains were obtained from CSIR Institute of Microbial Technology (IMTech, Staines, UK), Chandigarh and School of Microbiology, Shoolini University, Solan, India, respectively.

2.5.2. Antibacterial Activity

Disc Diffusion Method

The antibacterial activity of fruits' aqueous extract and ZnO-NPs were observed by using the disc diffusion method [31]. The bacterial culture (100 µL), with the help of sterile cotton swabs, was uniformly spread on the surface of the nutrient agar plates. A stock solution of ZnO-NPs and fruits aqueous extract was prepared by dissolving 10 mg ZnO-NPs and crude aqueous extract of fruits to 1 mL of dimethyl sulfoxide (DMSO) separately. The 300 µg/mL of ZnO-NPs and fruits aqueous extract were applied separately to each 6 mm sterilized paper disc. Discs were placed into Petri plates with bacterial culture and then placed in an incubator set to 37 °C for 24 h. After 24 h of incubation, an antibiotic zone scale was used to record zones of inhibition data. Each antibacterial assay was performed in triplicate. Dimethyl sulfoxide (solvent) and ampicillin (5 mg/mL) were used as negative and positive controls, respectively.

2.5.3. Antifungal Assay

Poison Food Technique

For the antifungal assay, the poison food technique was used with minor modifications [32]. The two fungal strains (*F. oxysporum* and *R. necatrix*) were cultured on potato dextrose agar (PDA) at 25 °C for 7 days before use. The 300 µg/mL of fruits aqueous extract and ZnO-NPs were mixed in 24 mL of PDA per plate separately. After the solidification of the media, a 6 mm diameter of fungal disc was cut with flame sterilized cork borer and then placed at the center of each petri dish. The dish was incubated at 25 °C for seven days. The colony diameter of fungus was measured on the seventh day of incubation. The tests were performed in triplicate. The antifungal activity of the fruit aqueous extract and ZnO-NPs were further compared to the control (dish without fruits aqueous extract and ZnO-NPs). The percentage inhibition of radial mycelial growth over the control was calculated by using the following formula:

$$\text{Inhibition (\%)} = \frac{C - T}{C} \times 100$$

where C is the diametric growth of the colony in control, T is the diametric growth in the nanoparticles extract.

2.5.4. Minimum Inhibitory Concentration (MIC)

For the determination of antimicrobial activity, the fruit aqueous extract and ZnO-NPs 96 well microtitre plate method were used [33]. The 12 wells of each row of microtiter plates were filled with 0.1 mL of sterilized NA and PDA broth for bacteria and fungus, respectively. Sequentially, wells 2–11 received an additional 5 mg/mL of plant extract and ZnO-NPs in separate columns. The serial dilution was done by transferring 100 µL of testing samples (500 µg/mL) from the first row of the subsequent wells in the next row of the same column. Finally, a volume of 10 µL was taken from bacterial or fungal suspension and then added to each well to achieve a final concentration of 5×10^6 CFU/mL. The well plates were incubated for 24 h at 37 °C and then 15 µL of resazurin solution, as an indicator, was added to each well. The color change from purple to pink or colorless in the well was then observed visually. The lowest concentration of extract at which color change occurred was recorded as the MIC value.

2.6. Anticancer Analysis

2.6.1. Cell Culturing and Maintenance

Cell lines were purchased from the National Centre for Cell Sciences (NCCS), Pune, India, A549 (Human lung adenocarcinoma) to evaluate the anticancer effect of ZnO-NPs by using the 3-(4,5-dimethylthiazol-2-yl)-2,5-diphenyl tetrazolium bromide (MTT) assay. The stock cells were cultured in DMEM (Dulbecco's Modified Eagle's Medium) supplemented with inactivated fetal bovine serum (FBS) 10%, penicillin-streptomycin (1%) in a humidified atmosphere of CO_2 (5%) at 37 °C until confluent. To take out dead cells and debris, the used media and debris were deposed and washed with phosphate-buffered saline (PBS). The trypsin (0.25%) was used for cell dissociation and the pallet was cultivated after the centrifugation process at 3000 rpm for 10 min. For sub-culturing, fresh aliquots were made and transferred to new culture dishes.

2.6.2. In Vitro Anticancer Assay (MTT Assay on A549)

To evaluate the anticancer effect through MTT assay, the method reported by Kumari et al. [34] was used. The lung cancer cells (A549) at a density of 1×10^4 cells/mL were seeded in a 96-well plate and incubated at 37 °C for 24 h under 5% CO_2. The next day, cancer cells were treated with different concentrations (1.56–200 µg/mL) of ZnO-NPs for 24 h, followed by an MTT assay (4 h incubation) to assess the cell viability. For positive and negative control, Paclitaxel and DMSO were used, respectively. After 4 h, DMSO (100 µL) was added to each well for dissolving the purple-colored complex. Using a microplate reader, the optical density was noted at 595 nm. IC_{50} values of the ZnO-NPs (in triplicate) were further recorded, and cell viability was analyzed using the following formula:

$$\text{Cell cytotoxicity} = \frac{(\text{Abs control} - \text{Abs sample})}{\text{Abs control}} \times 100$$

2.7. Dye Degradation Study

To study the dye degradation capability of ZnO-NPs, the method reported by Mansi et al. [35] was used. ZnO-NPs were used as probe catalysts to investigate the degradation of methylene blue dye in synthetic wastewater in the presence of sunlight. The experimental methodology includes the dispersion of 100 mg of ZnO-NPs in 200 mL of dye tainted wastewater at a concentration of 10 ppm. To achieve adsorption-desorption equilibrium with dye and ZnO-NPs, the solution was set up in the dark with continuous stirring for 30 min. at 100 rpm. The equilibrated dye and nanoparticles solution was, afterward, subjected to sunlight to study the photocatalysis of ZnO-NPs towards dye degradation. The systematic activity of ZnO-NPs throughout the bulk phase was achieved by continuous stirring at 100 rpm during the process. Thereafter, the reaction mixture was withdrawn after 15 min. intervals for dye concentration evolution. The collected samples were centrifuged at 12,000 rpm to separate the nanoparticles, and finally, the resulting supernatant was subjected to absorption measures using a UV-vis spectrophotometer (Shimadzu corporation, Japan). After a while, the dye absorbance decreased (λ_{max} = 665 nm) gradually, providing the decolorization rate, as well as photocatalytic efficiency of ZnO-NPs, as calculated by using the following equation

$$\eta = [(A_0 - A_t)/A_0] \times 100$$

where A_0 and A_t were initial and final absorbance after a certain reaction time, respectively.

2.8. Statistical Analysis

All the results were analyzed on MS excel (Microsoft, Redmond, WA, USA), and data were expressed as mean ± SEM. The results were compared with the control group and $p < 0.05$ was considered statistically significant. All the statistical analysis was carried out by SPSS software using paired sample t-test.

3. Results and Discussion

3.1. Characterization of ZnO-NPs

3.1.1. X-ray Diffraction Spectroscopy (XRD)

XRD spectroscopy is a technique used for the determination of as crystalline or amorphous nature of the material. Figure 2 represents the X-ray diffraction pattern of the synthesized ZnO-NPs. The characteristic peaks observed at 31.86°, 34.56°, 36.32°, 47.57°, 56.61°, 62.95°, 66.47°, 68.05°, and 69.16° belongs to (100), (002), (101), (102), (110), (103), (200), (112), and (201) *hkl* planes of ZnO-NPs with hexagonal wurtzite structure [36,37]. A corresponding peak matched well with the standard JCPDS card number: 36-1451 without the presence of any impurity phase. The phase purity was also determined by Rietveld refinement of the given pattern, and structural parameters were obtained as given in Table 1. The Rietveld refinement was performed using FullProf software by describing peak patterns using a pseudo-Voigt profile. Initially, global parameters such as background, scale factors, and cell parameters were refined. Further, the FWHM and atomic orientations were refined consecutively [38]. Figure 2b represents the Rietveld refined patterns of the ZnO-NPs. The broad XRD peaks reveal the lower particle size of ZnO-NPs, and the crystallite size of 20 nm was calculated using the Scherer formula given by [27]:

$$D = \frac{0.9\lambda}{\beta \cos \theta} \quad (1)$$

where λ is the wavelength (1.54Å), θ is the diffraction angle, and β is the full-width half maximum of the XRD peaks. The lattice strain ε was determined by employing the UDM (Universal Deformation Model) and William Hall equation [39]:

$$\beta \cos \theta = \frac{K\lambda}{D} + \varepsilon(4 \sin \theta) \quad (2)$$

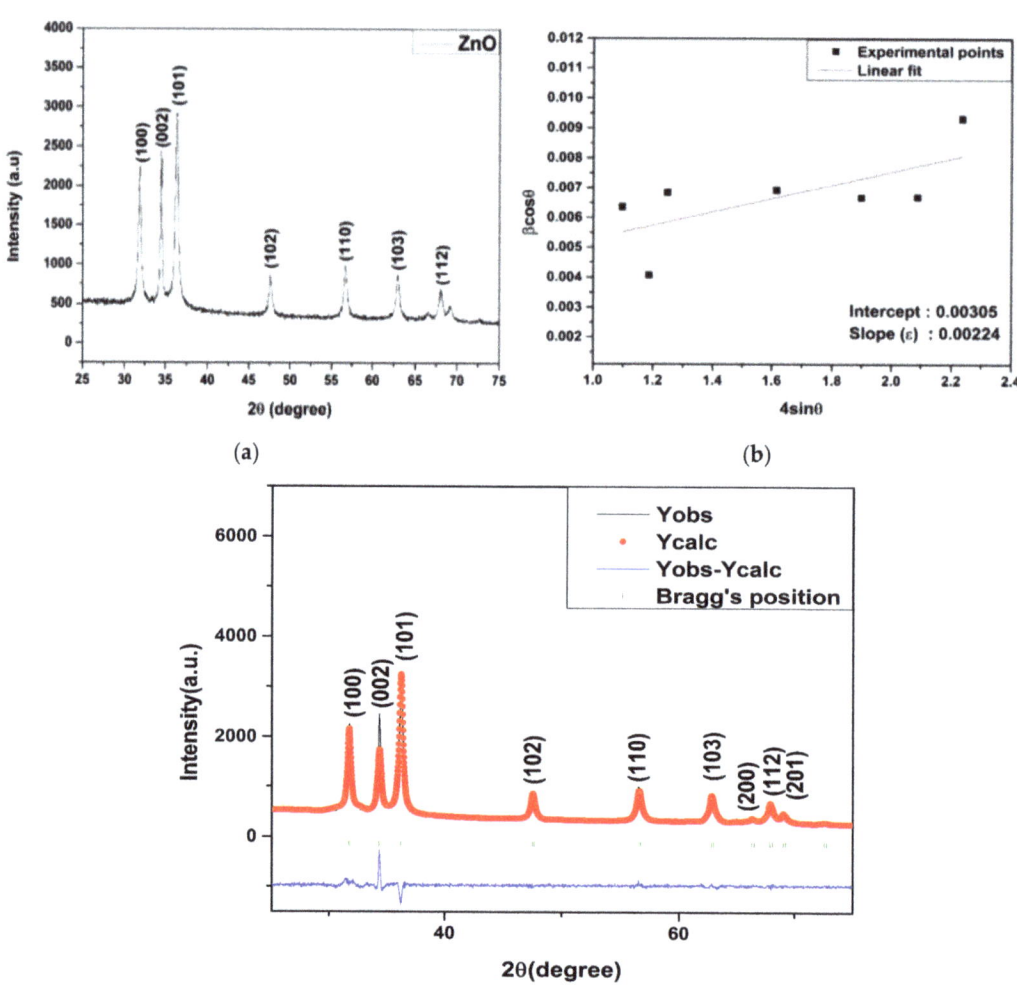

Figure 2. (a) XRD pattern, (b) W-H plot, and (c) Rietveld refined pattern of ZnO-NPs.

Table 1. Structural and Rietveld refine parameters of ZnO-NPs.

Sample	ZnO-NPs
Structure	Hexagonal-wurtize
Space group	P-63-mc
a (Å)	3.251287
b (Å)	3.251287
c (Å)	5.208760
Volume (Å)^3	47.684
χ^2	2.15
R_p	14.7
R_{wp}	21.5
R_e	25.4
Average Crystallite size (Scherer method) (nm)	20
Strain (ε)	0.00224

The UDM model assumes isotropic strain in the crystal in all directions and is independent of the direction of property measurement. The strain generated within the crystal is a result of dislocations and crystal imperfections. The strain was calculated from the slope of the plot between $\beta\cos\theta$ along the y-axis and $4\sin\theta$ along the x-axis corresponding to each peak of the XRD pattern as given in Figure 2c.

3.1.2. X-ray Photoelectron Spectroscopy (XPS)

The chemical bonding states of the elements of ZnO-NPs were determined using the XPS spectra. Figure 3 shows the O1s and Zn2p core-level XPS spectra of ZnO-NPs. The asymmetric O1s peak was deconvoluted into three different peaks at 531.17 eV, 532.26 eV, and 533.13 eV, which corresponds to the oxygen ions associated with the lattice oxygen of ZnO structure, oxygen vacancies, and chemisorbed oxygen species [40,41]. The chemisorbed species plays a major role in enhancing the antimicrobial and photocatalytic degradation properties [27,39]. In Zn2p core-level XPS spectra shows two symmetrical peaks at 1022.2 eV and 1045.35 eV are ascribed to Zn2p$_{1/2}$ and Zn2p$_{3/2}$. The given value of binding energy is in good agreement with the previously reported data for the Zn^{2+} oxidation state [42].

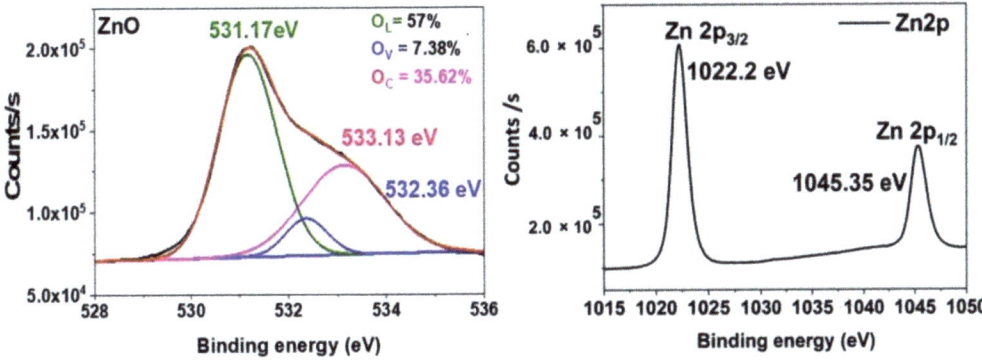

Figure 3. O1s and Zn2p core-shell XPS spectra of ZnO-NPs.

3.1.3. UV-Visible Spectroscopy

Figure 4 shows the UV absorbance spectra and Tauc's plot for the determination of optical bandgap. It can be observed that the absorption spectra of the synthesized ZnO-NPs exhibit a strong absorption band at 353 nm. The excitonic absorption peak can be observed at 257 nm, which is a result of the lower bandgap wavelength of ZnO-NPs than 358 nm [43].

The direct energy bandgap (Eg) that turned out to be 3.21 eV for the synthesized ZnO-NPs was determined using the Tau's relation [44]:

$$(\alpha h\nu)^2 = A(h\nu - Eg) \qquad (3)$$

where α is the absorption coefficient and $h\nu$ corresponds to the photon energy. Hence, the plot between $(\alpha h\nu)^2$ and $(h\nu)$ gives the energy bandgap (Eg).

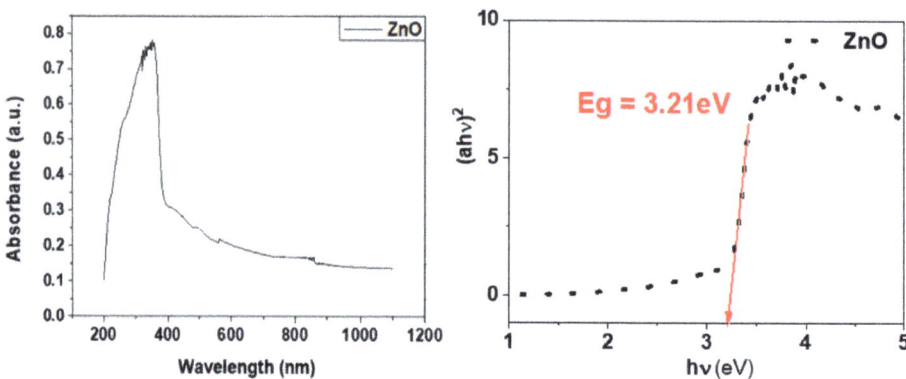

Figure 4. UV-vis Absorbance and Tauc's plot of ZnO-NPs.

3.1.4. FE-SEM and Elemental Mapping

The FE-SEM images given in Figure 5a reveal the self-assembly of spherical particles to form a flower-like structure and are arranged uniformly. The agglomeration of larger assembled structures can be observed. Figure 5b shows the elemental distribution of ZnO-NPs, which reveals the uniform distribution of zinc and oxygen in ZnO-NPs. The surface area of nanoparticles is much higher than their bulk counterparts, which allows them to have more molecules on the surface of the nanoparticles. This provides the nanoparticles with their special properties due to their surface chemistry as a photocatalyst and an antimicrobial agent. The plant extract could have played a relevant role as a capping agent and prevented the agglomeration of nanoparticles [39,45].

3.1.5. Transmission Electron Microscopy (TEM) Study

The TEM images reveal the formation of self-assembled grains with irregular shapes and size. It can be observed that the smaller grains diffused to form larger grains. The particle size distribution given in Figure 6b with average particle size of 19.12 ± 0.77 nm was obtained using ImageJ software and origin software. The d-spacing of 0.5906 nm was calculated using ImageJ software from the observed lattice spacing corresponding to (002) *hkl* plane (Figure 6d). The SAED pattern showed the presence of circular rings with bright spots which correspond to the polycrystalline nature of the ZnO-NPs (Figure 6).

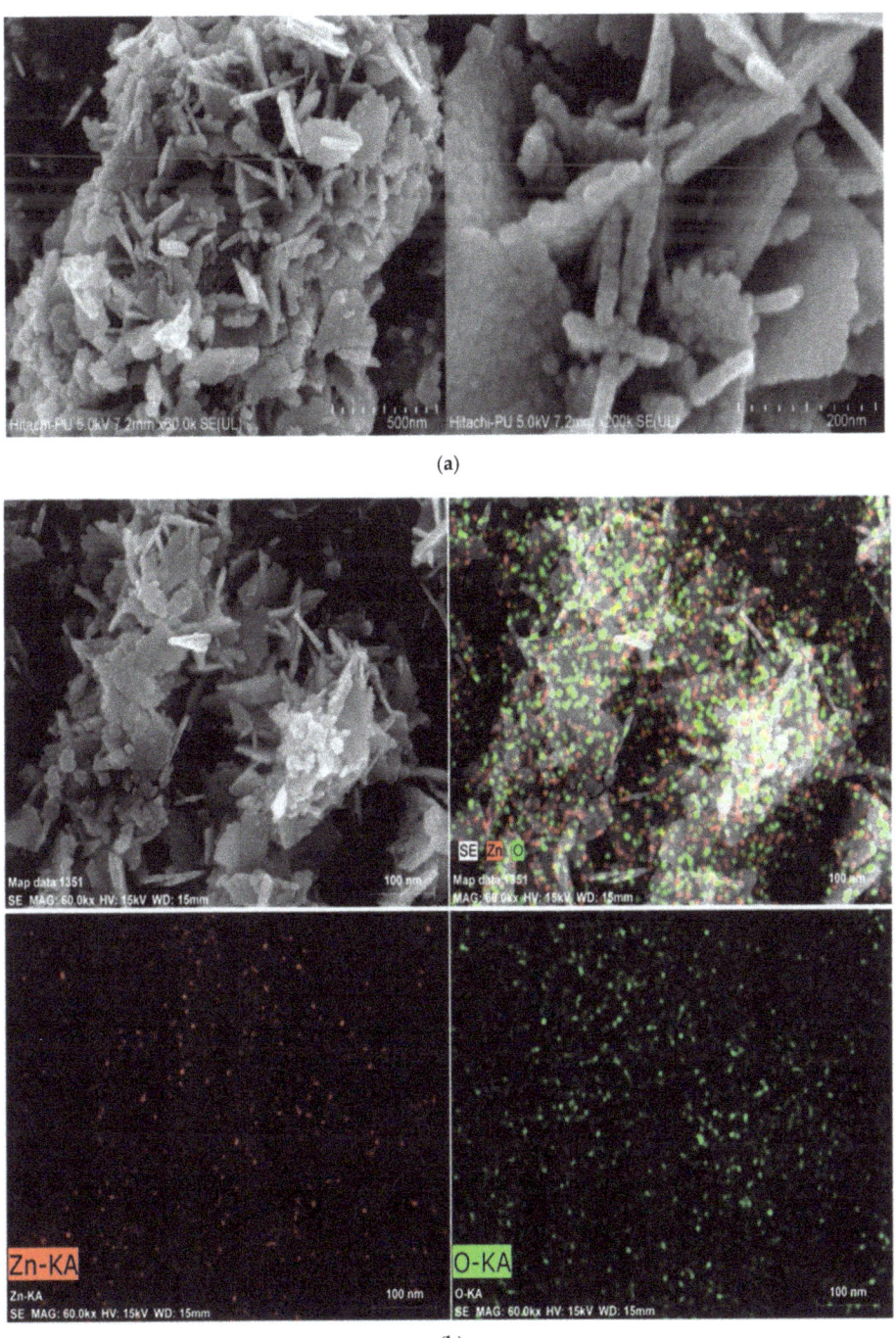

Figure 5. FE-SEM micrographs (**a**) and Elemental mapping (**b**) of ZnO-NPs.

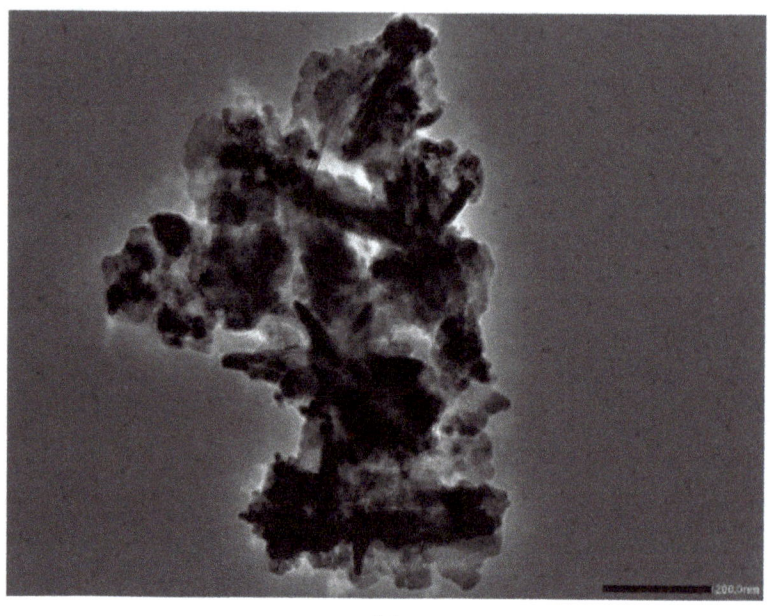

(a)

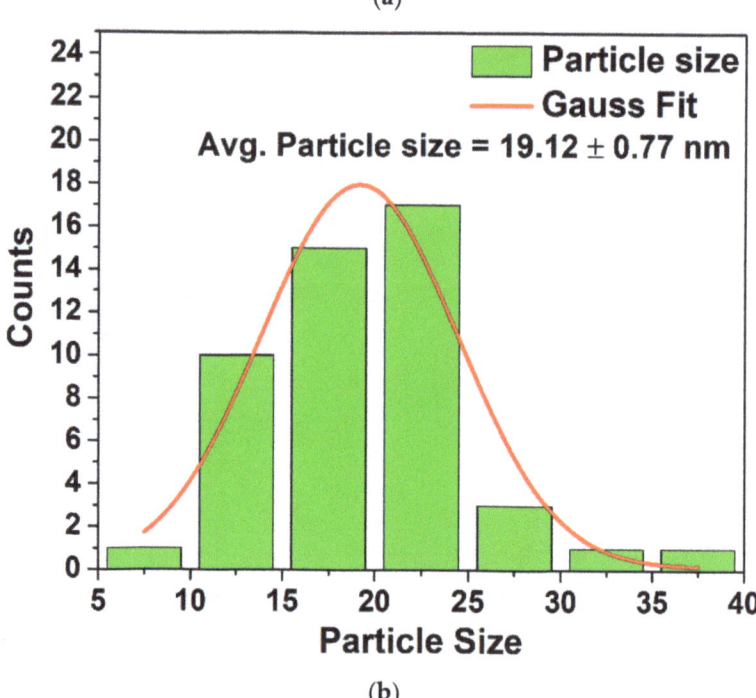

(b)

Figure 6. *Cont.*

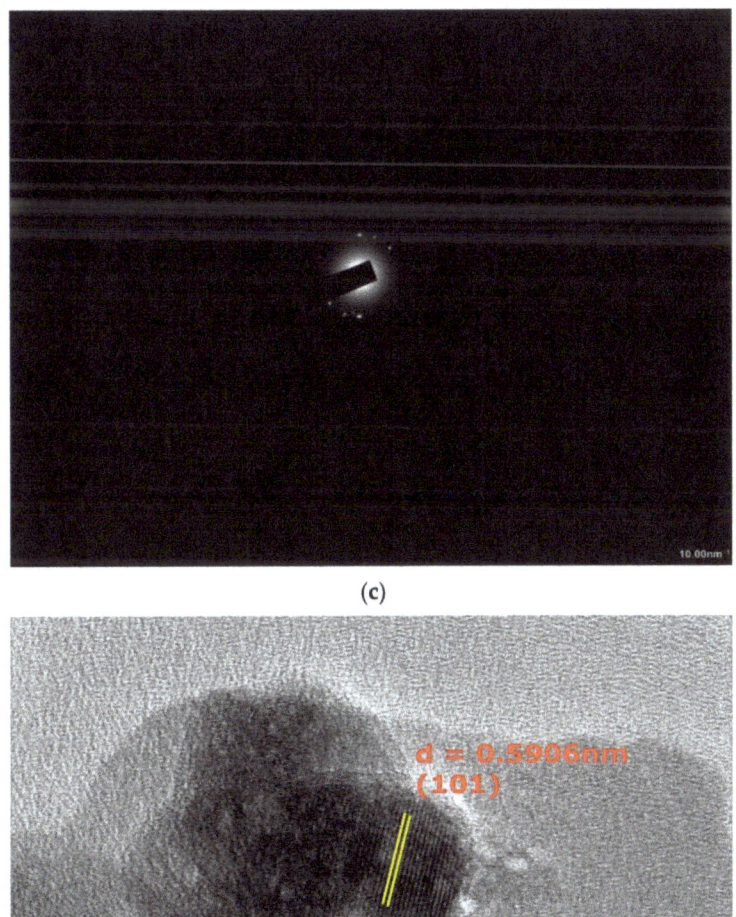

Figure 6. TEM micrograph (**a**), particle size distribution (**b**), SAED pattern (**c**) and d—spacing (given at 20 nm) (**d**) of ZnO-NPs.

3.1.6. FTIR

The Fourier transform infrared spectroscopy (FTIR) is carried out to study the chemical bonds in a molecule by generating the infrared transmittance spectrum. The molecular fingerprint design was used to investigate the different compounds present in the sample. The fingerprint region in the range of 1500–500 cm^{-1} is distinctive for every sample and the region in the range between 4000 to 1500 cm^{-1} is the functional group region. Figure 7 represents the FTIR spectrum of fruit extract and green-synthesized ZnO-NPs to identify the functional groups of phytochemical molecules. The synthesized ZnO-NPs exhibited strong bands at 3432, 2926, corresponding to O–H, N–H stretching vibrations of carboxylic acids, alcohols, amide, respectively. The absorption peaks at 1626 and 1395 cm^{-1} owing to the –C=C– and C–N stretching vibrations of aromatics and aromatic amines [46]. The absorption bands at 898 cm^{-1} and 563 cm^{-1} confirmed the formation of metal oxide [47]. The strong band stretching absorption peaks around 3200–3300 cm^{-1} was assigned to strong band interrelated overlapping stretching vibration of the amide (N–H) and hydroxyl group (OH) in the fruit extract. The OH group stretching indicated the presence of alcohol and phenol and also probably includes the N-H group of proteins [15,48,49]. The fruit extract showed the presence of phytochemical compounds (phenols, glycosides, tannins, saponins, and flavonoids) in *R. ellipticus* [25,26] that facilitate the formation of ZnO-NPs by acting as a capping, reducing, and stabilizing agent [15,27]. The analysis of peaks was in accordance with an earlier study of green synthesis of ZnO-NPs using different extracts [27,46,47].

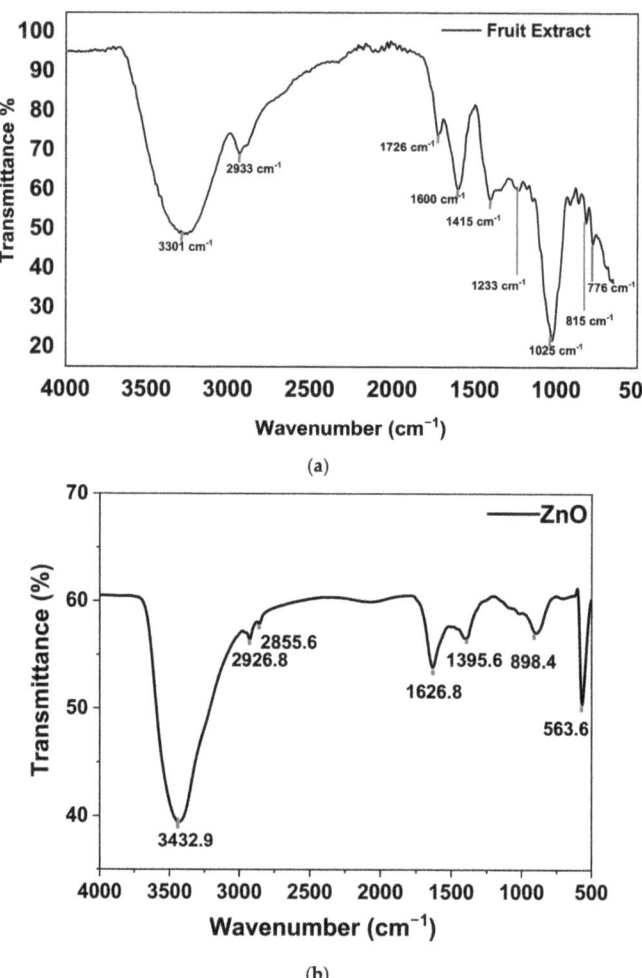

Figure 7. FTIR of (a) fruit extract and (b) ZnO-NPs.

3.2. Antioxidant Activity

The results of antioxidant potential of fruit extract and ZnO-NPs analyzed through DPPH, ABTS, and FRAP assays are presented in Table S1 and Figure 8. The results indicated that as the concentration of fruit extract, ZnO-NPs, and ascorbic acid increased from 20–100 µg/mL, the percentage inhibition was also increased (Table S1). In DPPH assay, the IC_{50} values for fruit extract, ZnO-NPs, and ascorbic acid were observed as 32.8 ± 0.5 µg/mL, 72.9 ± 0.7 µg/mL, and 19.5 µg/mL, respectively, whereas for FRAP assays, values were 73.1 ± 0.5 µM Fe^{2+} equivalents, 149.4 ± 0.9 µM Fe^{2+} equivalent, and 45.4 ± 2.4 µM Fe^{2+} equivalents for fruit extract, ZnO-NPs, and ascorbic acid, respectively. In the case of ABTS assay, the IC_{50} values for fruit extract, Ru-ZnO-NPs, and ascorbic acid were observed as 39.2 ± 1.1 µg/mL, 87.48 ± 0.3 µg/mL, and 27.73 ± 1.494 µg/mL, respectively. The paired sample t-test revealed a significant variation ($p > 0.05$) between the fruit extract and ZnO-NPs results.

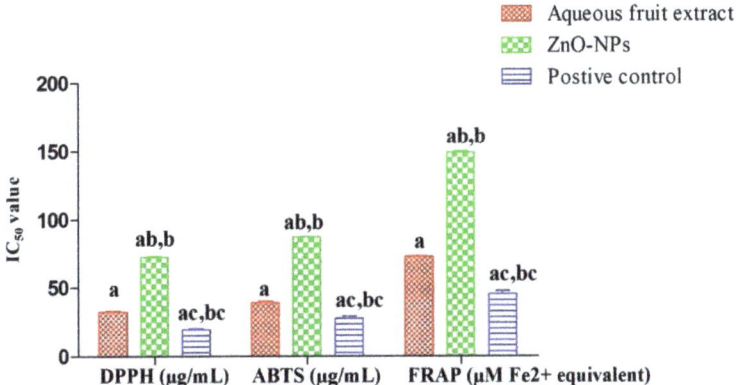

Figure 8. In vitro antioxidant potential (IC_{50} value) of zinc oxide nanoparticles and fruit extract [Different superscript showed significant ($p < 0.05$) difference between (a) fruit extract, (b) ZnO-NPs and (c) positive control between plant extract and nanoparticles].

The lower value of IC_{50} indicates a higher antioxidant potential [50]. In DPPH, ABTS, and FRAP assays, ascorbic acid showed the lower IC_{50} followed by fruit extract and ZnO-NPs (Figure 8). This means fruit extract has higher antioxidant potential than ZnO-NPs which could be due to the presence of higher phenolic components in the crude fruit extract of *R. ellipticus*. Mahendran and Kumari [51] and Reddy et al. [52] also observed similar observations in the fruit extract of *Nothapodytes nimmoniana* and *Piper longum* where fruit extracts were observed with higher antioxidant potential than the green-synthesized Ag-NPs. Previous studies also discussed the antioxidant potential of ZnO-NPs synthesized from the plant parts of *Azadirachta indica* [49], *Carica papaya* [53], and *Luffa acutangula* [54]. In all these studies, ZnO-NPs showed a broad spectrum of antioxidant activity by effectively inhibiting the reactive oxygen species.

Nanoparticles exposure causes oxidative stress, which raises concerns regarding their application in humans. However, the green synthesized nanoparticles of this study showed antioxidant activity and might be used to combat oxidative stress. The phytoconstituents of *R. ellipticus* fabricated on the surface of nanoparticles are responsible for this action. The hydroxyl group proved the presence of phenols that are supposed to be attached to the surface of nanoparticles based on FTIR results. These can be used as a substitute for nanoparticles synthesized using other methods.

3.3. Antimicrobial Activity

The antibacterial activity of *R. ellipticus* fruit extract and ZnO-NPs were tested against Gram-positive (*S. aureus* and *B. subtilis*) and Gram-negative (*E. coli* and *P. aeruginosa*) bacteria using the disc diffusion method and MIC value. Among all bacteria, the maximum inhibition zone was observed against B. subtilis (16 ± 0.5 mm) followed by S. aureus (14 ± 0.5 mm) and minimum in P. aeruginosa (12 ± 0.5 mm) (Figure 9 and Figure S1). Whereas, in the case of fruits aqueous extract, the maximum inhibition zone was achieved for B. subtilis (10 ± 1 mm) and lowest for E. coli (8 ± 0.5 mm). The results of the disc diffusion method showed higher efficacy of synthesized ZnO-NPs against Gram-positive bacterial cells as compared to Gram-negative bacterial cells. The paired sample t-test revealed a significant variation ($p > 0.05$) between the fruit extract and ZnO-NPs results. These results were similar to previous reports on the antibacterial effect of ZnO-NPs [6,49,55,56].

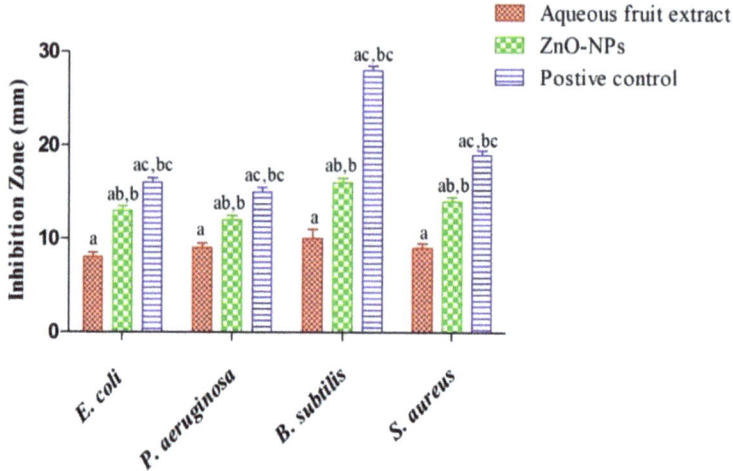

Figure 9. Antibacterial activity of aqueous extract and ZnO-NPs against pathogenic bacteria [Different superscript showed significant ($p < 0.05$) difference between (a) fruit extract, (b) ZnO-NPs, and (c) positive control].

The MIC values of ZnO-NPs against Gram-positive and Gram-negative bacteria ranged from 31.2 to 125 µg/mL, as shown in Table 2. The lowest MIC value was observed against B. subtilis (31.2 µg/mL) therefore, indicating the higher effectiveness of the nanoparticles to this microorganism. Whereas for E. coli and S. aureus MIC value was found to be 62.5 µg/mL and for P. aeruginosa MIC value was 125 µg/mL (Table 2). Literature also shows the lower MIC value of ZnO-NPs for B. subtilis as compared to other bacteria [57,58]. Whereas Alekish et al. [59] observed the lower MIC value for S. aureus as compared to E. coli. Ezealisiji and Siwe–Noundou [60] observed the MIC concentration of ZnO-NPs against bacterial pathogens in the range of 0.140 to 6.420 µg/mL.

Table 2. MIC of fruit extract and ZnO-NPs against bacterial and fungal strains.

Microbial Strains	MIC (µg/mL)		
	Fruit Extract	ZnO-NPs	Ampicillin
E. coli	250	62.5	6.25
P. aeruginosa	250	125	6.25
B. subtilis	500	31.2	1.56
S. aureus	500	62.5	3.12
F. oxysporum	125	62.5	–
R. necatrix	62.5	15.6	–

The antifungal activity of fruit aqueous extract and ZnO-NPs was determined by the food poison technique against two fungal strains (F. oxysporum and R. necatrix) and further confirmed with MIC value. As shown in Figure 10, ZnO-NPs showed a significantly ($p > 0.05$) higher percent of inhibition of R. necatrix (70 ± 1.4%) and F. oxysporum (47 ± 1.6%) compared to the fruits aqueous extract (40.8 ± 1.4% and 35 ± 1.4%, respectively) (Figure 10 and Figure S2). Chauhan et al. [27] observed 47% and 26% percentage inhibition of ZnO-NPs for F. oxysporum and R. necatrix, respectively. On the other hand, Yehia and Ahmed [61] in their study also observed good antifungal activity of ZnO-NPs against F. oxysporum and Penicillium expansum.

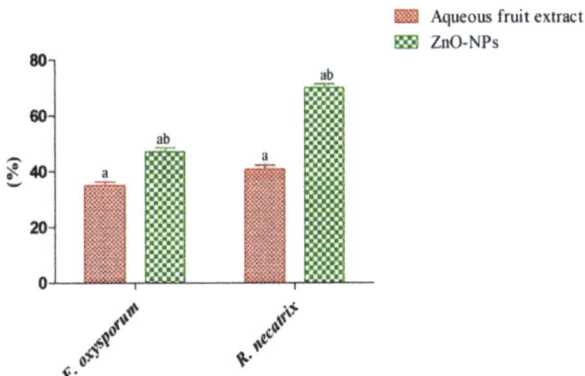

Figure 10. Graphical representation of the antifungal activity of plant extract and ZnO-NPs against plant pathogenic fungi [Different superscript showed significant ($p < 0.05$) difference between (a) fruit extract, (b) ZnO-NPs].

The results of MIC value also showed higher antifungal activity of the ZnO-NPs as compared to fruit's aqueous extract. Among these two fungi, the ZnO-NPs showed a minimum MIC value against R. necatrix (15.62 µg/mL) as compared to F. oxysporum (62.5 µg/mL) (Table 2). According to Rosa–Garcia et al. [62], the ZnO-NPs can be used as a strong antifungal agent against plant pathogens.

According to Mahamuni et al. [63], the antimicrobial activity of the ZnO-NPs is due to the release of zinc ions (Zn^+) in aerobic conditions, which shows toxicity to the microorganism. The released Zn^+ ions bind to the cell wall and result in disruption of the cell membrane to the production of reactive oxygen species (ROS), including hydrogen peroxide (H_2O_2) [63,64]. The generation of hydrogen peroxide is the main factor behind the antimicrobial property. The membrane lipid peroxidation damages the DNA of the cell and also inhibits the plasmid DNA replication. The revealed hydrogen peroxides also disrupt protein structure and reducing sugars, and therefore reduce overall cell viability [65–67]. The probable mechanism of antimicrobial activity of ZnO-NPs is presented in Figure 11.

3.4. Anticancer Activity

The anticancer effect of ZnO-NPs on A549 cells was done by MTT assay at different concentrations (1.56–200 µg/mL). The anticancer activity of ZnO-NPs on the A549 cell line revealed that the ZnO-NPs exhibited potent inhibitory activities comparable to that of Paclitaxel, as shown in Figure 12 and Figure S3. ZnO-NPs showed a high percentage (52.41%) of A549 cells death in 200 µg/mL concentration, whereas the IC_{50} value of ZnO-NPs on A549 cells was observed to be 158.1 ± 1.14 µg/mL. Results showed that with increasing concentration (1.56–200 µg/mL) of ZnO-NPs, there was an increase in the death rate of cancer cells. Rajeswaran et al. [68] also observed the potential anticancer activity (IC_{50} 120 µg/mL) of ZnO-NPs synthesized from Cymodocea serrulata extract against A549 cell lines. On the other hand, Selim et al. [12] also observed remarkable anticancer activity of ZnO-NPs of aqueous extract of *Deverra tortuosa* on human lung adenocarcinoma (A549) cells. According to Rasmussen et al. [69], the possible mechanism of the anticancer activity of ZnO-NPs is their semiconductor nature, which induces oxidative stress in cancer cells by generation of ROS and therefore leads to the death of cancer cells [69–72].

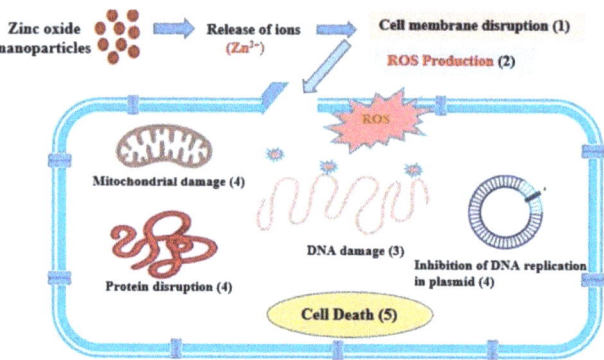

Figure 11. Antimicrobial mechanism of ZnO-NPs.

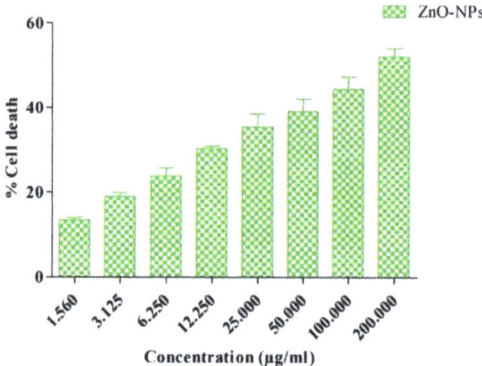

Figure 12. Anticancer effect of ZnO-NPs at a concentration (1.56–200 µg/mL) on A549 cells.

3.5. Photocatalytic Effectiveness

Methylene blue is a highly toxic cationic dye that is used to dye paper, leather, and textiles. The industrial effluent mixes this dye with river and other water sources, therefore preventing the penetration of solar radiation which eventually affects the water-based photosynthesis, resulting in harm to water ecology, causing environmental pollution, and poisoning the food chain [73]. In Figure 13, the decrease in peak absorbance of methylene blue at 665 nm as a function of irradiation time depicts the photocatalytic efficiency of applied ZnO-NPs, as synthesized ZnO-NPs utilizing *Rubus ellipticus* fruit extract has shown remarkable photodegradation activity against methylene blue contaminated water samples at various degradation intervals. Within the first 15 min. of photo degradation, there was a considerable decline in absorption intensity, which eventually disappeared completely at the end of 180 min. Figure 13b presents the time profile of MB dye degradation efficiency. Within 60 min. of photocatalysis, approximately 17.5% of the total methylene blue concentration was degraded, yielding a final MB degradation efficiency of 72.7% within 180 min. In the early stage of photodegradation, the availability of numerous catalytic sites and a high concentration gradient encouraged the fast colour deterioration [74]. The photocatalytic degradation of methylene blue confirms the amphoteric nature of ZnO-NPs.

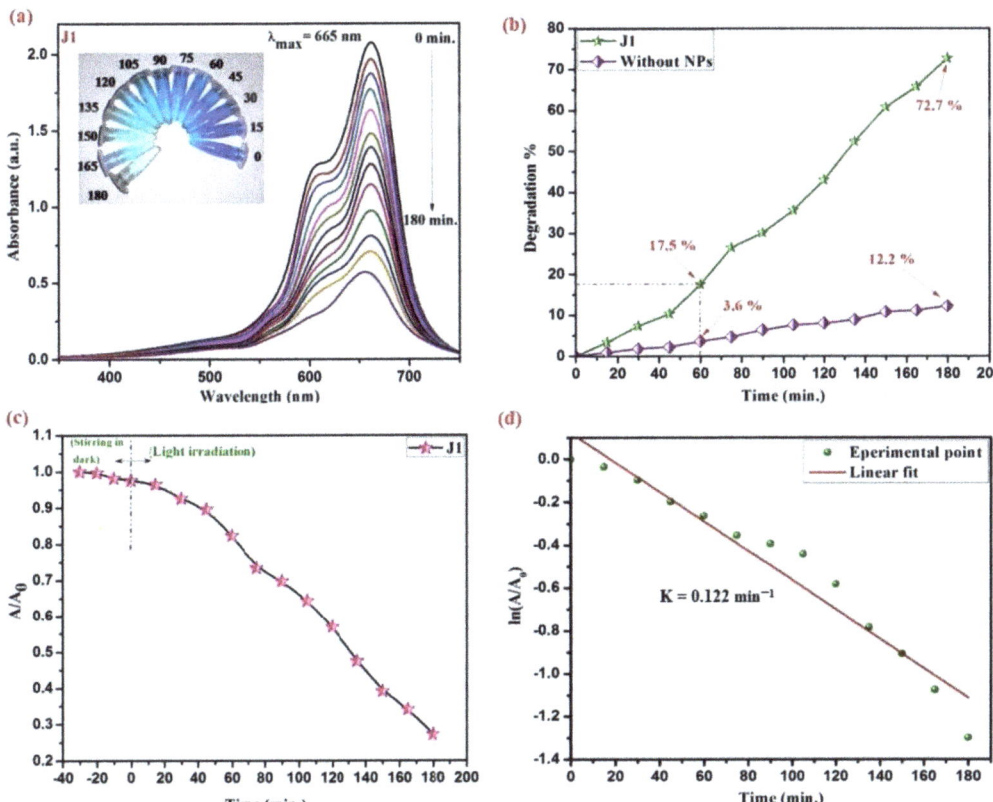

Figure 13. Photocatalytic dye degradation studies by ZnO-NPs (a) Absorption spectrum of MB dye samples along with decoloration of dye from dark blue to a nearly clear solution, as shown in the inset; (b) Time profile of degradation efficiency for MB dye (c) Normalized absorbance of dye in dark and sunlight in the presence of a catalyst (d) Kinetic study and rate constant calculated for the dye degradation reaction.

In the presence of sunlight, the photodegradation ability of ZnO-NPs proposed a three-step mechanism that involves the generation and transfer of electron-hole pairs, radical generation, and then degradation of dye. The electron and hole pair from ZnO conduction and valance bands are generated in the first step by irradiating the material with sunlight, as explained in Equation (4).

$$ZnO + h\nu \rightarrow e^-_{cb} + h^+_{vb} \tag{4}$$

In the second phase, oxygen molecules interact with surface electrons to generate superoxides, which are then converted to peroxide molecules. In addition, the surface holes oxidize the water molecules, resulting in the formation of hydroxyl ions as shown in Equations (5)–(7) [75].

$$O_2 + e^- \rightarrow O^\circ_2 + H^+ \rightarrow HOO \tag{5}$$

$$HOO^\circ + e^- + H^+ \rightarrow H_2O_2 \tag{6}$$

$$H_2O + h^+ \rightarrow OH^\circ + H^+ \quad (7)$$

The above-formed intermediates are exceedingly unstable, and when they react with the dye substituent, they lead to disintegration into mineralized products, as presented in Equation (8) [76].

$$\text{MB Dye molecule} + (OH^\circ, HOO^\circ, H^+ \text{ or } H_2O_2) \rightarrow \text{Degraded/mineralized products.} \quad (8)$$

A large surface area was provided by the spherical shape of green synthesized ZnO-NPs to facilitate electron-hole pair separation for distinct radical species that readily helps to disintegrate the dye molecule. However, the dye degradation efficiency of ZnO-NPs is dramatically lowered due to severe aggregation. The findings of photodegradation investigations revealed significant photocatalytic activity of the synthesized ZnO-NPs. The almost total disintegration of methylene blue, with around 17.5% photodegradation within 1 h, indicates that as-prepared ZnO-NPs had substantial photocatalytic efficiency.

4. Conclusions

Nanotechnology is evolving rapidly, yet research into the toxicological effects of nanoparticles on public health and the environment is still in its infancy. In the present study, the ZnO-NPs were synthesized using fruit extract of *R. ellipticus* for the first time via a simple, low-cost, and environment-friendly method. The fruit's aqueous extract acts as a capping and stabilizing agent. The X-ray diffraction patterns revealed the formation of the pure phase ZnO-NPs the hexagonal wurtzite structure with a crystallite size of 20 nm. The Rietveld refinement confirmed the phase purity of ZnO-NPs with the p-63-mc space group and a lower chi^2 value (χ) of 2.15. FTIR spectroscopy was utilized to determine the functional groups associated with the ZnO-NPs and the Z-O stretching band was observed at 536 cm^{-1}. XPS revealed the presence of oxygen chemisorbed species on the surface of ZnO-NPs. UV-vis spectroscopy showed the characteristic absorption peak of ZnO-NPs at 353 nm with an optical bandgap of 3.21 eV obtained using Tau's plot. FE-SEM and elemental mapping revealed the flowerlike morphology of self-assembled grains with a uniform distribution of elements. The TEM and SAED showed the formation of self-assembled grains with irregular shapes, size, and polycrystalline nature of the ZnO-NPs.

ZnO-NPs have shown antioxidant, antibacterial (against Gram-positive and Gram-negative bacteria), antifungal (plant pathogenic fungi), and anticancer activity (cancer cell (A549) lines through in vitro assays. The antioxidant activity exhibited could be due to the attachment of the -OH group of polyphenolic compounds on the surface of ZnO-NPs. The highest antibacterial activity of ZnO-NPs was observed against *B. subtilis* followed by *S. aureus*, and *E. coli* whereas, the maximum antifungal activity was observed against *R. necatrix* then *F. oxysporum*. The anticancer assay showed an increased death rate of cancer cells with increasing concentration (1.56–200 µg/mL) of ZnO-NPs. The ZnO-NPs have shown significant photocatalytic activity by degrading nearly 17.5% of methylene blue dye within 1 h.

In conclusion, the green synthesis technique is beneficial for plant-based fabrication. Due to less toxicity and chemical-free nature, and significant biological activities, the ZnO-NPs synthesized from fruits aqueous extract of *R. ellipticus* can be considered as biocompatible nanomaterials. Hence, these ZnO-NPs are safe and can be used for wastewater treatment and for medicinal purposes by pharmacological industries. From a future perspective, the in vivo study is required for their biological activities and phytoconstituents present in plants will develop a novel platform for green synthesis of nanoparticles for their biomedical purpose.

Supplementary Materials: The following supporting information can be downloaded at: https://www.mdpi.com/article/10.3390/ma15103470/s1, Table S1: Antioxidant activity of fruit extract and ZnO-NPs; Figure S1: Antibacterial activity of (a—300 µg; aqueous extract, b—300 µg; ZnO-NPs, c—50 µg; Ampicillin, d—negative control (DMSO—10 µl) fruit extract and ZnO-NPs against *E. coli*

(A), *P. aeruginosa* (B), *B. subtilis* (C), *S. aureus* (D). Figure S2: Antifungal activity against *F. oxysporum* and *R. necatrix*; A and D—*F. oxysporum* and *R. necatrix* Control (Without extract and nanoparticles), B and E—Aqueous extract (300 µg), C and F—ZnO-NPs (300 µg); Figure S3: Microscopic images of A549 cell before and after treatment [Cells alone (A); Cells treated with Paclitaxel (B), 25 µg ZnO-NPs (C), 50 µg ZnO-NPs (D), 100 µg ZnO-NPs (E), and 200 µg ZnO-NPs (F)].

Author Contributions: Conceptualization, J.D. and A.K. (Ashwani Kumar); Data curation, J.D. and I.G.; Synthesis and material characterization analysis, A.C. and K.M.B.; Analysis of biological activities, J.D., A.K. (Ashwani Kumar) and R.V.S.; Analysis of photocatalytic activity, K.M.; Materials Characterization and data interpretation, J.D., A.K. (Amita Kumari) and A.C.; Supervision, A.K. (Amita Kumari) and R.K.; Writing—original draft, J.D., S.T. and S.L.; Writing—review & editing, B.H.C. and A.-L.E.M. All authors have read and agreed to the published version of the manuscript.

Funding: Author A.-L.E.M. wants to thank Tshwane University of Technology for financial support. Author K.M.B. is thankful to the Researchers Supporting Project number (RSP-2021/148) at King Saud University for the financial support.

Institutional Review Board Statement: Not applicable.

Informed Consent Statement: Not applicable.

Data Availability Statement: All data are available in the manuscript.

Conflicts of Interest: Authors claim that they have no conflict of interests among them.

References

1. Iravani, S.; Varma, R.S. Plant-derived edible nanoparticles and miRNAs: Emerging frontier for therapeutics and targeted drug-delivery. *ACS Sustain. Chem. Eng.* **2019**, *7*, 8055–8069. [CrossRef]
2. Aritonang, R.S.; Koleangan, H.; Wuntu, H. Synthesis of silver nanoparticles using aqueous extract of medicinal plants' (*Impatiens balsamina* and *Lantana camara*) fresh leaves and analysis of antimicrobial activity. *Int. J. Microbiol.* **2019**, *2019*, 8642303. [CrossRef] [PubMed]
3. Piccinno, F.; Gottschalk, F.; Seeger, S.; Nowack, B. Industrial production quantities and uses of ten engineered nanomaterials in Europe and the world. *J. Nanopart. Res.* **2012**, *14*, 1109. [CrossRef]
4. Jiang, J.; Pi, J.; Cai, J. The advancing of zinc oxide nanoparticles for biomedical applications. *Bioinorgan. Chem. Appl.* **2018**, *2018*, 1062562. [CrossRef]
5. Siddiqi, K.S.; Rahman, A.; Husen, A. Properties of zinc oxide nanoparticles and their activity against microbes. *Nanoscale Res. Lett.* **2018**, *13*, 141. [CrossRef]
6. El-Belely, E.F.; Farag, M.; Said, H.A.; Amin, A.S.; Azab, E.; Gobouri, A.A.; Fouda, A. Green Synthesis of Zinc Oxide Nanoparticles (ZnO-NPs) Using *Arthrospira platensis* (Class: Cyanophyceae) and Evaluation of their Biomedical Activities. *Nanomaterials* **2021**, *11*, 95. [CrossRef]
7. Wojnarowicz, J.; Chudoba, T.; Lojkowski, W. A review of microwave synthesis of zinc oxide nanomaterials: Reactants, process parameters and morphologies. *Nanomaterials* **2020**, *10*, 1086. [CrossRef]
8. Divya, B.; Karthikeyan, C.; Rajasimman, M. Chemical synthesis of zinc oxide nanoparticles and its application of dye decolourization. *Int. J. Nanosci. Nanotechnol.* **2018**, *14*, 267–275.
9. Weldegebrieal, G.K. Synthesis method, antibacterial and photocatalytic activity of ZnO nanoparticles for azo dyes in wastewater treatment: A review. *Inorg. Chem. Commun.* **2020**, *120*, 108140. [CrossRef]
10. Noman, M.T.; Amor, N.; Petru, M. Synthesis and applications of ZnO nanostructures (ZONSs): A review. *Crit. Rev. Solid State Mater. Sci.* **2021**, 1–43. [CrossRef]
11. Pillai, A.M.; Sivasankarapillai, V.S.; Rahdar, A.; Joseph, J.; Sadeghfar, F.; Rajesh, K.; Kyzas, G.Z. Green synthesis and characterization of zinc oxide nanoparticles with antibacterial and antifungal activity. *J. Mol. Struct.* **2020**, *1211*, 128107. [CrossRef]
12. Selim, Y.A.; Azb, M.A.; Ragab, I.; Abd El-Azim, M.H. Green synthesis of zinc oxide nanoparticles using aqueous extract of *Deverra tortuosa* and their cytotoxic activities. *Sci. Rep.* **2020**, *10*, 3445. [CrossRef] [PubMed]
13. Marslin, G.; Siram, K.; Selvakesavan, R.K.; Kruszka, D.; Kachlicki, P.; Franklin, G. Secondary Metabolites in the Green Synthesis of Metallic Nanoparticles. *Materials* **2019**, *12*, 940. [CrossRef]
14. Helmy, E.T.; Abouellef, E.M.; Soliman, U.A.; Pan, J.H. Novel green synthesis of S-doped TiO2 nanoparticles using *Malva parviflora* plant extract and their photocatalytic, antimicrobial and antioxidant activities under sunlight illumination. *Chemosphere* **2021**, *271*, 129524. [CrossRef] [PubMed]
15. Rambabu, K.; Bharath, G.; Banat, F.; Show, P.L. Green synthesis of zinc oxide nanoparticles using *Phoenix dactylifera* waste as bioreductant for effective dye degradation and antibacterial performance in wastewater treatment. *J. Hazard. Mater.* **2021**, *402*, 123560. [CrossRef]

16. Tiwari, N.; Pandit, R.; Gaikwad, S.; Gade, A.; Rai, M. Biosynthesis of zinc oxide nanoparticles by petals extract of *Rosa indica* L., its formulation as nail paint and evaluation of antifungal activity against fungi causing onychomycosis. *IET Nanobiotechnol.* **2016**, *11*, 205–211. [CrossRef]
17. Pachaiappan, R.; Rajendran, S.; Ramalingam, G.; Vo, D.V.N.; Priya, P.M.; Soto-Moscoso, M. Green Synthesis of Zinc Oxide Nanoparticles by *Justicia adhatoda* Leaves and Their Antimicrobial Activity. *Chem. Eng. Technol.* **2021**, *44*, 551–558. [CrossRef]
18. Ambika, S.; Sundrarajan, M. Antibacterial behaviour of *Vitex negundo* extract assisted ZnO nanoparticles against pathogenic bacteria. *J. Photochem. Photobiol. B* **2015**, *146*, 52–57. [CrossRef]
19. Rad, S.S.; Sani, A.M.; Mohseni, S. Biosynthesis, characterization and antimicrobial activities of zinc oxide nanoparticles from leaf extract of *Mentha pulegium* (L.). *Microb. Pathog.* **2019**, *131*, 239–245. [CrossRef]
20. Suresh, D.; Nethravathi, P.C.; Rajanaika, H.; Nagabhushana, H.; Sharma, S.C. Green synthesis of multifunctional zinc oxide (ZnO) nanoparticles using *Cassia fistula* plant extract and their photodegradative, antioxidant and antibacterial activities. *Mater. Sci. Semicon. Proc.* **2015**, *31*, 446–454. [CrossRef]
21. Nilavukkarasi, M.; Vijayakumar, S.; Prathipkumar, S. *Capparis zeylanica* mediated bio-synthesized ZnO nanoparticles as antimicrobial, photocatalytic and anti-cancer applications. *Mater. Sci. Energy Technol.* **2020**, *3*, 335–343. [CrossRef]
22. Shabaani, M.; Rahaiee, S.; Zare, M.; Jafari, S.M. Green synthesis of ZnO nanoparticles using loquat seed extract; Biological functions and photocatalytic degradation properties. *LWT* **2020**, *134*, 110133. [CrossRef]
23. Qasim, S.; Zafar, A.; Saif, M.S.; Ali, Z.; Nazar, M.; Waqas, M.; Haq, A.U.; Tariq, T.; Hassan, S.G.; Iqbal, F.; et al. Green synthesis of iron oxide nanorods using *Withania coagulans* extract improved photocatalytic degradation and antimicrobial activity. *J. Photochem. Photobiol. B* **2020**, *204*, 111784. [CrossRef] [PubMed]
24. Lalla, R.; Cheek, M.D.; Nxumalo, M.M.; Renteria, J.L. First assessment of naturalised *Rubus ellipticus* Sm. populations in South Africa—A potential invasion risk? *S. Afr. J. Bot.* **2018**, *114*, 111–116. [CrossRef]
25. Saklani, S.; Chandra, S.; Badoni, P.P.; Dogra, S. Antimicrobial activity, nutritional profile and phytochemical screening of wild edible fruit of *Rubus ellipticus*. *Int. J. Med. Aromat. Plants* **2021**, *2*, 269–274.
26. Muniyandi, K.; George, E.; Sathyanarayanan, S.; George, B.P.; Abrahamse, H.; Thamburaj, S.; Thangaraj, P. Phenolics, tannins, flavonoids and anthocyanins contents influenced antioxidant and anticancer activities of *Rubus* fruits from Western Ghats. *India Food Sci. Hum. Wellness* **2019**, *8*, 73–81. [CrossRef]
27. Chauhan, A.; Verma, R.; Kumari, S.; Sharma, A.; Shandilya, P.; Li, X.; Batoo, K.M.; Imran, A.; Kulshrestha, S.; Kumar, R. Photocatalytic dye degradation and antimicrobial activities of Pure and Ag-doped ZnO using *Cannabis sativa* leaf extract. *Sci. Rep.* **2020**, *10*, 7881. [CrossRef]
28. Blois, M.S. Antioxidant determinations by the use of a stable free radical. *Nature* **1958**, *181*, 1199–2000. [CrossRef]
29. Oyaizu, M. Studies on products of browning reaction antioxidative activities of products of browning reaction prepared from glucosamine. *Jpn. J. Nutr. Diet.* **1986**, *44*, 307–315. [CrossRef]
30. Banerjee, D.; Chakrabarti, S.; Hazra, A.K.; Banerjee, S.; Ray, J.; Mukherjee, B. Antioxidant activity and total phenolics of some mangroves in Sundarbans. *Afr. J. Biotechnol.* **2008**, *7*, 805–810.
31. Bauer, A.W. Antibiotic susceptibility testing by a standardized single disc method. *Am. J. Clin. Pathol.* **1966**, *45*, 149–158. [CrossRef]
32. Grover, R.K.; Moore, J.D. Toxicometric studies of fungicides against brown rot organisms *Sclerotinia fructicola* and *S. laxa*. *Phytopathology* **1962**, *52*, 876–888.
33. CLSI. *Methods for Dilution Antimicrobial Susceptibility Tests for Bacteria that Grow Aerobically—Approved Standard*, 9th ed.; CLSI document M07–A9; Clinical and Laboratory Standards Institute: Wayne, PA, USA, 2012.
34. Kumari, R.; Saini, A.K.; Chhillar, A.K.; Saini, V.; Saini, R.V. Antitumor Effect of Bio-Fabricated Silver Nanoparticles Towards Ehrlich Ascites Carcinoma. *Biointerface Res. Appl. Chem.* **2021**, *11*, 12958–12972. [CrossRef]
35. Mansi, K.; Kumar, R.; Kaur, J.; Mehta, S.K.; Pandey, S.K.; Kumar, D.; Dash, A.K.; Gupta, N. DL-Valine assisted fabrication of quercetin loaded CuO nanoleaves through microwave irradiation method: Augmentation in its catalytic and antimicrobial efficiencies. *Environ. Nanotechnol. Monit. Manag.* **2020**, *14*, 100306. [CrossRef]
36. Zak, A.K.; Abrishami, M.E.; Majid, W.A.; Yousefi, R.; Hosseini, S.M. Effects of annealing temperature on some structural and optical properties of ZnO nanoparticles prepared by a modified sol–gel combustion method. *Ceram. Int.* **2011**, *37*, 393–398. [CrossRef]
37. Khatana, C.; Kumar, A.; Alruways, M.W.; Khan, N.; Thakur, N.; Kumar, D.; Kumari, A. Antibacterial potential of zinc oxide nanoparticles synthesized using *Aloe vera* (L.) Burm. f.: A Green approach to combat drug resistance. *J. Pure Appl. Microbiol.* **2021**, *15*, 1907–1914. [CrossRef]
38. Chauhan, A.; Verma, R.; Batoo, K.M.; Kumari, S.; Kalia, R.; Kumar, R.; Hadi, M.; Raslan, E.H.; Imran, A. Structural and optical properties of copper oxide nanoparticles: A study of variation in structure and antibiotic activity. *J. Mater. Res.* **2021**, *36*, 1496–1509. [CrossRef]
39. Verma, R.; Chauhan, A.; Batoo, K.M.; Kumar, R.; Hadhi, M.; Raslan, E.H. Effect of calcination temperature on structural and morphological properties of bismuth ferrite nanoparticles. *Ceram. Int.* **2021**, *47*, 3680–3691. [CrossRef]
40. Soto-Robles, C.A.; Luque, P.A.; Gómez-Gutiérrez, C.M.; Nava, O.; Vilchis-Nestor, A.R.; Lugo-Medina, E.; Ranjithkumar, R.; Castro-Beltrán, A. Study on the effect of the concentration of *Hibiscus sabdariffa* extract on the green synthesis of ZnO nanoparticles. *Results Phys.* **2019**, *15*, 102807. [CrossRef]

41. Diallo, A.; Ngom, B.D.; Park, E.; Maaza, M. Green synthesis of ZnO nanoparticles by *Aspalathus linearis*: Structural & optical properties. *J. Alloys Compd.* **2015**, *646*, 425–430. [CrossRef]
42. Das, J.; Pradhan, S.K.; Sahu, D.R.; Mishra, D.K.; Sarangi, S.N.; Nayak, B.B.; Verma, S.; Roul, B.K. Micro-Raman and XPS studies of pure ZnO ceramics. *Phys. B Condens.* **2010**, *405*, 2492–2497. [CrossRef]
43. Talam, S.; Karumuri, S.R.; Gunnam, N. Synthesis, characterization, and spectroscopic properties of ZnO nanoparticles. *Int. Sch. Res. Not.* **2012**, *2012*, 372505. [CrossRef]
44. Makuła, P.; Pacia, M.; Macyk, W. How to correctly determine the band gap energy of modified semiconductor photocatalysts based on UV-Vis spectra. *J. Phys. Chem. Lett.* **2018**, *9*, 6814–6817. [CrossRef] [PubMed]
45. Janaki, A.C.; Sailatha, E.; Gunasekaran, S. Synthesis, characteristics and antimicrobial activity of ZnO nanoparticles, Spectrochim. *Acta Part A Mol. Biomol. Spectros.* **2015**, *144*, 17–22. [CrossRef] [PubMed]
46. Vennila, S.; Jesurani, S.S. Eco-friendly green synthesis and characterization of stable ZnO Nanoparticle using small Gooseberry fruits extracts. *Int. J. ChemTech Res.* **2017**, *10*, 271–275.
47. Handore, K.; Bhavsar, S.; Horne, A.; Chhattise, P.; Mohite, K.; Ambekar, J.; Pande, N.; Chabukswar, V. Novel green route of synthesis of ZnO nanoparticles by using natural biodegradable polymer and its application as a catalyst for oxidation of aldehydes. *J. Macromol. Sci. A* **2014**, *51*, 941–947. [CrossRef]
48. Kumar, I.; Mondal, M.; Meyappan, V.; Sakthivel, N. Green one-pot synthesis of gold nanoparticles using *Sansevieria roxburghiana* leaf extract for the catalytic degradation of toxic organic pollutants. *Mater. Res. Bull.* **2016**, *117*, 18–27. [CrossRef]
49. Sohail, M.F.; Rehman, M.; Hussain, S.Z.; Huma, Z.E.; Shahnaz, G.; Qureshi, O.S.; Khalid, Q.; Mirza, S.; Hussain, I.; Webster, T.J. Green synthesis of zinc oxide nanoparticles by Neem extract as multi-facet therapeutic agents. *J. Drug Deliv. Sci. Technol.* **2020**, *59*, 101911. [CrossRef]
50. Dhatwalia, J.; Kumari, A.; Verma, R.; Upadhyay, N.; Guleria, I.; Lal, S.; Thakur, S.; Gudeta, K.; Kumar, V.; Chao, J.C.; et al. Phytochemistry, pharmacology, and nutraceutical profile of *Carissa* species: An updated review. *Molecules* **2021**, *26*, 7010. [CrossRef]
51. Mahendran, G.; Kumari, B.R. Biological activities of silver nanoparticles from *Nothapodytes nimmoniana* (Graham) Mabb. fruit extracts. *Food Sci. Hum. Wellness* **2016**, *5*, 207–218. [CrossRef]
52. Reddy, N.J.; Vali, D.N.; Rani, M.; Rani, S.S. Evaluation of antioxidant, antibacterial and cytotoxic effects of green synthesized silver nanoparticles by *Piper longum* fruit. *Mater. Sci. Eng.* **2014**, *34*, 115–122. [CrossRef] [PubMed]
53. Dulta, K.; Ağçeli, G.K.; Chauhan, P.; Jasrotia, R.; Chauhan, P.K. Ecofriendly Synthesis of Zinc Oxide Nanoparticles by *Carica papaya* leaf extract and their applications. *J. Clust. Sci.* **2021**, *27*, 603–617. [CrossRef]
54. Ananthalakshmi, R.; Rajarathinam, S.R.; Sadiq, A.M. Antioxidant activity of ZnO Nanoparticles synthesized using *Luffa acutangula* peel extract. *Res. J. Pharm. Technol.* **2016**, *12*, 1569–1572. [CrossRef]
55. Saif, S.; Tahir, A.; Asim, T.; Chen, Y.; Khan, M.; Adil, S.F. Green synthesis of ZnO hierarchical microstructures by *Cordia myxa* and their antibacterial activity, Saudi. *J. Biol. Sci.* **2019**, *26*, 364–1371. [CrossRef]
56. Demissie, M.G.; Sabir, F.K.; Edossa, G.D.; Gonfa, B.A. Synthesis of zinc oxide nanoparticles using leaf extract of *lippia adoensis* (koseret) and evaluation of its antibacterial activity. *J. Chem.* **2020**, *2020*, 7459042. [CrossRef]
57. Begum, J.S.; Manjunath, K.; Pratibha, S.; Dhananjaya, N.; Sahu, P.; Kashaw, S. Bioreduction synthesis of zinc oxide nanoparticles using Delonix regia leaf extract (Gul Mohar) and its agromedicinal applications. *J. Sci.-Adv. Mater. Dev.* **2020**, *5*, 468–475. [CrossRef]
58. Sharma, S.; Kumar, K.; Thakur, N.; Chauhan, S.; Chauhan, M.S. The effect of shape and size of ZnO nanoparticles on their antimicrobial and photocatalytic activities: A green approach. *Bull. Mater. Sci.* **2020**, *43*, 20. [CrossRef]
59. Alekish, M.; Ismail, Z.B.; Albiss, B.; Nawasrah, S. In vitro antibacterial effects of zinc oxide nanoparticles on multiple drug-resistant strains of *Staphylococcus aureus* and *Escherichia coli*: An alternative approach for antibacterial therapy for mastitis in sheep. *Vet. World* **2018**, *11*, 1428. [CrossRef]
60. Ezealisiji, K.M.; Siwe-Noundou, X. Green synthesis of zinc oxide nanoparticles and their antibiotic-potentiation activities of mucin against pathogenic bacteria. *Res. J. Nanosci. Nanotechnol.* **2020**, *10*, 9–14.
61. Yehia, R.S.; Ahmed, O.F. In vitro study of the antifungal efficacy of zinc oxide nanoparticles against *Fusarium oxysporum* and *Penicilium expansum*. *Afr. J. Microbiol. Res.* **2013**, *7*, 1917–1923. [CrossRef]
62. La Rosa-García, D.; Susana, C.; Martínez-Torres, P.; Gómez-Cornelio, S.; Corral-Aguado, M.A.; Quintana, P.; Gómez-Ortíz, N.M. Antifungal activity of ZnO and MgO nanomaterials and their mixtures against *Colletotrichum gloeosporioides* strains from tropical fruit. *J. Nanomater.* **2018**, *2018*, 3498527. [CrossRef]
63. Mahamuni, P.P.; Patil, P.M.; Dhanavade, M.J.; Badiger, M.V.; Shadija, P.G.; Lokhande, A.C.; Bohara, R.A. Synthesis and characterization of zinc oxide nanoparticles by using polyol chemistry for their antimicrobial and antibiofilm activity. *Biochem. Biophys. Rep.* **2019**, *17*, 71–80. [CrossRef] [PubMed]
64. He, L.; Liu, Y.; Mustapha, A.; Lin, M. Antifungal activity of zinc oxide nanoparticles against *Botrytis cinerea* and *Penicillium expansum*. *Microbiol. Res.* **2011**, *166*, 207–215. [CrossRef] [PubMed]
65. Sirelkhatim, A.; Mahmud, S.; Seeni, A.; Kaus, N.H.M.; Ann, L.C.; Bakhori, S.K.M.; Hasan, H.; Mohamad, D. Review on zinc oxide nanoparticles: Antibacterial activity and toxicity mechanism. *Nano-Micro Lett.* **2015**, *7*, 219–242. [CrossRef] [PubMed]
66. Tiwari, V.; Mishra, N.; Gadani, K.; Solanki, P.S.; Shah, N.A.; Tiwari, M. Mechanism of anti-bacterial activity of zinc oxide nanoparticle against carbapenem-resistant *Acinetobacter baumannii*. *Front. Microbiol.* **2018**, *9*, 1218. [CrossRef]

67. Madhumitha, G.; Fowsiya, J.; Gupta, N.; Kumar, A.; Singh, M. Green synthesis, characterization and antifungal and photocatalytic activity of *Pithecellobium dulce* peel–mediated ZnO nanoparticles. *J. Phys. Chem. Solids* **2019**, *127*, 43–51. [CrossRef]
68. Rajeswaran, S.; Thirugnanasambandan, S.S.; Subramaniyan, S.R.; Kandasamy, S.; Vilwanathan, R. Synthesis of eco-friendly facile nano-sized zinc oxide particles using aqueous extract of *Cymodocea serrulata* and its potential biological applications. *Appl. Phys.* **2019**, *125*, 105. [CrossRef]
69. Rasmussen, J.W.; Martinez, E.; Louka, P.; Wingett, D.G. Zinc oxide nanoparticles for selective destruction of tumor cells and potential for drug delivery applications. *Expert Opin. Drug Deliv.* **2010**, *7*, 1063–1077. [CrossRef]
70. Bisht, G.; Rayamajhi, S. ZnO nanoparticles: A promising anticancer agent. *Nanobiomedicine* **2010**, *3*, 3–9. [CrossRef]
71. Gupta, S.C.; Hevia, D.; Patchva, S.; Park, B.; Koh, W.; Aggarwal, B.B. Upsides and downsides of reactive oxygen species for cancer: The roles of reactive oxygen species in tumorigenesis, prevention, and therapy. *Antioxid. Redox Signal.* **2012**, *16*, 1295–1322. [CrossRef]
72. Amuthavalli, P.; Hwang, J.S.; Dahms, H.U.; Wang, L.; Anitha, J.; Vasanthakumaran, M.; Gandhi, A.D.; Murugan, K.; Subramaniam, J.; Paulpandi, M.; et al. Zinc oxide nanoparticles using plant *Lawsonia inermis* and their mosquitocidal, antimicrobial, anticancer applications showing moderate side effects. *Sci. Rep.* **2021**, *11*, 8837. [CrossRef] [PubMed]
73. Rambabu, K.; Bharath, G.; Monash, P.; Velu, S.; Banat, F.; Naushad, M.; Arthanareeswaran, G.; Show, P.L. Effective treatment of dye polluted wastewater using nanoporous CaCl2 modified polyethersulfone membrane. *Process Saf. Environ. Prot.* **2019**, *124*, 266–278. [CrossRef]
74. Nirmala, G.; Murugesan, T.; Rambabu, K.; Sathiyanarayanan, K.; Show, P.L. Adsorptive removal of phenol using banyan root activated carbon. *Chem. Eng. Commun.* **2021**, *208*, 831–842. [CrossRef]
75. Nagaraju, G.; Nagabhushana, H.; Suresh, D.; Anupama, C.; Raghu, G.K.; Sharma, S.C. *Vitis labruska* skin extract assisted green synthesis of ZnO super structures for multifunctional applications. *Ceram. Int.* **2017**, *43*, 11656–11667. [CrossRef]
76. Rupa, E.J.; Anandapadmanaban, G.; Chokkalingam, M.; Li, J.F.; Markus, J.; Soshnikova, V.; Perez, Z.E.J.; Yang, D.C. Cationic and anionic dye degradation activity of Zinc oxide nanoparticles from *Hippophae rhamnoides* leaves as potential water treatment resource. *Optik* **2019**, *181*, 1091–1098. [CrossRef]

MDPI AG
Grosspeteranlage 5
4052 Basel
Switzerland
Tel.: +41 61 683 77 34

Materials Editorial Office
E-mail: materials@mdpi.com
www.mdpi.com/journal/materials

Disclaimer/Publisher's Note: The statements, opinions and data contained in all publications are solely those of the individual author(s) and contributor(s) and not of MDPI and/or the editor(s). MDPI and/or the editor(s) disclaim responsibility for any injury to people or property resulting from any ideas, methods, instructions or products referred to in the content.

www.ingramcontent.com/pod-product-compliance
Lightning Source LLC
LaVergne TN
LVHW070421100526
838202LV00014B/1499